The Carbonic Anhydrases

Cellular Physiology and Molecular Genetics

The Carbonic Anhydrases

Cellular Physiology and Molecular Genetics

Edited by

SUSANNA J. DODGSON

University of Pennsylvania School of Medicine
Philadelphia, Pennsylvania

RICHARD E. TASHIAN

University of Michigan Medical School
Ann Arbor, Michigan

GEROLF GROS

Medizinische Hochschule
Hannover, Germany

and

NICHOLAS D. CARTER

University of London
London, England

Springer Science+Business Media, LLC

Library of Congress Cataloging-in-Publication Data

The Carbonic anhydrases : cellular physiology and molecular genetics / edited by Susanna J. Dodgson ... [et al.].
p. cm.
Includes bibliographical references.
Includes index.
ISBN 978-1-4899-0752-3 ISBN 978-1-4899-0750-9 (eBook)
DOI 10.1007/978-1-4899-0750-9
1. Carbonic anhydrase. I. Dodgson, Susanna J.
[DNLM: 1. Carbonic Anhydrase--analysis. 2. Carbonic Anhydrase--genetics. 3. Carbonic Anhydrase--physiology. QU 139 C264]
QP613.C37C37 1991
599'.012--dc20
DNLM/DLC
for Library of Congress 91-2475
CIP

ISBN 978-1-4899-0752-3

Originally published by Plenum Press, New York in 1991
Softcover reprint of the hardcover 1st edition 1991

Dedication

The contributions of Robert Elder Forster II to the understanding of the physiological functions of the carbonic anhydrases are many. He has always been a gifted tinkerer and delights in theoretical derivations of complicated mathematical models. During his tenure as Chairman of the Department of Physiology at the University of Pennsylvania (1959 – 1990), he provided a stimulating laboratory environment for junior scientists as well as for collaborations with colleagues. His interest in carbon dioxide homeostasis and pH continues unabated.

He was the first of three sons of a Philadelphia lawyer, born in time to remember Prohibition and his father's futile attempts at circumventing those dry days. The Spanish Civil War fired his ideological zeal, and perhaps this dedication could have ended there but for the refusal of his parents to let their teenager sign up. He was sent to Yale in 1937 and did well enough to be accepted into medical school at the University of Pennsylvania in 1940. The war effort left the nation's supply of physicians depleted: the university responded by offering a three-year M.D. program. After graduating in 1943, he interned in Boston at the Peter Bent Brigham Hospital, spent two years as a captain in the Quartermaster Corps in the Climatic Research Laboratory, and returned to Boston for a residency. He then settled down to become a physiologist, starting at Harvard under Eugene Landis and then returning to the University of Pennsylvania under the guidance of Julius J. Comroe, Jr.

Since his first publication in 1943, Dr. Forster has produced many reports concerning blood flow and gas exchange. It was his interest in gas exchange that brought him renown in the field of alveolar ventilation and also took him to his long-abiding interest in carbon dioxide and the carbonic anhydrases. He was the senior editor on the first book of collected papers about carbon dioxide and the carbonic anhydrases, which came out of a conference he organized at Haverford College in 1968. A collaboration and friendship with Professor F. J. W. Roughton, who is listed as a coeditor on the first carbonic anhydrase book, lasted long before and until Professor Roughton's death in 1971. Dr. Forster's many papers reflect his interest in developing new techniques. His review article in this book discusses every technique he could find which has been used to measure carbonic anhydrase activity; many of these techniques he developed himself.

This book includes papers from former and present students and colleagues: G. Gros, R. P. Henry, S. F. Silverton, B. T. Storey, S. Lahiri, and S. J. Dodgson. Others still in the field include E. D. Crandall and A. Bidani.

This dedication would not be complete without the acknowledgment of the love and comfort provided by Elizabeth Hilbert Forster (née Day) since their wedding in 1947. His beloved Betsy has also considerably aided quite a lot of us, both in her capacity as a psychiatric social worker and as a friend. What a great team.

Susanna J. Dodgson

No single individual, with the possible exception of F. J. W. Roughton, is more closely linked with carbonic anhydrase than Thomas H. Maren. Indeed, the affinity is so strong that even he himself has found the bond difficult to break. Although in dedicating this book to him, it is his many contributions toward our understanding of the enzyme that we celebrate, he has also figured prominently in other areas of pharmacology and medical education.

Surveying his three and a half decades of research in carbonic anhydrase, one readily appreciates both the depth and breadth of his investigative mind. His contributions include many fundamental studies into the physical chemistry, biochemistry, physiology, and pharmacology of the enzyme and its inhibitors. There is hardly a facet of carbonic anhydrase that has escaped his curiosity or not yielded to his theoretical and experimental skills. He came to focus on the enzyme relatively later on in his scientific career, after formal training in chemistry and English literature at Princeton and then in pharmacology, physiology, and medicine at Johns Hopkins. In the war years, between his undergraduate and medical education, there was further practical research experience gained first as a chemist in a small cosmetics firm and then in parasitology at the School of Public Hygiene at Hopkins. Such a broad grounding in the sciences and humanities set the stage and surely helps to explain his unmatched productivity and wide-ranging investigations on the enzyme. Since his earliest work in the 1950s that led to the clinical introduction of acetazolamide, carbonic anhydrase has been for him a scientific passion and portal into the joys of experimental chemistry, physiology, pharmacology, and medicine. He has elucidated the quantitative relationship between the chemical reaction rates of the enzyme and the physiological rates of CO_2 and acid–base transport in a host of tissues. This has led to a much better understanding of the physiological roles of carbonic anhydrase in gas exchange and fluid secretion, to a fuller understanding of the clinical consequences of enzyme inhibition, and in some cases to the possibility of important carbonic anhydrase-independent modes of acid-base movement. His pioneering work on acetazolamide, which found its greatest application in the treatment of glaucoma, has kept him in the forefront of basic research in ophthalmology. Thus, it is somewhat ironic but entirely fitting that his most recent work has closed a full circle. His laboratory has successfully synthesized new and effective topical carbonic anhydrase inhibitors with intraocular pressure-lowering effects equal to those of acetazolamide, whose systemic side effects preclude its use in many patients. By his own hand, the drug which he so successfully launched at the start of his career will likely be retired into relative clinical obscurity.

To conclude this brief tribute to Tom, I wish to express my deepest gratitude and affection to him. As mentor, collaborator, and friend, he has by his wisdom, integrity, humor, curiosity, and compassion provided an inspiring example of the best human synthesis of science and the arts. I believe that all who have had the honor and pleasure of working with him would share these sentiments.

Erik Swenson

Contributors

JONATHAN H. BARLOW • Department of Biochemistry, University College London, London WC1E 6BT, United Kingdom
HUGH J. M. BRADY • Department of Biochemistry, University College London, London WC1E 6BT, United Kingdom
WOLFGANG BRUNS • Zentrum Physiologie, Medizinische Hochschule Hannover, 3000 Hannover 61, Federal Republic of Germany
PETER H. W. BUTTERWORTH • University of Surrey, Guilford, Surrey GU2 5XH, United Kingdom
WENDY CAMMER • Department of Neurology, Albert Einstein College of Medicine, Bronx, New York 10461
NICHOLAS D. CARTER • Department of Child Health, St. George's Hospital Medical School, University of London, London SW17 ORE, United Kingdom
PIERRE M. CHAREST • Muscle Biology Research Group, Laval University, Ste-Foy, Quebec, Canada G1V 4G2
W. RICHARD CHEGWIDDEN • Division of Biomedical Sciences, Sheffield City Polytechnic, Sheffield S1 1WB, United Kingdom
JOEL L. COHEN • Biomedical Sciences Program and Department of Anatomy, School of Medicine, Wright State University, Dayton, Ohio 45435
CLAUDE CÔTÉ • Muscle Biology Research Group, Laval University, Ste-Foy, Quebec, Canada G1V 4G2
PATRICK CUMMINGS • The Wistar Institute of Anatomy and Biology, Philadelphia, Pennsylvania 19104-4268
PETER J. CURTIS • The Wistar Institute of Anatomy and Biology, Philadelphia, Pennsylvania 19104-4268
SUSANNA J. DODGSON • Department of Physiology, University of Pennsylvania School of Medicine, Philadelphia, Pennsylvania 19104-6085
MINA EDWARDS • Department of Biochemistry, University College London, London WC1E 6BT, United Kingdom
YVONNE H. EDWARDS • MRC Human Biochemical Genetics Unit, The Galton Laboratory, University College London, London NW1 2HE, United Kingdom

A. ELISABETH ERIKSSON • Institute of Molecular Biology, University of Oregon, Eugene, Oregon 97403-1229

MARIANNE E. FEITL • Department of Ophthalmology, Geisinger Medical Center, Danville, Pennsylvania 17822

ROSS T. FERNLEY • Howard Florey Institute of Experimental Physiology and Medicine, Parkville, Victoria 3052, Australia

ROBERT E. FORSTER II • Department of Physiology, University of Pennsylvania School of Medicine, Philadelphia, Pennsylvania 19104-6085

PETER FRASER • The Wistar Institute of Anatomy and Biology, Philadelphia, Pennsylvania 19104-4268

PIERRE FRÉMONT • Muscle Biology Research Group, Laval University, Ste-Foy, Quebec, Canada G1V 4G2

CAROL V. GAY • Departments of Molecular and Cell Biology and Poultry Science, The Pennsylvania State University, University Park, Pennsylvania 16802

CORNELIA GEERS • Zentrum Physiologie, Medizinische Hochschule Hannover, 3000 Hannover 61, Federal Republic of Germany

GEROLF GROS • Zentrum Physiologie, Medizinische Hochschule Hannover, 3000 Hannover 61, Federal Republic of Germany

RAYMOND P. HENRY • Department of Zoology and Wildlife Science, Auburn University, Auburn, Alabama 36849-5414

DAVID HEWETT-EMMETT • Center for Demographics and Population Genetics, Graduate School of Biomedical Sciences, The University of Texas Health Science Center at Houston, Houston, Texas 77225

STEPHEN JEFFERY • Department of Child Health, St. George's Hospital Medical School, University of London, London SW17 ORE, United Kingdom

RAJA G. KHALIFAH • Biochemistry Department, Kansas University Medical School, and Research Service, Veterans Administration Medical Center, Kansas City, Missouri 64128

THEODORE KRUPIN • Department of Ophthalmology, Northwestern University Medical School, Chicago, Illinois 60611

SUKHAMAY LAHIRI • Department of Physiology, University of Pennsylvania School of Medicine, Philadelphia, Pennsylvania 19104-6085

ANDERS LILJAS • Department of Molecular Biophysics, Chemical Center, University of Lund, S-221 00, Sweden

PAUL J. LINSER • Whitney Laboratory and the Department of Anatomy and Cell Biology, University of Florida, St. Augustine, Florida 32086-9604, and Mount Desert Island Biological Laboratory, Salsbury Cove, Maine 04672

ERIK A. LIPPA • Clinical Research, Merck Sharp & Dohme Research Laboratories, West Point, Pennsylvania 19486, Department of Ophthalmology, Jefferson Medical College of Thomas Jefferson University, Philadelphia, Pennsylvania 19107, and Department of Ophthalmology, University of Pennsylvania, Philadelphia, Pennsylvania 19104

NICHOLAS LOWE • Department of Biochemistry, University College London, London WC1E 6BT, United Kingdom
JUDITH A. NEUBAUER • Department of Medicine, Pulmonary Division, University of Medicine and Dentistry, Robert Wood Johnson Medical School, New Brunswick, New Jersey 08903
SHOKO NIOKA • Department of Biochemistry and Biophysics, University of Pennsylvania School of Medicine, Philadelphia, Pennsylvania 19104-6085
EEVA-KAISA PARVINEN • Department of Anatomy, University of Oulu, Kajaanintie 52 A, SF-90220 Oulu, Finland
YVONNE RIDDERSTRÅLE • Department of Animal Physiology, Swedish University of Agricultural Sciences, S-750 07 Uppsala, Sweden
PETER A. ROGERS • Muscle Biology Research Group, Laval University, Ste-Foy, Quebec, Canada G1V 4G2
DAVID N. SILVERMAN • Department of Pharmacology, University of Florida College of Medicine, Gainesville, Florida 32610
SUSAN F. SILVERTON • Department of Oral Medicine, University of Pennsylvania School of Dental Medicine, Philadelphia, Pennsylvania 19104-6003
WILLIAM S. SLY • Department of Biochemistry and Molecular Biology, St. Louis University School of Medicine, St. Louis, Missouri 63104
JANE C. SOWDEN • Department of Biochemistry, University College London, London WC1E 6BT, United Kingdom
BAYARD T. STOREY • Department of Obstetrics and Gynecology and Department of Physiology, University of Pennsylvania School of Medicine, Philadephia, Pennsylvania 19104-6085
ERIK R. SWENSON • Department of Medicine, University of Washington and Veterans Administration Medical Center, Seattle, Washington 98108
RICHARD E. TASHIAN • Department of Human Genetics, University of Michigan Medical School, Ann Arbor, Michigan 48109-0618
WILLIAM THIERFELDER • The Wistar Institute of Anatomy and Biology, Philadelphia, Pennsylvania 19104-4268
H. KALERVO VÄÄNÄNEN • Department of Anatomy, University of Oulu, Kajaanintie 52 A, SF-90220 Oulu, Finland
PATRICK J. VENTA • Department of Human Genetics, University of Michigan Medical School, Ann Arbor, Michigan 48109-0618

Preface

As we approach the twenty-first century the problems of industrialization are evident: we find there is a greenhouse effect, the ozone layer is being depleted, the rain is acidified, and there is a terrible problem of increasing CO_2 concentrations in the atmosphere. The carbonic anhydrases are a unique family of enzymes that solve these problems in the human body: they are responsible for converting CO_2 (a gas) to HCO_3^-, which is the biggest intracellular buffer, with a concomitant decrease in a hydroxyl ion. Globally, the functions of the carbonic anhydrases in photosynthesis in rain forests and in the algae and plankton that cover our oceans indicate that they are also of utmost importance in the maintenance of the acid–base balance on our planet. Although the whole field of CO_2 metabolism is enormous and still rapidly expanding, because of the research interests of the editors this book is mainly concerned with mammalian carbonic anhydrases. However, if the interested reader intends to purify carbonic anhydrases from nonmammalian sources, Dr. Chegwidden has provided the necessary information in Chapter 7.

The carbonic anhydrases were first discovered in 1933; until 1976 there were thought to be only two isozymes. Since then CA III, IV, V, VI, and VII have been discovered and well characterized. There is, of course, no reason to believe that we have found them all. Certainly, a carbonic anhydrase sitting on a plasma membrane (CA IV) would be expected to function under entirely different conditions than one floating freely in an erythrocyte (CA II) or one in a mitochondrial matrix (CA V); however, this certainty has only come with hindsight and with the huge advance in knowledge of cellular physiology. The idea of copious quantities of a carbonic anhydrase being produced just to be secreted in the saliva (CA VI) sounds absurd; Dr. Fernley has isolated the cDNA clone (Chapter 34). The carbonic anhydrases, particularly CA II, are well known as being those enzymes with the highest turnover numbers of any found in nature. Perhaps there are a plethora of low-turnover-number carbonic anhydrase isozymes which have so far remained undetected, of which CA III is the first example, or a whole group of acetazolamide-insensitive isozymes, of which, again, CA III is so far the only mammalian example.

It was first discovered in 1940 that some sulfonamide drugs are specific carbonic anhydrase inhibitors. Acetazolamide has been used successfully to treat glaucoma since the early 1950s: the importance of carbonic anhydrase inhibitors in ophthalmology is described by Drs. Feitl and Krupin (Chapter 13) and the discovery and pharmacology of topical carbonic anhydrase inhibitors is described by Dr. Lippa (Chapter 14). Studies with other organs described in this book led to the conclusion that mostly activation, rather than inhibition, of carbonic anhydrases is desirable. This conclusion is certainly reached after consideration of the enormous medical problems of patients lacking just one isozyme, CA II, as described by the discoverer of the CA II-deficiency syndrome, Dr. Sly (Chapter 15).

When I explain to colleagues that there are several carbonic anhydrase isozymes eyes light up. They glitter when I explain that the different isozymes are mainly products of different genes. Nature could not spend so much energy on a family of unnecessary enzymes. A major aim of this book is to bring together information about the individual isozymes, from gene sequencing, crystallographic structures and kinetics of the isolated organs, to their functioning in intact organs. We know a great deal more about this family of isozymes than in 1984 when Dr. Tashian and Dr. Hewett-Emmett edited the previous compendium of reviews (*Annals New York Academy of Sciences*, volume 429, 1984). Another major aim of this book is to present details of common techniques used by workers in the field so that any interested scientist can easily set up assays for carbonic anhydrase activity, and immunohistochemical or histochemical visualization. The special problems of carbonic anhydrases in cultured cells are addressed by Dr. Venta (Chapter 5).

To those understanding only the meaning of "the" and "and" in the title of this book, I explain this is an anthology of the wonderful things that can happen throughout the body when a gas becomes a salt, with recipes. I explain that this process is so fundamental that, without it, life is probably impossible. To these intrepid and persevering readers I refer especially to Chapters 1 and 2, which give a general outline of the field from physiological and genetic perspectives.

This book has the distinction of being the first exclusively dedicated to the carbonic anhydrases which is not the proceedings of a conference. This has not been an easy task and there are many people to thank. My older sons Angus Z. Dodgson Pekala and Miles C. Dodgson Pekala have helped immensely by amusing my youngest son, Allister M. Dodgson Blossfeld while I was glued to the computer. Miss Karen Lapps and Dr. Fernley helped with the design of the front cover, which ultimately was executed by Dr. Tashian and his staff. The staff at Plenum has been helpful and pleasant; without Mrs. Mary Phillips Born there would not be a book. Each chapter was reviewed by up to four colleagues; Dr. Storey's help was particularly commendable. I would also like to thank the co-editors, who are all my good friends and colleagues, for their input. Since it is cheaper to call Ann Arbor than Germany or England, Dr. Tashian has been a most frequent listener and arbiter of disputes.

Susanna J. Dodgson

Philadelphia

Contents

VII. The Carbonic Anhydrases of Other Organs

PART I

The Carbonic Anhydrase Isozymes

CHAPTER 1

The Carbonic Anhydrases

Overview of Their Importance in Cellular Physiology and in Molecular Genetics

SUSANNA J. DODGSON

1. The Discovery of Carbonic Anhydrase Simultaneously in Philadelphia and Cambridge

Carbonic anhydrase (CA) (EC 4.2.1.1.) was first characterized in erythrocytes in 1933 directly as a result of a search by several laboratories for a catalytic factor in the erythrocytes that had been theoretically determined as necessary for rapid transit of the HCO_3^- from the erythrocyte to the pulmonary capillary. Two laboratories simultaneously published papers describing a catalytic factor. From Dr. Roughton's laboratory at Cambridge University came a paper with an elegant title including a name for what they thought was a single enzyme ("Carbonic Anhydrase: Its Preparation and Properties")[61]; from Dr. Stadie's laboratory in Philadelphia came another paper with a less concise title ("The Catalysis of the Hydration of Carbon Dioxide and Dehydration of Carbonic Acid by the Enzyme Isolated from Red Blood Cells"; the other author was Dr. Helen O'Brien).[73] Perhaps the more frequent citation of the Cambridge paper is due to the shorter title with scholarly colon as well as the confident naming of this newly discovered enzyme. The adjective "late" before the senior author's name describes the tragically premature end of N. M. Meldrum's life by his own hand not long after a crippling accident. Dr. Roughton remained in Cambridge until his death in 1971; he worked with many stellar white male scientists from both sides of the Atlantic Ocean (e.g., Sir Joseph Barcroft, Sir Hans Krebs, P. F. Scholander, Britton Chance, Quentin Gibson, Robert Forster, and J. W. Severinghaus). Dr. Roughton continued his

---

SUSANNA J. DODGSON • Department of Physiology, University of Pennsylvania School of Medicine, Philadelphia, Pennsylvania 19104-6085.

interest in CO_2 transport until his death; his life and career have been discussed and his list of publications documented by Dr. Quentin Gibson, a fellow member of the Royal Society and one-time colleague.[31] Dr. Stadie continued in the field of metabolism, and by his death in the 1940s he was a noted researcher in diabetes. Dr. O'Brien also left the CA field.

CA has not been a fashionable enzyme since 1933. Researchers having finally found and named the catalytic factor "carbonic anhydrase," little excitement was generated by its contemplation for some years. The CA field stayed in the erythrocytes, with a diversion to the kidney and acidification of the urine after the discovery of the enzyme in the kidney in the 1940s (reviewed in reference 10). Meanwhile, the world changed. Nazi Germany forced the removal of Sir Hans Krebs from his idyllically located laboratory in Freiburg-im-Breisgau, from whence he conveniently (for us) located in England and wrote henceforth in English, even a few papers concerning CA and its sulfonamide inhibitors.[43,45,46] He had already reported that the synthesis of urea was by the intracellular cycling of ornithine[44]; after this time more biological cycles were discovered, and enzymes with cofactors and effectors that act like real enzymes were studied. CA became an enzyme relegated to one paragraph on gas transport in physiology textbooks. Meanwhile, sulfonamide CA inhibitors were developed; some were discovered to be specific CA inhibitors[41,53] and used in high concentrations on whole animals and organ preparations, with extraordinary claims as to their effects. Combined with the unfashionable subunitless, effectorless position that the CAs (still thought to be one enzyme) had assumed was an additional blow from the pharmacologists, who concluded that since rats and dogs survived after massive doses of acetazolamide, the CAs were nonessential enzymes.[54,56] The CAs are thus in the peculiar position of having to justify their existence in the body more than a half century after the frenetic search for them on both sides of the Atlantic. It must also be remembered that CA II has, arguably, the highest turnover number of any enzyme.[42] It is hoped that this chapter and those following will lead researchers in the field into extreme caution when concluding that a sulfonamide-induced effect is due to CA inhibition and when concluding that lack of an effect results from inhibited CAs being uninvolved in a process rather than the inability of a CA inhibitor to reach one or more CAs.

The first CA review appeared in 1935 with a description of the presence and absence of CA activity in most of the body's organs.[69] Dr. Roughton emphatically denied there was activity in the muscle and just did not find any in the liver. He reasoned that activity in the muscle would be deleterious, probably because of the assumption that the physiological function of CA was the same everywhere else in the body as in the erythrocytes. Since 1935, several CA isozymes have been found in muscle and liver (reviewed in reference 33); throughout the human body are distributed at least seven different CA isozymes.[37,76] Papers in the *Annual Reviews of Physiology* of 1988 demonstrate the diversity of physiological functions of the isozymes.[2,16,33,58]

2. The Carbonic Anhydrases and Molecular Biology

Since 1968, conferences focusing on CO_2 and the CA isozymes have resulted in three books[1,26,78] The first two[1,26] are compilations of reviews from the invited speakers. The third[78] is the first and only tome entirely dedicated to the CA isozymes; it includes short papers as well as reviews. The major advances since the 1983 conference and its subsequent 1984 book have largely been due to the great changes in biological research that have occurred within the past decade because of techniques in molecular biology. These techniques have resulted in the identification and preparation of cDNA clones to CA I, CA II, CA III, CA VII[7,20,83] (reviewed in reference 76) and hence the finding that the genes to CA I, CA II, and CA III are contiguous on chromosome 8.[19,20] The technique by which a single amino acid can be altered in a molecule, site-directed mutagenesis, is being used in Umeå, Sweden, in the laboratory of Dr. Sven Lindskog and in Gainesville, Florida, in the laboratory of Dr. David Silverman.[24,79] Dr. Lindskog believed that his relatively recent beginnings with this technique precluded his providing a review for this book. One chapter in this book covers reaction mechanisms and reviews the sparse reports in which this exciting technique has been used.[42] Dr. Lindskog's previous elucidations of reaction mechanisms of the CA isozymes are noted.[48,49] The understanding of the localization of chicken brain CA isozymes has been greatly helped by a recent exhaustive study using the technique of *in situ* hybridization.[68]

The constant development of new techniques is necessary for growth in any field; at the same time, old techniques can also be extremely useful. The early techniques of CA assay were reviewed in 1948[80]; some can still be easily used when quantitation is not critical. The Philpot technique[65] was modified for analysis of small quantities in the laboratory of Dr. Thomas Maren[59]; a further modification of this technique is detailed in this book.[4] Other old and new techniques for quantitation of CA activity are reviewed in this volume.[25,36] These techniques include a sophisticated isotope ratio mass spectrometric micromethod that enables detection of CA activity when there is only 10^{-12} M present.[25]

3. Carbon Dioxide

The basic physical principles controlling CO_2, HCO_3^- and OH^- must always be taken into account in any discussion of the physiological importance of the CA isozymes.

When CO_2 is bubbled into distilled water, with no freely dissociable ionic species present, CO_2 remains in the gaseous form, with negligible HCO_3^- formation.[18] However, when dry $NaHCO_3$ is added to ice-cold aqueous solution with alkaline pH, there is negligible CO_2 present. Because of this it is possible to determine which species crosses membranes, which species is the product, and which is the reactant; the application of this to liver mitochondria has been discussed.[11] The Henderson-Hasselbach equation describing the equilibrium between these species only becomes important if there really is an equilibrium:

$$pH = pK' + \log_{10}\{[HCO_3^-]/[CO_2]\}$$

When pK′ is plotted as a function of ionic strength, the linear relationships are as follows[35] (Itada and Forster; unpublished date): at 25°C, $pK' = 6.35 - 0.38\sqrt{(\text{ionic strength})}$; at 37°C, $pK' = 6.30 - 0.59\sqrt{(\text{ionic strength})}$. There is a marked increase in pK′ with increasing ionic strength and with increasing temperature, and thus the ratio of CO_2 to HCO_3^- changes greatly with pK′. At pH 7.4 and 37°C, changes in ionic strength throughout the various extracellular, intracellular, and intraorganellular compartments can result in a large difference in the concentration of CO_2. The rate of uncatalyzed hydration of CO_2 is dependent on CO_2 concentration, pH, and ionic strength; this observation forms the basis of the Philpot CA assay,[65] which depends on a high CO_2 to maximize the time in which an indicator changes color.

4. Overview of the Carbonic Anhydrase Isozymes and Their Functions

There is now good evidence for at least seven distinct CA isozymes[76]; the evolution of the genes encoding these isozymes is presented elsewhere in this volume.[37]

4.1. CA I

Until the early 1960s, CA was thought to exist in one form. The differences between the two major erythrocyte isozymes CA I (= CA B) and CA II (=CA C) were reviewed in 1976.[77] It was soon apparent that the isozyme with the higher concentration in the erythrocyte (CA I) had the lower turnover number. The isozymes have different Cl^- sensitivities, and at the concentration of Cl^- in the erythrocyte, 75 mM, CA I has been predicted as being 85% inhibited[60] and thus was thought to contribute little to total erythrocyte CA activity. A recent report provides evidence that at 37°C there is less inhibition of erythrocyte CA I than was predicted,[14] since CA I was less inhibited by Cl^- at 37°C than previously reported and erythrocytes from CA II-deficient patients had 50% normal CA activity. These data suggest that CA I can function effectively in the erythrocyte.

4.2. CA II in Bone Metabolism

CA II is the isozyme most commonly referred to in discussion of the physiological functions of CA isozymes. This is the high-activity isozyme that has been localized in more tissues than any other isozyme.[42,54,76,78] CA II is present in certain cells in the brain,[6,63] is needed for secretion of fluid into the aqueous humor of the eye,[21,50] and is in the cytosol of cells from many organs such as liver, kidney, and lungs.

The importance of CA II in the osteoclast was understood from the work of Dr. Carol Gay before the discovery that CA II deficiency resulted in osteopetrosis[72] (reviewed in references 71, 72, and 80); however, the implication of a CA isozyme in a serious disease certainly focused attention on this function of CA II in the bone. If it is true that lack of CA II leads to a denser matrix in the bone, then perhaps the contrary is also true, and CA II in some way is linked with osteoporosis. The answer is unknown.

4.3. CA III: The Muscle Isozyme

In 1974, the existence of sulfonamide-resistant CA activity was noted in male rat liver homogenates but not in female rat liver homogenates.[28,29] Two years later, a sulfonamide-resistant CA isozyme was isolated from chicken muscle and subsequently designated CA III.[39] Although CA III is known as the muscle isozyme, there is a high concentration of CA III in male rat livers. This localization is not universal; human liver does not have CA III.[40,83] The regulation of rat liver CA III by growth hormone is a fascinating story.[40]

Determination of serum concentration of CA III is reported to be a more sensitive skeletal muscle marker than creatine kinase in both myogenic and neurogenic muscle.[27,82]

4.4. CA IV: The Membrane-Bound Isozyme

The limited discussion of the membrane-bound carbonic anhydrase, CA IV, in this book reflects the paucity of experimenters interested in this enzyme rather than its importance. The purification of CA IV from the brush border membranes of human kidney proximal tubules has been recently been reported[87] with some of the kinetic properties. The majority of reports about this isozyme have come from Dr. Wistrand's laboratory and have been concerned with CA IV isolated from the kidney. Extensive washing of cellular membranes of the brain glial cells (reviewed in references 6 and 63) and the lung[85] (reviewed in reference 64) have resulted in residual CA activity believed to be due to CA IV. Indicator dilution experiments with perfused livers have provided evidence for its presence on liver parenchymal cell membranes (reviewed in reference 11).

4.5. CA V: The Mitrochondrial Isozyme

This unique mitochondrial enzyme has been found in the mitochondria of liver and kidney of rat, in the liver of guinea pigs, and in the skeletal muscle of guinea pigs but not rabbits (reviewed in references 11, 33, and 34). This is the first isozyme determined to have a function in intermediary metabolism: that of supplying HCO_3^- for the mitochondrial carboxylases in urea and glucose synthesis. There has since been evidence that CA II in the glial cells similarly supplies HCO_3^- for acetyl-CoA carboxylase for fatty acid metabolism in the brain,[6] so this function is not confined to the mitochondria nor to CA V.

A recent finding from this laboratory was that CA V activity was doubled in 6-day streptozotocin-administered male rats, whereas CA III activity was halved and the total liver CA activity remained constant.[15] Regulation of kidney CA V has also been reported in mildly acidotic rats; CA V was doubled compared with control rats (reviewed in reference 12).

4.6. CA VI: The Isozyme Secreted in Saliva

CA VI isozyme is secreted into the saliva. The finding that a CA isozyme secreted into the saliva in humans is a unique isozyme was reported in 1987. CA VI concentration was reported to be much lower in CA-II deficient patients, suggesting some link in the genetic regulation of these isozymes.[62] The sheer size of the CA VI isozyme from sheep parotid gland is unique to CA isozymes: molecular weight (MW) 540,000.[22,23] It is not yet known whether CA VI is located anywhere else in the body; recent evidence from Dr. Ross Fernley's group in Australia suggests that it is confined entirely to the salivary glands.[23] Properties of this isozyme are reviewed in this book.[22]

4.7. CA VII: The Salivary Gland Isozyme

CA VII is the latest isozyme, which was discovered in Dr. Richard Tashian's laboratory.[76] It appears to be specifically expressed in salivary glands.

5. Facilitated Diffusion

In the 1980 book on CO_2 and CA, there are several chapters devoted to the subject of facilitated diffusion.[1] This phrase describes a function of CA first observed in Dr. Forster's laboratory[51]; CO_2 diffuses faster through solutions if CA is present. It has been concluded that the reason for the accelerated diffusion is the concentration gradient, since there is a higher concentration of HCO_3^- than CO_2. The studies with skeletal muscle CA isozymes from Dr. Gros's laboratory have indicated that CA-catalyzed facilitated diffusion is possibly of great importance in the functioning of the skeletal muscle.[30]

6. Carbonic Anhydrase Inhibitors

6.1. Sulfonamide Carbonic Anhydrase Inhibitors

The sulfonamide CA inhibitors have been very important in determining the physiological function of CA isozymes in cells, organelles, and solutions. The kidney is an organ in which this is particularly true (reviewed in references 54, 65, and 86). Since 1940, when sulfanilamide was first shown to inhibit CA activity,[53] there has been a huge literature devoted to studies with these drugs. Many of these studies have come from the laboratory of Dr. Thomas Maren, and the researcher intending to use these drugs for the first time is strongly advised to obtain three of his reviews for constant reference.[55–57] A point that Dr. Maren makes is that if greater than 10^{-4} M of one of the highly specific drugs is needed for an *in vivo* physiological effect, then the effect may not be due to CA inhibition. The large concentrations of base needed to dissolve acetazolamide in millimolar quantities may cause physiological effects by themselves. A requirement for sulfonamide inhibition is that it must cross the various membranes in the organ to reach the CAs. The considerable amount of evidence that ions such as Na^+, Cl^-, and K^+ cross membranes through channels may suggest that large molecules such as acetazolamide (MW 222.2) and ethoxzolamide (MW 258.3) also require channels to cross membranes, which may well be every bit as tightly regulated as those for smaller ions. It is generally believed that the sulfonamides dissolve in the lipid bilayers to cross the membranes.[54]

Although it is fortuitous that very specific CA inhibitors exist, their limitations must be recognized. Plant CA isozymes[32] and mammalian CA III[27,29,30,39] are peculiarly insensitive to acetazolamide.

A major problem with physiological studies conducted with sulfonamide CA inhibitors is that of compartmentation. It is conceivable that even at the massive doses given to dogs and rats described by Dr. Maren, not all CA was inhibited. CA inhibitors can be bound by intracellular proteins such as albumin[53] or even perhaps glutathione S transferase.[13] It is most likely that the highest concentration of these small drugs (MW 200–350) is not retained by the body but is filtered through the glomerulus very quickly. In this laboratory, acetazolamide, 50 mg/kg body weight, was given daily intraperitoneally for 5 weeks to two rabbits. Liver mitochondria were isolated from the excised livers; the CA activity was identical to that of control rabbits. This limited study suggested that the liver CA V was never inhibited; certainly isolation of mitochondria would not wash away the acetazolamide, a drug with K_i for CA V of 10^{-8} M.[13]

In the 1950s and 1960s, there was an enormous literature describing the various uses of the sulfonamide CA inhibitors throughout the body. A perusal of *Index Medicus* reveals uses of acetazolamide in congestive heart failure, prevention of pancreatic fat necrosis, depression of gastric acid secretion, decrease in blood sugars, and amelioration of epileptic seizures (*Index Medicus*, 1955). As other drugs were developed which were much better for each use, the use of the sulfonamide CA inhibitors decreased; until the 1980s, they were used consistently only in the field of ophthalmology.

In 1939, a careful study of the CA activity of each part of the stomach was published by Dr. Horace Davenport.[8] Working in Cambridge, Dr. Davenport (who is an American) found that the CA activity varied depending on the area of the stomach and that this variation corresponded closely to the areas where acid secretion was the highest. In 1940, the anti-CA effect of the sulfonamides was discovered, which led to a physiological study in which animals administered sulfonamides were examined for decreased parietal acid secretion. It was found that acid secretion was unaffected by much higher concentrations of sulfonamides than were needed to inhibit the CA activity. Back in America, Dr. Davenport subsequently published a retraction of his so-called carbonic anhydrase theory of gastric acid secretion in 1942,[9] and there the matter rested for four decades. In the 1984 New York Academy of Sciences meeting "Biology and Chemistry of the Carbonic Anhydrases," Dr. Davenport presented a historical perspective of his work. Also present were several Romanian gastrointestinal physicians, who showed a film of documenting the cures of patients with stomach ulcers who were given acetazolamide orally. Why did sulfonamides inhibit acid secretion in the 1980s when they had not in the 1940s? There are serious problems with the Romanian studies; the principal one is their claim that four times the usual clinical dose did not result in acidosis.[67] However, perhaps this whole field should be reexamined. The location of CA isozymes in the entire gastrointestinal tract plus studies with the sulfonamide inhibitors are exhaustively reviewed in this book by Dr. Swenson.[75]

An extremely useful technique for quantitation of CA inhibitor concentration, modified by Dr. Maren from a previously published technique,[17,54] is detailed in Chapter 9.

6.2. Ionic Carbonic Anhydrase Inhibitors

It is well established that certain ions are extremely good inhibitors of CA isozymes and that measurement of CA activity in the presence of certain ions may be used to differentiate between CA isozymes. CA II activity of a dilute hemolysate solution is completely inhibited by 1 μM Cu^{2+}, which has no effect on CA I.[52] CA I is much more susceptible than CA II to inhibition by chloride and imidazole (reviewed in reference 42).

6.3. Naturally Occurring Plasma Carbonic Anhydrase Inhibitors

There have been reports published and then forgotten over the years about the existence in plasma of naturally occurring CA inhibitors (reviewed in reference 80). A recent paper has provided evidence for this in the plasma of dogs.[38]

Assuming that the inhibitor is not an artifact, e.g., because the dogs were given sulfonamide antibiotics in their food, this finding is of physiological importance. These substances would prevent loss of the Zn^{2+} prosthetic group through the kidneys. An unresolved question is the body's disposal of CA isozymes released from erythrocytes. Their molecular weight of 30,000 precludes filtration through the glomerulus and excretion into the urine. The possibility that CA isozymes are in some way metabolized by the lung after becoming stuck to the endothelial cells has been considered.[64]

6.4. Altitude Sickness and Carbonic Anhydrase Inhibitors

A major problem with mountain climbing and otherwise ascending quickly to altitude is mountain sickness; sudden onset of this illness at its severest can lead to convulsions and death within a matter of hours if one does not immediately descend. This illness can strike the first time one is at altitude or not at all for many years. The first non-Nepalese conqueror of Mount Everest, Sir Edmund Hilary, now suffers from mountain sickness. It appears to be a problem of water and electrolyte balance in the body, and so, almost inevitably, the sulfonamide CA inhibitor acetazolamide was given to mountain climbers to determine whether this compound alleviates onset of the symptoms. Acetazolamide does alleviate the symptoms of acute mountain sickness.[3,5,74]

7. The Carbonic Anhydrases: A Family of Essential Enzymes?

In 1983, the groups of Dr. William Sly in St. Louis and Dr. Richard Tashian in Ann Arbor together reported that three sisters severely afflicted with osteopetrosis with renal tubular acidosis were also entirely lacking in erythrocyte CA II[71,72]; it was concluded that the lack of CA II in the bone osteoclasts led to the inability of the osteoclasts to resorb bone and that the acidosis was due to lack of CA II in the kidneys. Dr. Sly's brilliant detective work in discovering the genetic reason for this extremely debilitating disease arose from his reading an entry in a popular pharmacology textbook that sulfonamide CA inhibitor administration resulted in thickening of the bones.

The finding that a severe disease arose when a CA isozyme was lacking greatly increased the morale of the CA physiologists, especially this one. Those who had believed *a priori* that CA is an essential enzyme now argued that even though only one isozyme was missing (although it is acknowledged that CA II is the most active and ubiquitous isozyme), this disease interfered greatly with the ability to enjoy a normal life. There also are scientists who believe that CA is not an essential enzyme (see Section 6.1) but study it anyway; they have argued that the fact that the CA II-deficient patients can exist at all is proof.[57]

There is not available at this time an animal model completely lacking all CA isozymes. Perhaps a total lack would be fatal, and thus the fetus would be aborted. The CA II-deficient mouse is runted and acidotic but does not have osteopetrosis.[47] Humans totally lacking erythrocyte CA I are without clinical abnormalities; it is not known whether these humans also lack CA I elsewhere in their bodies.[76]

ACKNOWLEDGMENTS. The work from this laboratory over the past decade has been entirely supported by grants from the National Institutes of Health; currently these are HL-19737, Program Project with Director R. E. Forster II, and DK-38041 to S. J. Dodgson. Over this time, L. Lin maintained the mass spectrometer; D. Rivers occasionally assisted with experiments; D. A. Schwed, S. Fox, J. M. Kamerling, and C. Miller each worked for one or two summers; J. A. Krawiec, M. Connelly, L. C. Contino, and K. Cherian provided technical assistance for between 1 and 2 years. Special thanks to Angus Z. Dodgson Pekala, Miles C. Dodgson Pekala, and Allister M. Dodgson Blossfeld.

References

1. Bauer, C., Gros, G., and Bartels, H. (eds.), 1980, *Biophysics and Physiology of CO2*, Springer-Verlag, New York.
2. Bidani, A., and Crandall, E., 1988, *Annu. Rev. Physiol.* **50**:639–652.
3. Bradwell, A. R., Dykes, P. W., and Coote, J. H. 1987, *Sports Med.* **4**:157–163.
4. Bruns, W., and Gros, G. this volume, Chapter 9.
5. Cain, S. M., and Dunn, J. E., II, 1966, *J. Appl. Physiol.* **21**:1195–1200.
6. Cammer, W., this volume, Chapter 28.
7. Curtis, P. 1983, *J. Biol. Chem.* **258**:4459–4463.
8. Davenport, H. W., 1939, *J Physiol.* **97**:3–43.
9. Davenport, H. W., 1946, *Gastroenterology* **7**:374–375.
10. Davenport, H. 1946, *Physiol. Rev.* **26**:560–573.
11. Dodgson, S. J., this volume, Chapter 25.
12. Dodgson, S. J., this volume, Chapter 36.
13. Dodgson, S. J., 1987, *J. Appl. Physiol.* **63**:2134–2141.
14. Dodgson, S. J., Forster, R. E., II, Sly, W. S., and Tashian, R. E., 1988, *J. Appl. Physiol.* **65**:1472–1480.
15. Dodgson, S. J., and Watford, M., 1990, *Arch. Biochem. Biophys.* **277**:410–414.
16. Dubrose, T., and Bidani, A. 1988, *Annu. Rev. Physiol.* **50**:653–667.
17. Easson, L. H., and Stedman, E., 1936–7, *Proc. R. Soc. London Ser. B.* **121**:142–164.
18. Edsall, J. T., 1969, in: *CO_2: Chemical, Biochemical, and Physiological Aspects* (E. E. Forster, II, J. T. Edsall, A. B. Otis, and F. J. W. Roughton, eds.) NASA SP-188, pp. 15–34.
19. Edwards, Y. H., Charlton, J., and Brownson, S., 1988, *Gene* **7**:473–481.
20. Edwards, Y. H., Barlow, J. H., Konialis, C. P., Povey, S., and Butterworth, P. H., 1986, *Ann. Hum. Genet.* **50**:123–129.
21. Feitl, M., and Krupin, T. this volume, Chapter 13.
22. Fernley, R. T., this volume, Chapter 34.
23. Fernley, R. T., Coghlan, J. P., and Wright, R. D., 1988, *Biochem. J.* **249**:201–207.

24. Forsman, C., Behravan, G., Jonsson, B. H., Liang, Z. W., Lindskog, S., Ren, X. L., Sandstrom, J., and Wallgren, K., 1988, *FEBS Lett.* **229**;360–362.
25. Forster, R. E., II, this volume, Chapter 6.
26. Forster, R. E., II, Edsall, J. T., Otis, A. B., and Roughton, F. J. W. (eds.), 1969, *CO_2: Chemical, Biochemical, and Physiological Aspects* NASA SP-188.
27. Frémont, P., Charest, P. M., Côte, C., and Rogers, P. A., this volume, Chapter 20.
28. Garg, L. C., 1974, *J. Pharmacol. Exp. Ther.* **189**:557–562.
29. Garg, L. C., 1974, *Biochem. Pharmacol.* **23**:3153–3161.
30. Geers, C., and Gros, G., this volume, Chapter 19.
31. Gibson, Q. H., 1973, *Biogr. Mem. Fellows R. Soc.* **19**:1899–1972.
32. Graham, D., Reed, M. L. Patterson, B. D., Hockley, D. G., and M. R. Dwyer, 1984, *Ann. N. Y. Acad. Sci.* **429**:222–237.
33. Gros, G., and S. J. Dodgson. 1988. *Annu. Rev. Physiol.* **50**:669–694.
34. Gros, G., Forster, R. E., II, and Dodgson, S. J., 1988, (D. Häussinger, ed.), Academic Press, London, pp. 203–231.
35. Harned, H. S., and Bonner, F. T., 1945, *J. Am. Chem. Soc.* **67**:1026–1031.
36. Henry, R. P., this volume, Chapter 8.
37. Hewett-Emmett, D., and R. E. Tashian, this volume, Chapter 2.
38. Hill, E. P., 1986, *J. Appl. Physiol.* **60**:191–197.
39. Holmes, R. S., 1976, *J. Exp. Zool.* **197**:289–295.
40. Jeffery, S. this volume, Chapter 12.
41. Keilin, D., and Mann, T., 1940, *Biochem. J.* **34**:1163–1176.
42. Khalifah, R. G., and Silverman, D. N., this volume, Chapter 4.
43. Krebs, H. A., 1948, *Biochem. J.* **43**:525–528.
44. Krebs, H. A., and Henseleit, K., 1932, *Hoppe-Seyler's Z. Physiol. Chem.* **210**:33–66.
45. Krebs, H. A., and Roughton, F. J. W., 1948, *Biochem. J.* **43**:550–595.
46. Krebs, H. A., and Speakman, J. C. 1946. *Br. Med. J.* **1**:47–50.
47. Lewis, S. E., Erickson, R. P., Barnett, L. B., Venta, P. J., and Tashian, R. E., 1988, *Proc. Natl. Acad. Sci. USA* **85**:1361–1366.
48. Lindskog, S., Henderson, L. E., Kannan, K. K., Lijas, A., Nyman, P. O., and Strandberg, B., 1971, in: *The Enzymes*, Volume 5 (P. D. Boyer, ed.), Academic Press, New York, pp. 587–665.
49. Lindskog, S., Engberg, P., Forsman, C., Ibrahim, S. A., Jonsson, B.-H., Simonsson, I., and Tibell, L., 1984, *Ann. N. Y. Acad. Sci.* **429**:61–75.
50. Lippa, E. this volume, Chapter 14.
51. Longmuir, I. S., Forster, R. E., II, and Woo, C. Y., 1966, *Nature* (London) **209**:393–394.
52. Magid, E., 1967, *Scand. J. Haematol.* **4**:257–270.
53. Mann, T., and Keilin, D., 1940, *Nature* (London) **146**:164–165.
54. Maren, T. H., 1967, *Physiol. Rev.* **47**:595–781.
55. Maren, T. H., 1977, *Am. J. Physiol.* **232**:F291–F297.
56. Maren, T. H., 1984, *Ann. N.Y. Acad. Sci.* **429**:568–579.
57. Maren, T. H., 1985, *N. Engl. J. Med.* **313**:179–181.
58. Maren, T. H., 1988, *Annu. Rev. Physiol.* **50**:695–717.
59. Maren, T. H., and Couto, E. O., 1979, *Arch. Biochem. Biophys.* **196**:501–510.
60. Maren, T. H., Rayburn, C. S., and Liddell, N. E., 1976, *Science* **191**:469–472.
61. Meldrum, N. U., and Roughton, F. J. W., 1983, *J. Physiol.* **(London) 80**:113–142.
62. Murakami, H., and Sly, W. S., 1987, *J. Biol. Chem.* **262**:1382–1388.
63. Neubauer, J., this volume, Chapter 27.
64. Nioka, S., and Forster, R. E., II, this volume, Chapter 29.
65. Philpot, F. J., and J. St. L. Philpot, 1936–7, *Biochem. J.* **30**:2191–2193.
66. Preisig, P. A., Toto, R. D., and Alpern, R. J., 1987, *Renal Physiol. Basel* **10**:136–159.
67. Puscas, I., 1984, *Ann. N.Y. Acad. Sci.* **429**:587–591.

68. Rogers, J. H., and Hunt, S. P., 1987, *Neuroscience* **23**:343–361.
69. Roughton, F. J. W., 1935, *Physiol. Rev.* **15**:241–206.
70. Silverton, S. F., this volume, Chapter 33.
71. Sly, W. S., this volume, Chapter 15.
72. Sly, W. S., Hewett-Emmett, D., Whyte, M. P., Yu, Y.-S. L., and Tashian, R. E., 1983, *Proc. Natl. Acad. Sci. USA* **80**2752–2756.
73. Stadie, W. C., and O'Brien, H., 1933, *J. Biochem.* **103**:521–529.
74. Sutton, J. R., Houston, C. S., Mansell, A. L., McFadden, M. D., Hackett, P. M., Rigg, J. R. A., and Powles, A. C. P., 1979, N. Engl. J. Med. **301**:1329–1331.
75. Swenson, E. R., this volume, Chapter 23.
76. Tashian, R. E., 1989, *BioEssays* **10**:186–192.
77. Tashian, R. E., and Carter, N. D., 1976, *Adv. Hum. Genet.* **7**:1–56.
78. Tashian, R. E., and Hewett-Emmett, D. (eds.), 1984, *Ann. N.Y. Acad. Sci.* **429**.
79. Tu, C., Silverman, D. N., Forsman, C., Jonsson, D. H., and Lindskog, S., 1989, *Biochemistry*.
80. van Goor, H., 1948, *Enzymology* **13**:73–164.
81. Väänänen, H. K., and Parvinen, E.-K., this volume, Chapter 32.
82. Väänänen, H. K., Takala, T. E., Tolonen, V., Vuori, J., and Myllyla, V. V., 1988, *Arch. Neurol.* **45**:1254–1256.
83. Venta, P. J., Montgomery, J. C., Hewett-Emmett, D., Wiebauer, K., and Tashian, R. E., 1985, *J. Biol. Chem.* **260**:12130–12135.
84. Wade, R., Gunning, P., Eddy, R., Shows, T., and Kedes, L., 1986, *Proc. Nat. Sci. USA* **83**:9571–9575.
85. Whitney, P. L., and Briggie, T. V., 1982, *J. Biol. Chem.* **257**:12056–12059.
86. Wistrand, P., 1984, *Ann. N.Y. Acad. Sci.* **429**:609–619.
87. Wistrand, P., and Knuuttila, K.-G., 1989, *Kidney Int.* **35**:851–859.

CHAPTER 2

Structure and Evolutionary Origins of the Carbonic Anhydrase Multigene Family

DAVID HEWETT-EMMETT and RICHARD E. TASHIAN

1. Introduction

The purpose of this chapter is to provide a concise and useful resource detailing our current knowledge of the organization, structure, and evolution of the carbonic anhydrase (CA) multigene family. Much has been accomplished since the last such synthesis,[36] and aspects of these advances have been reviewed elsewhere in the interim.[23,66,76] In 1984, there was firm structural evidence, mostly at the protein sequence level, for three distinct CA genes (*CA1*, *CA2*, and *CA3*) encoding the CA I, CA II, and CA III isozymes. However, compelling evidence existed that membrane-bound CA from kidney brush border and lung membranes[47,81,83] was a distinct gene product as determined from its amino acid composition and other properties. Arguments could also be advanced that mitochondrial CA[15,16,77] and the high-molecular-weight CA from ovine parotid gland[24] and rat saliva[22] were distinct isozymes. However, alternative splicing of mRNA or use of different promoters associated with CA I, CA II, or CA III could not be ruled out as the basis for these additional isozymes. Earlier reviews[5,56,65] had raised the possibility of further tissue-specific isozymes.

DAVID HEWETT-EMMETT • Center for Demographics and Population Genetics, Graduate School of Biomedical Sciences, The University of Texas Health Science Center at Houston, Houston, Texas 77225. ***RICHARD E. TASHIAN*** • Department of Human Genetics, University of Michigan Medical School, Ann Arbor, Michigan 48109-0618.

2. The CA Gene Family

In the last 5 years, extensive characterization of the CA I, CA II, and CA III genes has taken place (Table I). Further characterization, including immunospecific detection and protein or cDNA sequence determination, has led to the conclusion that membrane-bound CA IV,[84,86] liver mitochondrial CA V,[37] and salivary (parotid) CA VI[23,25,51] are encoded by distinct genes. In addition, another apparently distinct functional gene, CA VII (formerly CA 'Z'[76]), was completely sequenced following isolation by cross-hybridization during screening of a human genomic DNA library.[48,73,76] We have also been made aware of a partial sequence (CA 'Y') from a mouse liver cDNA library (M. Amor and T. Meo, personal communication) which shows most similarity with the short stretch of protein

TABLE I
CA and Reportedly Homologous Genes That Have Been Cloned and Characterized

Gene	Species	Cloned sequence	References
CA1	Human	cDNA (reticulocyte)	1
		Genomic	4, 88
	Mouse[b] (BALB/c)	cDNA (anemic spleen)	28
		Genomic[a]	27
	Rabbit	cDNA[a] (reticulocyte)	41
	Pigtail macaque	Genomic[a]	P. H. Nicewander (unpublished)
CA2	Human	cDNA (liver)	50
		cDNA (kidney)	52
		Genomic[a]	60, 74; P. J. Venta (unpublished)
	Mouse[b] (BALB/c)	cDNA (anemic spleen)	9, 10
	(YBR)	Genomic	75
	(DBA/2J)	Genomic[a]	S. R. Diehl and P. J. Venta (unpublished)
	Chicken	cDNA (reticulocyte)	85
		cDNA (retina)	57
		Genomic	85
CA3	Human	cDNA (skeletal muscle)	45, 78
		Genomic	46
	Mouse[b]	cDNA (skeletal muscle)	71
	Rat	cDNA (skeletal muscle)	39
CA6	Human	cDNA[a] (parotid gland)	23; R. T. Fernley (unpublished)
CA7	Human	Genomic	48
CA'Y'	Mouse[b]	cDNA (liver)	87
D8	Vaccinia virus	Genomic	54, 55
v-*erb*A	Avian virus	Genomic	12, 33
c-*erb*A	Human	cDNA (placental)	79
	Chicken	cDNA (whole embryo)	59
Ubiquitin	Human	Genomic	82

[a]Sequences that are not yet complete.
[b]Formal designations for murine genes are: *Car-1*, *Car-2*, *Car-3*, and *Car-Y*.

sequence from guinea pig liver mitochondrial CA V.[37] While CA 'Y' might represent an additional (eighth) gene, it is now clear that there are at least seven CA genes in mammals.

Another advance of great interest has been the characterization of erythrocyte CA from the tiger shark,[3] which represents a lineage that diverged long enough ago to shed light on the gene duplication events leading to the evolution of the CA I, CA II, and CA III genes found in mammals, birds and reptiles (amniotes).

Besides the seven or eight genes that apparently encode functional CA isozymes, it has been claimed that vaccinia virus gene D8, which lacks introns and whose product is a transmembrane protein,[54,55] the avian viral oncogene v-*erb*A,[12] and the ubiquitin gene[14,82] show sequence homology with the CA gene family. In all of these cases, however, the gene product cannot possibly function as a CA because of major changes in or absence of the residues that participate in the active site.[37,66,76] While the evidence for shared ancestry is very convincing in the case of vaccinia virus D8, it is at best extremely doubtful for v-*erb*A and ubiquitin. It is now known that human and chicken c-*erb*A, the cellular homologs of v-*erb*A, are thyroid hormone receptors and are part of a multigene receptor family that includes the estrogen and glucocorticoid receptors.[30,59,79] No significant homology was noted between human estrogen receptor and the CAs, however,[31] casting further doubt on the relationship of the receptor gene superfamily to the CAs.

3. The CA Isozymes: Structure and Evolution

The evolution of the CA gene family was examined by using both the available protein sequence data and sequences inferred from cDNA and genomic DNA sequences. Two phylogenetic trees were constructed from these data, using the neighbor-joining method.[58]

The first tree (Fig. 1) was based on a set of 21 complete sequences from CA I, CA II, CA III, CA VI, and CA VII that span a 270-residue alignment with only very small insertions or deletions, as shown in Table II. The unpublished human CA VI sequence (R. T. Fernley, personal communication) was used but is not shown. The rabbit CA I, turtle CA I, and chicken CA III sequences are shown in Table II for information purposes but were not used to construct the tree, since they are not complete. The vaccinia virus D8 gene product lacks regions present in all other CAs and was also omitted from this tree. The main feature of the tree is the confirmation for the first time that CA I, CA II, and CA III result from gene duplications extremely close together in time. This finding is possible because of the presence in the tree of two more distantly related genes, CA VI and CA VII. Other features of the tree are the relatively close relationship of CA VII to CA I, CA II, and CA III and the conserved nature of this isozyme subsequent to its divergence. This is further confirmation that CA VII is undoubtedly a functionally important gene probably encoding a cytosolic isozyme, although Montgomery[48]

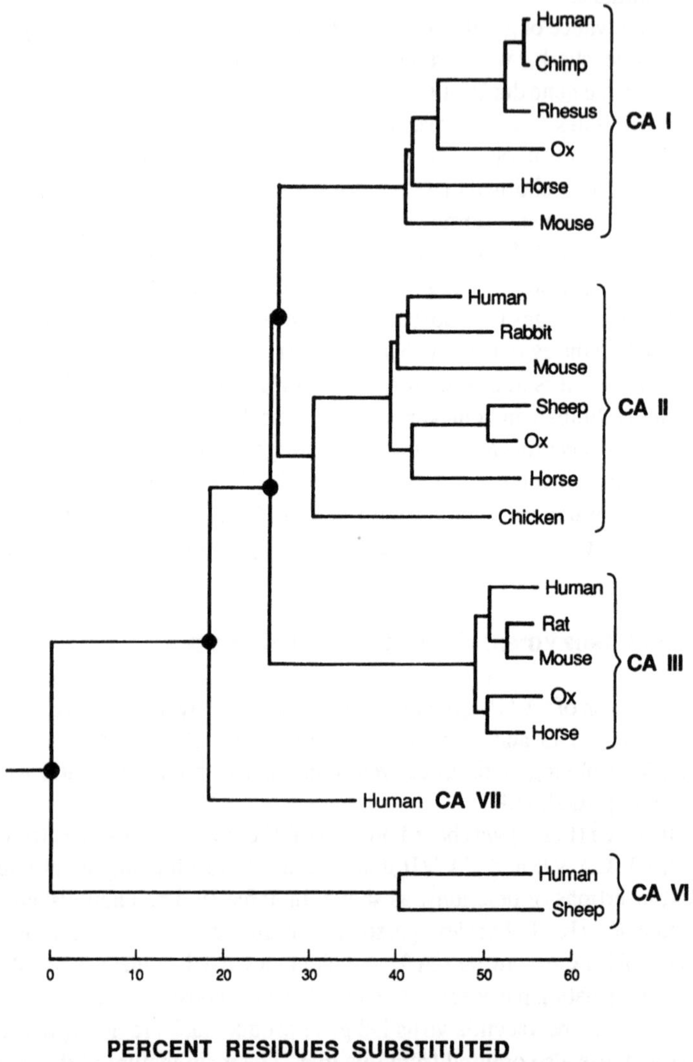

Figure 1. Phylogenetic tree of the CA isozymes. The tree was constructed by the neighbor-joining method[58] from a 270-residue alignment of 21 complete protein sequences, all of which are shown in Table II except for the unpublished human CA VI sequence inferred from a cDNA (R. T. Fernley, personal communication). Branch lengths reflect Poisson-corrected (for multiple hits) percent of 270 residues substituted. ●, Gene duplication event; other bifurcations are speciation events.

found evidence of mRNA only in primate salivary gland. The third finding is the distant relationship of CA VI to the cytosolic CAs. In the mammals, CA VI is evolving more rapidly (15.5%) than CA I (13.7%), which in turn is evolving more rapidly than CA II (12.4%) and considerably more rapidly than CA III (6.8%). These estimates, based on weighted averages of the Poisson-corrected percent divergences in the mammalian lineages in Fig. 1, should be regarded as approximate, since species representation and number are not identical for each isozyme. It is noteworthy that the estimates are in general agreement with those found earlier by using fewer sequence data and a different tree-building algorithm.[36] Then, it was found that whereas the active-site residues of CA II were evolving much less rapidly than CA I or CA III, the remaining regions of CA II were less conserved than CA I and considerably less than CA III. It was suggested that both CA I and CA III may interact with other molecules, resulting in conservation of external residues.

A second tree (Fig. 2) was constructed from a 174-residue subalignment (shown in Table II) that included 10 of the 21 sequences used to construct the first tree and an additional four sequences: tiger shark CA,[3] mouse CA 'Y'[87] (M. Amor and T. Meo, personal communication), and human CA IV[86] (W. S. Sly, personal communication), which are not shown in Table II, and vaccinia virus D8.[54] The reduction in size of the alignment (from 270 to 174 residues) results from removing regions that are absent or uncharacterized in CA IV, CA 'Y', tiger shark CA, and the vaccinia virus D8 gene product. This tree affords us an opportunity to examine a number of other interesting questions.

First, tiger shark CA clearly branches off prior to the gene duplications that resulted in CA I, CA II, and CA III. This firmly places the latter events between the divergence of the elasmobranchs [~ 450 million years ago (mya)] and the divergence of the amniotes (~300 mya).

Second, the duplications leading to CA VII and CA 'Y' (which may in fact be CA V) occurred very close together in time and very shortly before the divergence of the elasmobranch lineage. These events are likely to be coincident with the divergence of vertebrates (~500 mya), a time when many mutigene families developed, including the blood coagulation portion of the serine proteases.

Third, the gene duplication events resulting in membrane-bound CA IV and extracellular CA VI occurred much earlier, and it is possible that representatives of these lineages exist in plants for which there is evidence of multiple isozymes.[29]

4. The CA Gene Sequences: Organization and Evolution

All four CA genes so far characterized have the same general exon–intron structure; six introns interrupt genomic sequences encoding the CA I, CA II, CA III, and CA VII isozymes at homologous sites (Fig. 3). Originally, it was thought that the donor and acceptor sites of mouse CA II intron 4 were shifted 14 nucleotides 5′ relative to chicken CA II and the other CA genes. It now appears,

TABLE II

Alignment of CA and CA-Related Protein Sequences[a]

Species:Protein	Sequence
	10 20 30 40 50 60
	* * * * * * * * * * * * *
HUMAN:CAI	-ASPDWGYDD-KNGPEQWSKLYPIANGNNQSPVDIKTSETKHDTSLKPISVSYNPATAKE--IINVGHS
CHIMP:CAI	-ASPDWGYDD-KNGPEQWSKLYPIANGNNQSPVDIKTSETKHDTSLKPISVSYNPATAKE--IINVGHS
RHESUS:CAI	-ASPDWGYDD-KNGPEQWSKLYPIANGNNQSPVDIKTSEAKHDTSLKPISVSYNPATAKE--IINVGHS
MOUSE:CAI	-ASADWGYGS-ENGPDQWSKLYPIANGNNQSPIDIKTSEANHDSSLKPLSISYNPATAKE--IVNVGHS
OX:CAI	-ASPDWGYDG-ENGPEHWGKLYPIANGNNQSPIDIKTSETKYDPSLKPRSVSYNPATAKE--IVNVGHS
HORSE:CAI	-AHSDWGYDS-PBGPZEWVKLYPIABGBBQSPIDIKTSETKHDTSLKPFSVSYDPATAKE--IVNVGHS
RABBIT:CAI	- - GNKQSPVDIKSSEVKHDTSLKPFSVSYNPASAKE--IINVGHS
TURTLE:CAI	- RYQG-NNGPDQWHKLYPIADGNYQSPIDIK DVKKDPALGHLHISYN STSKE--IVNVGHS
HUMAN:CAII	--SHHWGYGK-HNGPEHWHKDFPIAKGERQSPVDIDTHTAKYDPSLKPLSVSYDQATSLR--ILNNGHA
MOUSE:CAII	--SHHWGYSK-HNGPENWHKDFPIANGDRQSPVDIDTATAHHDPALQPLLISYDKAASKS--IVNNGHS
SHEEP:CAII	--SHHWGYGE-HNGPEHWHKDFPIADGERQSPVDIDTKAVVPDPALKPLALLYEQAASRR--MVNNGHS
OX:CAII	--SHHWGYGK-HBGPZHWHKDFPIANGERQSPVNIDTKAVVQDPALKPLALVYGEATSRR--MVNNGHS
HORSE:CAII	--SHHWGYGE-HDGPKHWHKDFPIAKGERQSPVDIDTKAAVHDAALKPLAVHYQEATSRR--IVNNGHS
RABBIT:CAII	--SHHWGYGK-HNGPEHWHKDFPIADGERQSPIDIDTDAAKHDPSLKPLRVSYEHPISRR--IINNGHS
CHICK:CAII	--SHHWGYDS-HNGPAHWYEHFPIANGERQSPIAISTKAARYDPALKPLSFSYDSGTAKA--IVNNGHS
HUM:CAIII	--AKEWGYAS-HNGPDHWHELFPNAKGENQSPIELHTKDIRHDPSLQPWSVSYDGGSAKT--ILNNGKT
OX:CAIII	--AKEWGYAD-HNGPDHWHELFPNAKGENQSPIELNTKEINHDPSLKPWTASYDPGSAKT--ILNNGKT
HORSE:CAIII	--AKEWGYAD-HNGPDHWHEFYPIAKGDNQSPIELHTKDINHDPSLKAWTASYDPGSAKT--ILNNGRT
RAT:CAIII	--AKEWGYAS-HNGPEHWHELYPIAKGDNQSPIELHTKDIRHDPSLQPWSVSYDPGSAKT--ILNNGKT
MOUSE:CAIII	--AKEWGYAR-HNGPDHWHELYPIAKGDNQSPIELHTKDIKHDPSLQPWSASYDPGSAKT--ILNNGKT
CHICK:CAIII	-- - MAKGDKQSPIEINSKDV HDTEL P T--ILNNGR
TIGSHARK:CA	EGTRQSPI NQEAKFDSNLKPLDL YDPA
HUMAN:CAVII	-GHHGWGYGQ-DDGPSHWHKLYPIAQGDRQSPINIISSQAVYSPSLQPLELSYEACMSLS--ITNNGHS
SHEEP:CAVI	GHGVEWTYSEGMLDEAHWPLEYPKCGGRRQSPIDLQMKKVQYNPSLRALNLTGYGLWHGEFPVTNNGHT
VACCINIA:D8	--------------------------POOLSPINIETKKAISNARLKPLDIHYNESKPTT--IQNTGKL
	70 80 90 100 110 120 130
	* * * * * * * * * * * * *
HUMAN:CAI	FHVNFEDNDNRSVLKGGPFSDSYRLFQFHFHWG--STNEHGSEHTVDGVKYSAELHVAHWNSAKYSS
CHIMP:CAI	FHVNFEDNDNRSVLKGGPFSDSYRLFQFHFHWG--STNEHGSEHTVDGVKYSAELHIAHWNSAKYSN
RHESUS:CAI	FHVNFEDNDNRSVLKGGPFSDSYRLFQFHFHWG--SSNEYGSEHTVDGVKYSSELHIVHWNSAKYSS
MOUSE:CAI	FHVIFDDSSNQSVLKGGPLADSYRLTQFHFHWG--NSNDHGSEHTVDGTRYSGELHLVHWNSAKYSS
OX:CAI	FHVNFEDSDNRSVLKGGPLSESYRLFQFHFHWG--ITDEDGSEHLVDGAKFSAELHLVHWNSAKYPS
HORSE:CAI	FQVKFEDSDNRSVLKDGPLPGSYRLVQFHFHWG--STDDYGSEHTVDGVKYSAELHLVHWNSSKYSS
RABBIT:CAI	FHVNFED-DSQSVLKGGPLSDNYRLSQFHFHWG--KTDDYGSEHTVDGAKFSAELHLVHWNSGKYPN
TURTLE:CAI	FHVNFEDIDNRSVVTGGPLTGNYRLHQFHFHWG--QADDHGSEHTVDGAKYASELHLVHWNTLKY S
HUMAN:CAII	FNVEFDDSQDKAVLKGGPLDGTYRLIQFHFHWG--SLDGQGSEHTVDKKKYAAELHLVHWNT-KYGD
MOUSE:CAII	FNVEFDDSQDNAVLKGGPLSDSYRLIQFHFHWG--SSDGQGSEHTVNKKKYAAELHLVHWNT-KYGD
SHEEP:CAII	FNVEFDDSQDKAVLKDGPLTGTYRLVQFHFHWG--SSDDQGSEHTVDRKKYAAELHLVHWNT-KYGD
OX:CAII	FNVEYDDSQDKAVLKDGPLTGTYRLVQFHFHWG--SSBBQGSEHTVDRKKYAAELHLVHWNT-KYGD
HORSE:CAII	FNVEFDDSEDKAVLEGGPLTGTYRLIQFHFHWG--SSNGQGSEHTVDKKKYAAELHLVHWNT-KYGD
RABBIT:CAII	FNVEFDDSHDKSVLKEGPLEGTYRLIQFHFHWG--SSDGEGSEHTVNKKKYAAELHLVHWNT-KYGD
CHICK:CAII	FNVEFDDSSDKSVLQGGALDGVYRLVQFHIHWG--SCEGQGSEHTVDGVKYDAELHIVHRNV-KYGK
HUM:CAIII	CRVVFDDTYDRSMLRGGPLPGPYRLRQFHLHWG--SSDDHGSEHTVDGVKYAAELHLVHWNP-KYNT
OX:CAIII	CRVVFDDTYDRSMLRGGPLAAPYRLRQFHLHWG--SSDDHGSEHSVDGVKYAAELHLVHWNS-KYNS
HORSE:CAIII	CRVVFDDTYDRSMLRGGPLTAPYRLRQFHLHWG--SSDDHGSEHTVDGVKYA ELHLVHWNP-KYNT
RAT:CAIII	CRVVFDDTFDRSMLRGGPLSGPYRLRQFHLHWG--SSDDHGSEHTVDGVKYAAELHLVHWNP-KYNT
MOUSE:CAIII	CRVVFDDTYDRSMLRGGPLSRPYRLRQFHLHWG--SSDDHGSEHTVDGVKYAAELHLVHWNP-RYNT
CHICK:CAIII	VVFDDTFD -- -
TIGSHARK:CA	YVLTGGPLCG RQFHFHWG--ASDTHGSEHILDGKTYAAELHLVH E-KYSD
HUMAN:CAVII	VQVDFNDSDDRTVVTGGPLEGPYRLKQFHFHWG--KKHDVGSEHTVDGKSFPSELHLVHWNAKKYST
SHEEP:CAVI	VQISLPSTMSMTTS-DG---TQYLAKQMHFHWGGASSEISGSEHTVDGMRYVIEIHVVHYNS-KYNS
VACCINIA:D8	VRINFKGGY----ISGGFLPNEYVLSSLHIYWG--KEDDYGSNHLIDVYKYSGEINLVHWNKKKYSS

TABLE II *(Continued)*

Species:Protein	Sequence
	140 150 160 170 180 190
	* * * * * * * * * * * * *
HUMAN:CAI	LAEAASKADGLAVIGVLMKVGEANPK--LQKVLDALQAIKTKGKRAPFTN-FDPSTLLPSSLDFWTYP
CHIMP:CAI	LAEAASKADGLAVIGVLMKVGEANPK--LQKVLDALQAIKTKGKRAPFTN-FDPSTLLPSSLDFWTYP
RHESUS:CAI	LAEAVSKADGLAVIGVLMKVGEANPK--LQKVLDALHAIKTKGKRAPFTN-FDPSTLLPSSLDFWTYS
MOUSE:CAI	ASEAISKADGLAILGVLMKVGPANPS--LQKVLDALNSVKTKGKRAPFTN-FDPSSLLPSSLDYWTYF
OX:CAI	FADAASKADGLALIGLLVKVGQANPN--LQKVLDALKAVKNKNKKAPFTN-FDPSTLLPSSLDFWTYS
HORSE:CAI	FDEASSQADGLAILGVLMKVGEANPK--GQKVLDALNEVKTKGKKAPFKN-FDPSSLLPSSPDYWTYS
RABBIT:CAI	IADSVSKADGLAIVAVFLKVGQANPK--LQKVLDALSAVKTKGKKASFTN-FDPSTLLPSSLDYWTYS
TURTLE:CAI	FAEASDKPDGL LKVGPPNEH--VQDIVKALGSIKTKGKKAPFTN-FDPSTLLPGILDYGTYP
HUMAN:CAII	FGKAVQQPDGLAVLGIFLKVGSAKPG--LQKVVDVLDSIKTKGKSADFTN-FDPRGLLPESLDYWTYP
MOUSE:CAII	FGKAVQQPDGLAVLGIFLKIGPASQG--LQKVLEALHSIKTKGKRAAFAN-FDPCSLLPGNLDYWTYP
SHEEP:CAII	FGTAAQQPDGLAVVGVFLKVGDANPA--LQKVLDVLDSIKTKGKSADFPN-FDPSSLLKRALBYWTYP
OX:CAII	FGTAAQQPDGLAVVGVFLKVGDANPA--LQKVLDALDSIKTKGKSTDFPN-FDPGSLLPQVLDYWTYP
HORSE:CAII	FGKAVQEPDGLAVVGVFLKVGGAKPG--LQKVLDVLDSIKTKGKSADFTN-FDPRGLLPESLDYWTYP
RABBIT:CAII	FGKAVKHPDGLAVLGIFLKIGSATPG--LQKVVDTLSSIKTKGKSVDFTN-FDPRGLLPESLDYWTYP
CHICK:CAII	FAEALKHPDGLAVVGIFMKVGNAKPE--IQKVVDALNSIQTKGKQASFPT-FDPTELLPPCRDYWTYP
HUM:CAIII	FKEALKQRDGIAVIGIFLKIGHENGE--FQIFLDALDKIKTKGKEAPFTK-FDPSCLFPACRDYWTYQ
OX:CAIII	YATALKQADGIAVVGVFLKIGREKGE--FQLLLDALDKIKTKGKEAPFNN-FNPSCLFPACRDYWTYH
HORSE:CAIII	YGGALKQPDGIAVVGVFLKIGREKGE--FQLFLDALDKIKTKGKEAPFTN-FDPSCLFPTCRDYWTYR
RAT:CAIII	SEEALKQPDGIAVVGIFLKIGREKGE--FQILLDALDKIKTKGKEADFNH-FDPSCLFPACRDYWTYH
MOUSE:CAIII	FGEALKQPDGIAVVGILLKIGREKGE--FQILLDALDKIKTKGKEAPFTH-FDPSCLFPACRDYWTYH
CHICK:CAIII	GVFLKIGR --MQRILLEIDNIKTKGKEAPFQ -FDPS LFP S DY
TIGSHARK:CA	FAEALKAPDGLAV GVML VCDSNPA--L KIVCSLESLKAKGD EFKD-FSPCSLLPSCLQYWTYL
HUMAN:CAVII	FGEAASAPDGLAVVGVFLETGDEHPS--MNRLTDALYMVRFKGTKAQFSC-FNPKCLLPASRHYWTYP
SHEEP:CAVI	YEEAQKEPDGLAVLAALVEVKDYTENAYYSKFISHLEDIRYAGQSTVLRGLDIEDMLPGDLRYYYSYL
VACCINIA:D8	YEEAKKHDDGLIIISIFLQVLDHKNV-YFQKIVNQLDSIRSANTSAPFDSVFYLDNLLPSKLDYFTYL

	200 210 220 230 240 250 260
	* * * * * * * * * * * * *
HUMAN:CAI	GSLTHPPLYESVTWIICKESISVSSEQLAQFRSLLSNVEGDNAVPMQHNNRPTQPLKGRTVRASF--
CHIMP:CAI	GSLTHPPLYESVTWIICKESISVSSEQLAQFRSLLSNVEGDNAVPMQHNNRPTQPLKGRTVRASF--
RHESUS:CAI	GSLTHPPLYESVTWIICKESISVSSEQLAQFRSLLSNVEGSNPVPIQRNNRPTQPLKGRTVRASF--
MOUSE:CAI	GSLTHPPLHESVTWVICKDSISLSPEQLAQLRGLLSSAEGESAVPVLSNHRPPQPLKGRTVRASF--
OX:CAI	GSLTHPPLLESVTWIIFKETISASSEQLAQFRCLLANAEGDGELLIKQNHRPPQPLKGRTVKASF--
HORSE:CAI	GSLTHPPLYESVTWIVCKENISISSQQLSQFRSLLSNVEGGKAVPIQHNNRPPQPLKGRTVRAFF--
RABBIT:CAI	GSLTHPPLHESVTWLICKDSISISSEQLAQFRSLLSNAEGEAAVPILHNNRPPQPLKGRTVKASF--
TURTLE:CAI	GSLTHPPLFESVV IIYKEPTTISSEQLAQFR KSLILTNHRLPQPLKGRQVRT --
HUMAN:CAII	GSLTTPPLLECVTWIVLKEPISVSSEQVLKFRKLNFNGEGEPEELMVDNWRPAQPLKNRQIKASFK-
MOUSE:CAII	GSLTTPPLLECVTWIVLREPITVSSEQMSHFRTLNFNEEGDAEEAMVDNWRPAQPLKNRKIKASFK-
SHEEP:CAII	GSLTNPPLLESVTWVVLKEPTSVSSQQMLKFRSLNFNAEGEPELLMLANWRPAQPLKNRQVRVFPK-
OX:CAII	GSLTTPPLLESVTWIVLKEPISVSSQQMLKFRTLNFNAEGEPELLMLANWRPAQPLKNRQVRGFPK-
HORSE:CAII	GSLTTPPLLECVTWIVLREPISVSSEQLLKFRSLNFNAEGKPEDPMVDNWRPAQPLNNRQIRASFK-
RABBIT:CAII	GSLTTPPLLECVTWIVLKEPITVSSEQMLKFRNLNFNKEAEPEEPMVDNWRPTQPLKGRQVKASFV-
CHICK:CAII	GSLTTPPLHECVIWHVLKEPITVSSEQMCKLRELCFSAENEPVCRMVDNWRPCQPLKSREVRASFQ-
HUM:CAIII	GSFTTPPCEECIVWLLLKEPMTVSSDQMAKLRSLLSSAENEPPVPLVSNWRPPQPINNRVVRASFK-
OX:CAIII	GSFTTPPCEECIVWLLLKEPITVSSDQIAKLRTLYSSAENEPPVPLVRNWRPPQPIKGRIVKASFK-
HORSE:CAIII	GSFTTPPCEECIVWLLLKEPITVSSDQVAKLRSLFSSAENEPPVPLVRNWRPPQPLKGRVVRASFK-
RAT:CAIII	GSFTTPPCEECIVWLLLKEPMTVSSDQMANVRSLFASAENEPPVPLVGNWRPPQPVKGRVVRASFK-
MOUSE:CAIII	GSFTTPPCEECIVWLLLKEPMTVSSDQMAKLRSLFSSAENEPPVPLVGNWRPPQPVKGRVVRASFK-
CHICK:CAIII	EPIEVSPDQMARL L FNGENE MAPLVDNW PLQPV G V ASF -
TIGSHARK:CA	GSLTTPPL ESVIWLVIKEPISLGPDQIAKFRG RGLEFGPNEPMQNNYRPPQPLKGREIRKNFE-
HUMAN:CAVII	GSLTTPPLSESVTWIVLREPICISERQMGKFRSLLFTSEDDERIHMVNNFRPPQPLKGRVVKASFRA
SHEEP:CAVI	GSLTTPPCTENVHWFVVADTVKLSKTQVEKLENSLLNHQNKTIQ---NDYRRTQPLNHRVVEANFMS(+44)
VACCINIA:D8	G--TTINHSADAVWIIFPTPINIHSDQLSKFRTLLSSS-NHDGKPHYINYRNPYKLNDTQVYYSGEI(+67)

[a]Sequences are in the one-letter code and are taken from Hewett-Emmett *et al.*,[36] except for the following: mouse CA I[28]; rabbit CA I[41]; mouse CA II (39-His allele)[10,75] (S. R. Diehl and P. J. Venta, unpublished); chicken CA II[85]; human CA III (31-Ile allele)[35,45,78]; ox CA III corrections[21]; horse CA III[80]; rat CA III[39]; mouse CA III[71]; tiger shark CA[3]; human CA VII[48]; sheep CA VI[25]; vaccinia virus D8.[54,55] Sheep CA VI and vaccinia virus D8 extend 44 and 67 residues, respectively, beyond the alignment shown. Numbering is based on the human CA I sequence. Restricted alignment of 174 residues used to construct tree in Fig. 2 is denoted by a line over the top of the human CA I sequence. Unsequenced regions are shown as blank; a dash indicates a homologous residue not present in a gene product.

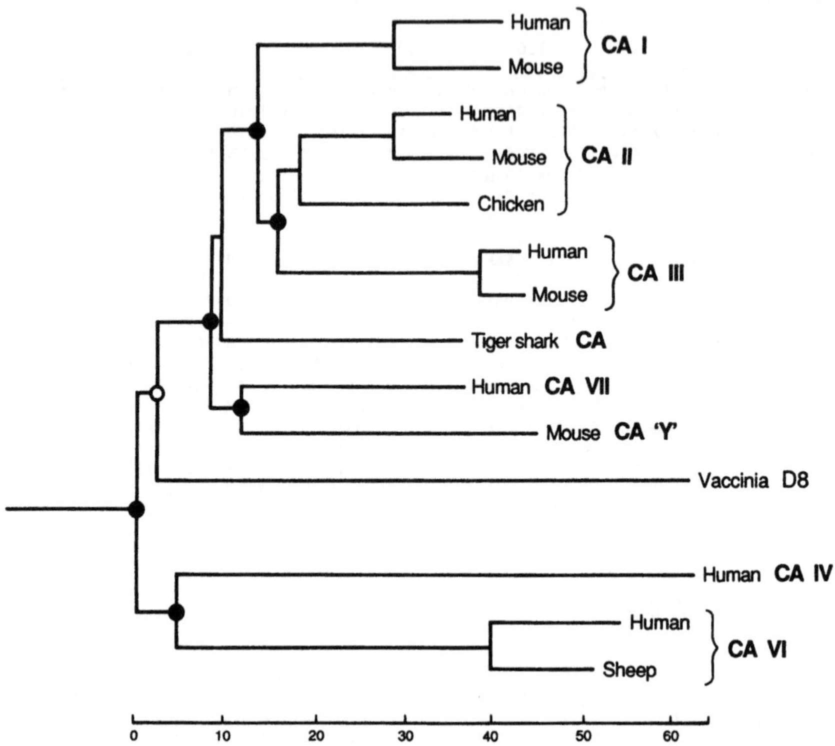

Figure 2. Phylogenetic tree of CA and CA-related gene products, using a 174-residue alignment. The restricted alignment is shown in Table II. Sequences are from Table II except for the following unpublished data: mouse CA 'Y,' inferred from cDNA[87] (M. Amor and T. Meo, personal communication); human CA IV partial protein sequence[86] (W. S. Sly, personal communication; human CA VI inferred from cDNA (R. T. Fernley, personal communication). Branch lengths reflect Poisson-corrected (for multiple hits) percent of 174 residues substituted. ●, Gene duplication event; ○, insertion of a CA gene into a vaccinia virus precursor or speciation event of the species from which the gene was derived; other bifurcations are speciation events.

however, that an error in the original mouse CA II cDNA sequence led to this conclusion. Resequencing of the original cDNA clone[10], (P. J. Venta, personal communication) and additional genomic sequencing of this region in mouse strain DBA/2J (S. R. Diehl and P. J. Venta, personal communication) has now demonstrated that the mouse CA II intron 4 donor follows condon 149 as in all other CA genes rather than interrupting codon 145.[75]

One unexpected finding has recently emerged from the work of Fraser *et al.*[27] and Brady *et al.*[4] They have shown that the promoter regions of mouse and human CA I are both extensive [spanning ~36 kb in the human gene (see Chapter 16)] and

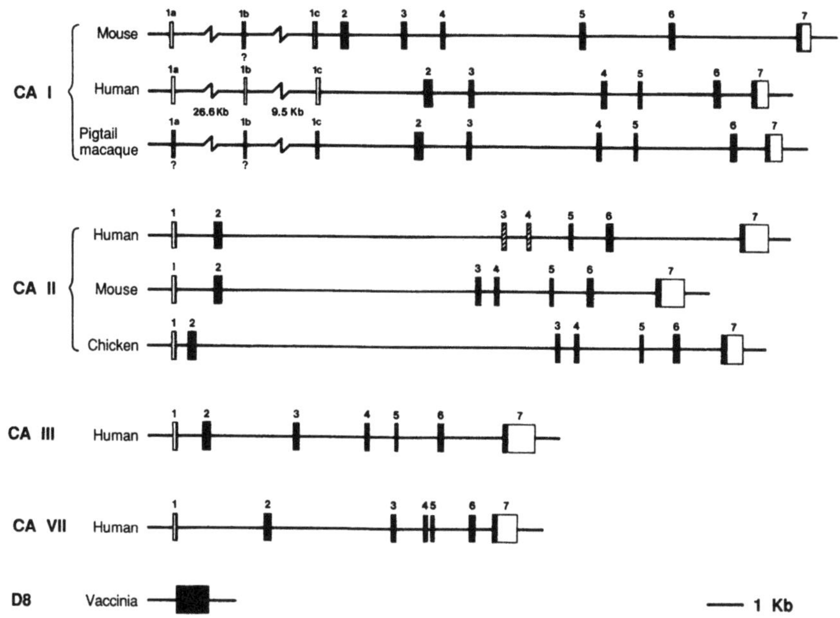

Figure 3. Organization of the CA isozyme and CA-related genes. Genes are drawn to scale. Vertical bars represent exons; solid bars are regions that are translated; open regions are the 5′ and 3′ untranslated regions; shaded bars represent exon regions that are not yet fully sequenced. The source of the exon–intron organization and sequence data is given in Table I. ? indicates absence of firm data about the sizes and precise locations of these exons. Note that vaccinia virus D8 lacks introns.

complex (Fig. 3). Exon 1 (henceforth 1c) is preceded by two additional small exons (1a and 1b) in the human gene, whereas in the mouse gene, Fraser *et al.*[27] have found evidence of two exons (1a and 1c) and two promoters. One promoter is upstream from exon 1a and is used in erythroid tissues; the second is upstream from exon 1c (i.e., at the 3′ end of the intron interrupting exons 1b and 1c) and is used in colon and perhaps other nonerythroid tissues. Thus, in mouse erythroid tissues the mature mRNA contains exons 1a, 1c, and 2–7. Whether the mouse gene has an exon corresponding to human 1b is not yet known. In their work on human erythroid tissue, Brady *et al.*[4] found two different mature CA I mRNAs; the predominant species contained exons 1a, 1c, and 2–7 (as in the mouse), while <5% contained 1a, 1b, 1c, and 2–7. An interesting observation is that the "optional" exon (1b) has the features of an intron, leading to the possibility that removal of 1b is the predominant primary event, although Brady *et al.*[4] discount this on the basis of the small size of the exon.

The tissue-specific use of two different promoters in the mouse may explain why erythrocyte CA I deficiencies in humans and pigtail macaques are not associated with clinical symptoms,[40,67] unlike deficiencies or erythrocyte CA II, which are reflected in deficiencies and attendant symptoms in other tissues.[61–63]

Interestingly, an induced mouse CA II deficiency mutant[43] does not precisely mimic the human counterparts, reminding us that animal models must account for physiological differences between species. This is further emphasized by the species differences in tissue expression of CA I,[65,66] CA III,[39,66,71] and CA V.[17,77]

It now becomes possible to speculate that human CA I deficiency (in which CA II levels are normal) results from a promoter or splicing defect typical of those found in the β^0 thalassemias.[8] By contrast, the deficiency of pigtail macaque CA I, which unlike the human deficiency is not complete and is always associated with a less complete deficiency of the *cis* CA II gene product, may be due to a lesion in an erythroid enhancer element that influences both linked CA genes. Of further interest is the complete absence of CA I in erythrocytes of all ruminants and felids.[65] It appears likely that separate mutations occurred early in these lineages and affected transcription at the exon la promoter. While turtle red cells express both CA I and CA II,[32,36,68] the red cells of birds, a separate amniote lineage, appear to lack CA I.[65,68] It is apparent that lack of erythrocyte CA I is not harmful and that numerous separate events have probably silenced use of the exon la promoter in amniote evolution.

There are now published DNA sequences corresponding to the coding regions of CA I, CA II, CA III, CA VII, and the CA-related D8 gene product in vaccinia virus (Table III). The CA sequences were used to construct a phylogenetic tree, using the neighbour-joining method.[58] Corrections for multiple hits are made by using the Jukes–Cantor approach as described elsewhere.[58] The resulting tree is shown in Fig. 4. The main feature of the tree is that, as in Fig. 2, CA II and CA III appear to have resulted from a gene duplication occurring slightly more recently than the gene duplication that led to the CA I lineage. The order of duplications is particularly sensitive to which sequences (protein or DNA) are included in the data set (D. Hewett-Emmett, unpublished results), and until more data are forthcoming, it seems most appropriate to regard the CA I, CA II, and CA III lineages as having resulted from a trifurcation.

5. Chromosomal Location and Linkage Relationships

The human CA genes are now known to be located on at least three chromosomes. *CA1*, *CA2*, and *CA3* have been mapped to chromosome 8,[18,19,72,78] with most probable regional assignment of all three genes to 8q22.[11,53] *CA6* is on the tip of the small arm of chromosome 1,[64] and *CA7* has been regionally assigned to chromosome 16q21–23.[48,49] *CA4* and *CA5* have yet to be mapped. These findings are consistent with the evolutionary tree (Fig. 1) inasmuch as CA I, CA II, and CA III result from the most recent of the gene duplications. The other four CA genes are derived from gene duplications at least 450 mya and almost all genes become dispersed to different chromosomes over such time periods. Indeed, the conservation of the CA I–CA II–CA III cluster over >300 million years, although not unique, is probably the exception rather than the rule. Other gene families of comparable antiquity display similar patterns; in the globin family, the α- and β-globin gene clusters have been dispersed in the mammals and birds but remain linked in the amphibians,[8,34] while myoglobin, which resulted from an earlier gene duplication, is nonsyntenic with either the α- or β-globin cluster in humans.

The organization of the three CA genes on human 8q22 is not yet fully understood. However, by use of pulsed-field gels capable of separating large DNA fragments, it is clear that the *CA1* and *CA3* genes (each in uncertain orientation) are located 27–180 kb 5′ to the *CA2* gene.[66,76]

The genes encoding CA I and CA II are also tightly linked in the guinea pig[6] and pigtail macaque[13] and are near the centromere of mouse chromosome 3,[20] as indicated by genetic evidence. Recently, Beechey *et al.*[2] have shown that mouse *Car-3* is tightly linked to *Car-1* and *Car-2*, thus demonstrating the conservation of this linkage in two mammalian lineages.

6. Genetic Variation at the CA Loci

Genetic variation of CA I and CA II at the protein level has been studied for 25 years but little new has been reported in the last 5 years, probably because of the greater attention being paid to DNA-level variability. The reader is referred to several comprehensive reviews for details of human polymorphisms and rare variants.[38,68,69] In contrast to hemoglobin,[8] the natural mutants of the CA isozymes have been rather disappointing with regard to shedding light on their structure–function relationships; however, progress is being made on creating and expressing tailored variants *in vitro*.[26,70] In addition, studies continue on active-site modification of residues thought to be responsible for the unusual properties of CA III,[7] although tailored mutants will surely provide more clear-cut results.

The characterization of inherited mammalian polymorphisms, found between different laboratory strains, between interbreeding subspecies, or within species, was particularly influential (as described above) in demonstrating linkage of the CA I and CA II genes[6,13,20] and in describing *cis*- effects in pigtail macaques.[67]

At the DNA level, several polymorphisms and strain differences have been reported. A *Taq*I polymorphism 1 kb upstream of the human CA II gene was discovered by Lee *et al.*[42] Murakami *et al.*[52] pointed out that several CA II variants, including the polymorphic CA II variant (Asn-252 → Asp) found in blacks, can be detected directly by use of restriction endonucleases. A polymorphism of human CA III (Ile-31 ↔ Val) detected at the protein level[35] was confirmed at the cDNA level when two groups cloned the different alleles.[45,78] Lloyd *et al.*[45] also described a *Taq*I polymorphism in human CA III. Such polymorphisms provide excellent markers for the 27–180 kb of DNA on chromosome 8q22 that encompasses the *CA1–CA3–CA2* cluster.

In mice, sequencing of cDNA and genomic DNA representing CA II in different inbred laboratory strains revealed four sites that differ,[75] one of them (His-39 ↔ Gln) explaining the well-known electrophoretic polymorphism that had

TABLE III

Alignment of CA and CA-Related cDNA and Genomic Sequences[a]

Gene	Sequence
	1 10 20
Hu1	ATGGCAAGTCCAGACTGGGGATATGATGACAAAAATG/GTCCTGAACAATGGAGCAAGCTGTATCCCATTGCC
Mo1	ATGGCAAGTGCAGACTGGGGATATGGAAGCGAAAATG GTCCTGACCAATGGAGCAAGCTGTATCCCATTGCC
Rb1	....
Hu2	ATGTCCCATCACTGGGGGTACGGCAAACACAACG/GACCTGAGCACTGGCATAAGGACTTCCCCATTGCC
Mo2	ATGTCCCACCACTGGGGATACAGCAAGCACAACG/GACCAGAGAACTGCCACAAGGACTTCCCCATTGCC
Ch2	ATGTCCCATCACTGGGGGTACGACAGCCACAACG/GACCCGCGCACTGCCACGAGCACTTCCCCATCGCC
Hu3	ATGGCCAAGGAGTGGGGCTACGCCAGTCACAACG/GTCCTGACCACTGGCATGAACTTTTCCCAAATGCC
Mo3	ATGGCTAAGGAGTGGGGCTACGCCAGGCACAATG GTCCTGATCACTGGCATGAACTTTATCCAATTGCA
Rt3	ATGGCTAAGGAGTGGGGTTACGGCAGCCACAATG GCCCTGAGCACTGGCATGAACTTTATCCAATTGCC
Hu7	ATGACCGGCCACCACGGCTGGGGCTACGGCCAGGACGACG/GCCCCTCGCATTGGCACAAGCTGTATCCCATTGCC
VD8	--
	30 40
Hu1	AATGGAAATAACCAATCCCCTGTTGATATTAAAACCAGTGAAACCAAACATGACACCTCTCTGAAACCTATTAGT
Mo1	AATGGTAACAACCAGTCTCCTATTGATATTAAAACCAGTGAAGCCAATCATGACTCCTCTCTGAAACCACTCAGC
Rb1	...GGAAATAAGCAGTCTCCAGTAGATATTAAAAGCAGCGAAGTGAAACATGACACCTCTCTGAAACCTTTCAGT
Hu2	AAGGGAGAGCGCCAGTCCCCTGTTGACATCGACACTCATACAGCCAAGTATGACCCTTCCCTGAAGCCCCTGTCT
Mo2	AATGGAGACCGGCAGTCCCCTGTGGATATTGACACAGCAACTGCCCACCATGACCCTGCCCTACAGCCTCTGCTC
Ch2	AATGGGGAGCGCCAGTCGCCCATCGCCATCAGCACCAAAGCCGCCCGCTACGACCCCGCGCTGAAGCCCCTCAGC
Hu3	AAGGGGGAAAACCAGTCGCCCGTTGAGCTGCATACTAAAGACATCAGGCATGACCCTTCTCTGCAGCCATGGTCT
Mo3	AAGGGGGACAACCAGTCACCCATTGAACTGCATACTAAAGACATCAAGCATGACCCCTCTCTGCAGCCCTGGTCA
Rt3	AAAGGGGACAACCAGTCACCCATTGAACTGCATACTAAAGACATCAGGCATGATCCTTCTCTGCAGCCTTGGTCA
Hu7	CAGGGAGATCGCCAATCACCCATCAATATCATCTCCAGCCAGGCTGTGTACTCTCCGAGCCTGCAACCACTGGAG
VD8	ATGCCGCAACAACTATCTCCTATTAATATAGAAACTAAAAAAGCAATTTCTAACGCGCGATTGAAGCCGTTAGAC
	50 60 70
Hu1	GTCTCCTACAACCCAGCCACAGCCAAAGAAATTATCAATGTGGGGCATTCTTTCCATGTAAATTTTGAGGACAAC
Mo1	ATCTCCTATAATCCTGCAACTGCCAAAGAAATTGTTAACGTGGGACATTCTTTCCATGTAATTTTTGATGACAGT
Rb1	CTCTCCTACAATCCAGCCTCTGCCAAAGAAATTATCAACGTGGGACATTCCTTCCATGTCAATTTTGAAGAT---
Hu2	GTTTCCTATGATCAAGCAACTTCCCTGAGGATCCTCAACAATGGTCATGCTTTCAACGTGGAGTTTGATGACTCT
Mo2	ATATCTTATGATAAAGCTGCGTCCAAGAGCATTGTCAACAACGGCCACTCCTTTAACGTTGAGTTTGATGACTCT
Ch2	TTCAGCTACGATGCCGGCACGGCCAAAGCCATCGTCAACAACGGGCACTCCTTCAACGTGGAGTTTGACGACTCC
Hu3	GTGTCTTATGATGGTGGCTCTGCCAAGACCATCCTGAATAATGGGAAGACCTGCCGAGTTGTATTTGATGATACT
Mo3	GCATCTTATGACCCTGGCTCTGCTAAGACCATCCTGAACAATGGGAAGACCTGCAGAGTTGTGTTTGATGATACT
Rt3	GTATCTTATGATCCTGGCTCTGCTAAGACCATCCTGAACAATGGGAAGACCTGCAGAGTTGTGTTTGATGATACC
Hu7	CTTTCCTATGAGGCCTGCATGTCCCTCAGCATCACCAACAATGGCCACTCTGTCCAGGTAGACTTCAATGACAGC
VD8	ATACATTATAATGAGTCGAAACCAACCACTATCCAGAACACTGGAAAACTAGTAAGGATTAATTTTAAAGGAGGA
	80 90
Hu1	GATAACCGATCAG/TGCTGAAAGGTGGTCCTTTCTCTGACAGCTACAGGCTCTTTCAGTTTCATTTTCACTGGGGC
Mo1	AGCAACCAATCTG TTCTGAAAGGTGGCCCTCTTGCTGATAGCTATCGGCTCACTCAGTTCCATTTTCACTGGGGC
Rb1	GACAGCCAATCAG TGCTGAAAGGCGGCCCTCTTTCTGACAACTACCGACTTTCTCAGTTCCATTTTCATTGGGGC
Hu2	CAGGACAAAGCAG/TGCTCAAGGGAGGACCCCTGGATGGCACTTACAGATTGATTCAGTTTCACTTTCACTGGGGT
Mo2	CAGGACAATGCAG/TGCTGAAAGGAGGACCCCTCAGTGACTCCTACAGATTGATCCAGTTTCACTTTCACTGGGGC
Ch2	TCCGACAAGTCAG/TGCTGCAAGGAGGAGCGCTGGATGGAGTCTACAGGTTGGTGCAGTTTCACATTCACTGGGGA
Hu3	TATGATAGGTCAA/TGCTGAGAGGGGGTCCTCTCCCTGGACCCTACCGACTTCGCCAGTTTCATCTTCACTGGGGC
Mo3	TATGACAGGTCTA TGCTGAGGGGTGGTCCTCTCTCTCGGCCCTACCGACTTCGCCAATTCCATCTTCACTGGGGC
Rt3	TTCGATAGGTCCA TGCTGAGAGGTGGGCCTCTCTCTGGACCCTACCGACTTCGCCAGTTCCATCTTCACTGGGGC
Hu7	GATGACCGAACCG/TGGTGACTGGGGGCCCCCTGGAAGGGCCCTACCGCCTCAAGCAGTTTCACTTCCACTGGGGC
VD8	TAT----------*--ATAAGTGCAGGGTTTCTCCCCAATGAATATGTGTTATCATCACTACATATATATTGGGGA
	100 110 120
Hu1	AGTACAAATGAGCATGGTTCAGAACATACAGTGGATGGAGTCAAATATTCTGCCGAG/CTTCACGTAGCTCACTGG
Mo1	AACTCAAACGACCATGGCTCTGAGCACACCGTGGATGGAACTAGATATTCTGGAGAG CTTCACTTAGTTCACTGG
Rb1	AAGACAGATGACTATGGTTCTGAACACACAGTGGATGGAGCCAAATTCTCTGCAGAG CTTCACCTTGTTCACTGG
Hu2	TCACTTGATGGACAAGGTTCAGAGCATACTGTGGATAAAAAGAAATATGCTGCAGAA/CTTCACTTGGTTCACTGG
Mo2	TCATCTGATGGCCAGGGCTCTGAGCACACTGTGAACAAAAAAAAATATGCTGCAGAG/CTTCACTTGGTTCACTGG
Ch2	TCCTGTGAGGGCCAGGGCTCTGAGCACACTGTGGATGGCGTGAAGTACGATGCAGAG/CTTCATATTGTTCACTGG
Hu3	TCTTCGGATGATCATGGCTCTGAGCACACCGTGGATGGAGTCAAGTATGCAGCGGAG/CTTCATTTGGTTCACTGG
Mo3	TCCTCTGATGACCACGGCTCTGAGCACACCGTGGACGGAGTAAAATACGCTGCTGAG CTTCACCTGGTTCACTGG
Rt3	TCCTCGGATGACCATGGCTCTGAGCACACAGTGGATGGAGTGAAGTATGCTGCTGAG CTTCACCTGGTTCACTGG
Hu7	AAGAAGCACGATGTGGGTTCTGAGCACACGGTGGACGGCAAGTCCTTCCCCAGCGAG/CTGCATCTGGTTCACTGG
VD8	AAGGAAGACGATTATGGATCCAATCACTTGATAGATGTGTACAAATACTCTGGAGAG*ATTAATCTTGTTCATTGG
	130 140
Hu1	AATTCTGCAAAGTACTCCAGCCTTGCTGAAGCTGCCTCAAAGGCTGATGGTTTGGCAGTTATTGGTGTTTTGATG
Mo1	AATTCTGCAAAGTACTCCAGTGCTTCTGAAGCCATCTCCAAGGCTGATGGCCTGGCAATCCTTGGCGTTTTGATG
Rb1	AATTCTGGGAAGTACCCCAACATCGCTGACTCTGTTTCCAAGGCTGATGGCTTGGCGATTGTTGCTGTTTTTCTG
Hu2	AACACC---AAATATGGGGATTTTGGGAAAGCTGTGCAGCAACCTGATGGACTGGCCGTTCTAGCTATTTTTTTG
Mo2	AACACC---AAATATGGGGACTTTGGAAAAGCTGTGCAGCAACCGGATGGATTGGCTGTTTTGGGTATTTTTTTG
Ch2	AATGTA---AAATATGGCAAATTTGCTGAAGCTCTGAAGCATCCTGATGGTTTGGCCGTCGTAGGCATCTTCATG
Hu3	AACCCG---AAGTATAACACTTTTAAAGAAGCCCTGAAGCAGCGCGATGGGATCGCTGTGATTGGCATTTTTCTG
Mo3	AATCCA---AGGTATAACACCTTTGGAGAGGCTCTGAAGCAGCCTGATGGCATCGCTGTGGTTGGCATTTTGCTG
Rt3	AACCCG---AAGTATAACACTTCGGAGGAGGCTCTGAAGCAGCCCGATGGGATTGCTGTGGTTGGCATTTTTCTG
Hu7	AATGCCAAGAAGTACAGCACTTTTGGGGAGGCGGCCTCAGCACCTGATGGCCTGGCTGTGGTTGGTGTTTTTTTG
VD8	AATAAGAAAAAATATAGTTCTTATGAAGAGGCAAAAAAACACGATGATGGACTTATCATTATTTCTATATTCTTA

TABLE III *(Continued)*

Gene	Sequence	
	150 · 156 A · 160 · 170	
Hu1	AAG/GTTGGTGAGGCCAACCCAAAG---CTGCAGAAAGTACTTGATGCCCTCCAAGCAATTAAAACCAAG/GGC	
Mo1	AAG GTTGGTCCAGCCAACCCAAGC---CTGCAGAAAGTACTTGATGCTCTAAACTCAGTTAAAACTAAG GGA	
Rb1	AAG GTTGGTCAGGCCAACCCCAAG---CTGCAGAAAGTTCTTGATGCACTGAGTGCAGTTAAAACTAAG GGC	
Hu2	AAG/GTTGGCAGCGCTAAACCGGGC---CTTCAGAAAGTTGTTGATGTGCTGGATTCCATTAAAACAAAG/GGC	
Mo2	AAG/ATTGGACCTGCCTCACAAGGC---CTTCAGAAAGTCCTTGAAGCACTGCATTCCAGGAAAACAAAG/GGG	
Ch2	AAG/GTAGGGAATGCCAAACCTGAA---ATACAGAAAGTTGTTGATGCTCTGAACTCCATTCAAACCAAG/GGG	
Hu3	AAG/ATAGGACATGAGAATGGCGAG---TTCCAGATTTTCCTTGATGCCATGGACAAGATTAAGACAAAG/GGC	
Mo3	AAG ATAGGACGGGAGAAAGGCGAG---TTCCAGATTCTTCTTGATGCCTTGGACAAAATTAAGACGAAG GGC	
Rt3	AAG ATAGGACGGGAGAAAGGCGAG---TTCCAGATTCTCCTTGATGCCCTGGACAAAATTAAGACTAAG GGC	
Hu7	GAG/ACAGGAGACGAGCACCCCAGC---ATGAATCGTCTGACAGATGCGCTCTACATGGTCCGGTTCAAG/GGC	
VD8	CAA*GTATTGGATCATAAAAATGTATATTTTCAAAAGATAGTTAATCAATTGGATTCCATTAGATCCGCC*AAT	
	178 A · 180 · 190	
Hu1	AAACGAGCCCCATTCACAAAT---TTTGACCCCTCTACTCTCCTTCCTTCATCCCTGGATTTCTGGACCTACCCT	
Mo1	AAACGAGCCCCATTCACAAAT---TTTGACCCATCCAGTCTGCTTCCTTCATCTCTGGATTACTGGACCTACTTT	
Rb1	AAAAAAGCCTCATTCACGAAC---TTTGACCCTTCCACTCTGCTTCCTCCGTCCTGGGATTACTGGACCTACTCT	
Hu2	AAGAGTGCTGACTTCACTAAC---TTCGATCCTCGTGGCCTCCTTCCTGAATCCCTGGATTACTGGACCTACCCA	
Mo2	AAGCGTGCGGCCTTTGCTAAC---TTCGACCCTTGCTCCCTTCTTCCTGGAAACTTGGACTACTGGACATACCCT	
Ch2	AAACAAGCTTCTTTCACAAAC---TTTGACCCTACTGGACTGCTGCCTCCATGCAGAGACTATTGGACGTACCCT	
Hu3	AAGGAGGCGCCCTTCACAAAG---TTTGACCCATCCTGCCTGTTCCCGGCATGCCGGGACTACTGGACCTACCAG	
Mo3	AAGGAGGCCCCTTTTACACAC---TTTGACCCATCATGCCTGTTCCCTGCTTGCCGGGACTATTGGACCTATCAC	
Rt3	AAGGAGGCTCCTTTTAATCAC---TTCGACCCATCGTGCCTGTTCCCTGCTTGCCGGGACTATTGGACCTACCAT	
Hu7	ACCAAAGCCCAGTTCAGCTGC---TTCAACCCCAAGTGCCTCCTGCCTGCCAGCCGGCACTACTGGACCTACCCG	
VD8	ACGTCTGCACCGTTTGATTCAGTATTTTATCTAGACAATTTGCTGCCTAGTAAGTTGGATTATTTTACATATCTA	
	200 · 210 · 220	
Hu1	GGCTCTCTGACTCATCCTCCTCTTTATGAGAGTGTAACTTGGATCATCTGTAAGGAGAGCATCAGTGTCAGCTCA	
Mo1	GGCTCTCTGACTCACCCTCCTCTTCATGAAAGTGTGACCTGGGTGATCTGCAAGGATAGCATCAGTCTAAGCCCA	
Rb1	GGCTCTCTGACTCATCCCCCTCTTCACGAGAGTGTGACCTGGCTCATCTGTAAGGACAGCATCAGCATCAGCTCA	
Hu2	GGCTCACTGACCACCCCTCCTCTTCTCGAATGTGTGACCTGGATTGTGCTCAAGGAACCCATCAGCGTCAGCAGC	
Mo2	GGCTCTCTGACCACTCCGCCTCTGCTGGAATGTGTGACCTGGATCGTGCTCAGGGAGCCCATTACTGTCAGCAGC	
Ch2	GGCTCCCTGACTACTCCACCACTGCATGAATGTGTGATTTGGCATGTTCTGAAGGAGCCCATCACTGTCAGCTCT	
Hu3	GGCTCATTCACCACGCCGCCCTGCGAGGAATGCATTGTGTGGCTGCTGCTGAAGGAGCCCATGACCGTGAGCTCT	
Mo3	GGCTCCTTCACCACGCCGCCCTGCGAGGAGTGCATTGTGTGGCTGCTGCTCAAAGAGCCCATGACTGTGAGCTCA	
Rt3	GGCTCCTTCACCACGCCACCCTGCGAGGAGTGCATTGTGTGGCTGCTACTGAAAGAGCCCATGACAGTGAGCTCA	
Hu7	GGCTCTCTGACGACTCCCCCACTCAGTGAGAGTGTCACCTGGATTGTGCTCCGGGAGCCCATCTGCATCTCTGAA	
VD8	GGA------ACAACTATCAACCACTCTGCAGACGCTGTATGGATAATTTTTCCAACXCCAATAAACATTCATTCT	
	230 · 240 · 242 A 242 B	
Hu1	GAGCAG/CTGGCACAATTCCGCAGCCTTCTTTCAAATGTTGAAGGTGATAACGCTGTCCCCATGCAG------CAC	
Mo1	GAGCAG CTGGCCCAGCTCCGTGGTCTTCTGTCAAGTGCAGAGGGAGAGTCTGCAGTTCCAGTTCTG------AGC	
Rb1	GAGCAG TTGGCACAATTCCTCAGTCTTCTATCGAATGCTGAAGGTGAGGCTGCTGTCCCCATACTG------CAC	
Hu2	GAGCAG/GTGTTGAAATTCCGTAAACTTAACTTCAATGGGGAGGGTGAACCCGAAGAACTCATGGTG------GAC	
Mo2	GAGCAG/ATGTCTCATTTCCGTACGCTGACCTTCAATGAGGAGGGGGATGCTGAAGAAGCGATGGTG------GAC	
Ch2	GAGCAG/ATGTGCAAACTCCGTGGCCTTTGCTTCAGTGCTGAGAATGAGCCGGTGTGCCGCATGGTG------GAC	
Hu3	GACCAG/ATGGCCAAGCTGCGGAGCCTCCTCTCCAGTGCTGAGAACGAGCCCCCAGTGCCTCTTGTG------AGC	
Mo3	GACCAG ATGGCCAAGCTGCGCAGCCTCTTCTCCAGCGCAGAGAATGAGCCCCCAGTGCCTCTGGTG------GGG	
Rt3	GACCAG ATGGCCAACGTGCGCAGCCTGTTCGCCAGTGCAGAGAATGAGCCCCCGGTGCCTCTGGTG------GGG	
Hu7	AGGCAG/ATGGGGAACTTCCGGAGCCTGCTTTTTACCTCGGAGGACGATGAGAGGATCCACATGGTG------AAC	
VD8	GATCAA*CTATCTAAATTCAGAACACTATTGTCGTCGTCT---AATCATGATGGAAAACCGCATTATATAACAGAG	
	250 · 252 A · 260	*References*
Hu1	AACAACCGCCCAACCCAACCTCTGAAG---GGCAGAACAGTGAGAGCTTCATTT*TGA*	1,4,87
Mo1	AACCACCGTCCACCCCAACCCCTGAAG---GGCAGAACAGTCAGAGCCTCATTT*TGA*	28
Rb1	AACAACCGACCACCCCAGCCTCTGAAG---GGCAGAACTGTGAAAGCTTCATTC*TGA*	41
Hu2	AACTGGCGCCCAGCTCAGCCACTGAAG---AACAGGCAAATCAAAGCTTCCTTCAAA*TAA*	50,52,74
Mo2	AACTGGCGTCCAGCTCAGCCGCTAAAG---AATAGAAAGATCAAAGCGTCCTTTAAG*TAA*	10,75
Ch2	AACTGGCGCCCATGCCAGCCTCTAAAG---AGCAGGGAAGTCAGAGCTTCCTTCCAG*TAA*	57,85
Hu3	AACTGGCGACCTCCACAGCCTATCAAT---AACAGGGTGGTGAGAGCTTCCTTCAAA*TGA*	45,78
Mo3	AATTGGCGCCCTCCTCAGCCTGTCAAG---GGCAGGGTGGTGAGGGCCTCCTTCAAG*TAA*	70
Rt3	AATTGGCGCCCTCCTCAGCCGATCAAG---GGCAGGGTGGTGAGGGCCTCCTTCAAG*TAA*	39
Hu7	AACTTCCTGCCACCACAGCCACTGAAG---GGCCGCGTGGTAAAGGCCTCCTTCCGGGCC*TGA*	48
VD8	AACTATAGAAATCCGTATAAATTGAACGACGACACGCAAGTATATTATTCTGGGGAGATTATA+	54,55

[a]Sequences shown are as follows: CA1—human (HuI), mouse (MoI), rabbit (RbI); CA2—human (Hu2), mouse (Mo2), chicken (Ch2); CA3—human (Hu3), mouse (Mo3), rat (Rt3); CA7—human (Hu7); D8—vaccinia virus (VD8). References are from Table I, but positions of introns in Hu2 and Mo2 are based in part on unpublished data of P. J. Venta and of S. R. Diehl and P. J. Venta, respectively. Codon triplets are numbered according to the CA I protein sequence (cf. Table II). Symbols: ---, homologous sequence absent in that gene; ..., sequence not yet determined; /, position of intron in genomic sequence; *, genomic sequence lacks intron at this position; +, sequence extends 201 bp (including termination codon).

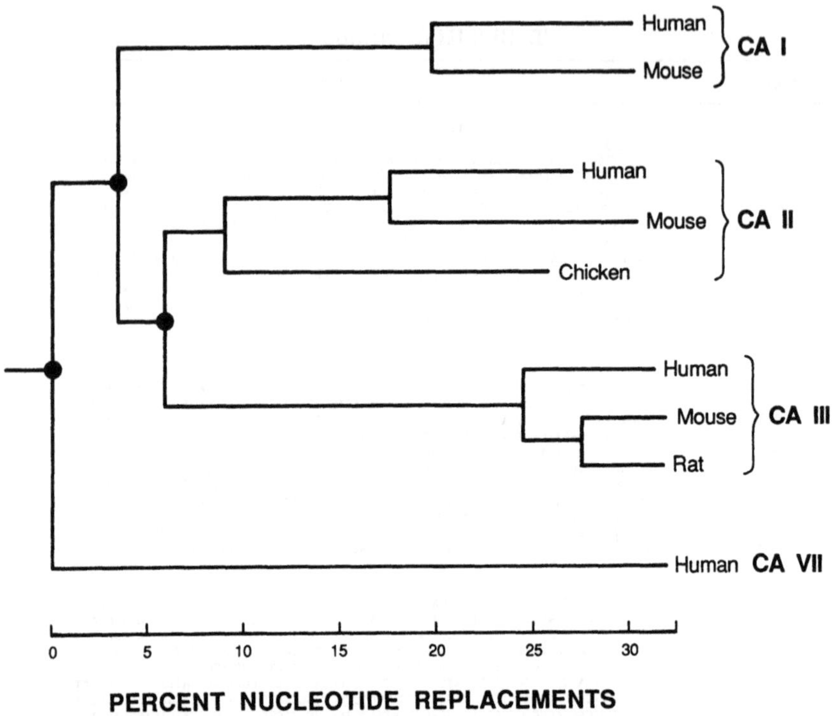

Figure 4. Phylogenetic tree of the CA gene sequences. Sequences are from the alignment in Table III. Rabbit CA I and vaccinia virus D8 were not included since they are incomplete or lack a segment of homologous DNA. Branch lengths reflect Jukes–Cantor-corrected (for multiple hits) percent of the nucleotide replacements. ●, Gene duplication event; other bifurcations are speciation events.

been used in demonstrating linkage of CA I and CA II.[20] Tweedie and Edwards[71] have recently described *Taq*I and *Pst*I polymorphisms in the CA III gene from different mouse species that can interbreed under laboratory conditions.

With the advent of *in vitro* techniques of DNA amplification and sequencing, we can anticipate that new genetic variation in all of the cloned CA genes will be discovered and that characterization of some of the protein variants, particularly the functionally interesting ones such as CA II Baniwa and the zinc-activated CA I Michigan-1,[69] will become feasible.

7. Summary

We are entering another exciting phase of research on the CA family. Five years from now we shall probably understand the full scope of the CA multigene family, its origins, more on the basis of tissue-specific expression, the active-site and kinetic differences between isozymes, and a great deal more about the isozymes less tractable at the protein level, i.e., CA IV, CA V, and CA VII. For now, we can report that CA I, CA II, and CA III resulted from gene duplications close together and shortly after the elasmobranch radiation ~450 mya. CA VII and CA 'Y' probably result from gene duplications at the time of the vertebrate–invertebrate split ~500 mya, and CA IV and CA VI result from much earlier duplications.

Note added in proof. Portions of the human CA IV protein sequence[86] and the entire mouse CA 'Y' cDNA sequence[87] have now been published. In addition, more complete reports are now available on the structure and expression of the human CA I gene.[88,89] A cDNA sequence from the green alga, *Chlamydomonas reinhardtii*,[90] encodes a periplasmic CA clearly related to the animal CAs. It retains 15 active-site residues invariant in animal CAs. The cDNA corresponds to one of two closely similar, linked genes.[91] We have learned that a bacterial CA, from *Neisseria sicca*, contains an active-site peptide (N. Bergenhem, U. Carlsson, and B.-H. Jonsson, personal communication) with sequence identical to human CA II and algal CA residues 198–202. However, two homologous plant chloroplast CA cDNA sequences from spinach[92,93] and pea[94] have been characterized and, surprisingly, are unrelated to the animal kingdom CAs. Kato[95] has now reported the structure of a mouse brain cDNA that encodes a new carbonic anhydrase-related polypeptide (CARP) displaying significant homology (33–41% identity) with CA I, II, III, and VI. It is also distinct from other mammalian CAs. If two radical active-site substitutions (94His → Arg; 92 Gln → Glu) allow CARP to function as a CO_2 hydratase, the murine *Carp* gene, transcribed in Purkinje cells of the cerebellum, will qualify as *Car-8*.

ACKNOWLEDGMENTS. We wish especially to thank the following who kindly provided us with sequence data and/or manuscripts ahead of publication: Mounira Amor, Nils Bergenhem, Nicholas Carter, Peter Curtis, Scott Diehl, Yvonne Edwards, Ross Fernley, Tommaso Meo, Jeffry Montgomery, Penelope Nicewander, William Sly, and Patrick Venta. We appreciate too the help of Li Jin and Manolo Gouy in Houston, who adapted the phylogenetic tree-building programs of N. Saitou and M. Nei for use on the personal computer. Preparation of this chapter was in part supported by grants RR 07148 (D.H.E.) and GM 24681 (R.E.T.) from the National Institutes of Health.

References

1. Barlow, J. H., Lowe, N., Edwards, Y. H., and Butterworth, P. H. W., 1987, *Nucleic Acids Res.* **15**:2386.
2. Beechey, C., Tweedie, S., Spurr, N., Peters, J., and Edwards, Y., 1989, *Genomics* **6**:692–696.
3. Bergenhem, N., and Carlsson, U., 1990, *Comp. Biochem. Physiol.* **95B**:205–213.
4. Brady, H. J. M., Lowe, N., Sowden, J. C., Barlow, J. H., and Butterworth, P. H. W., 1989, *Biochem. Soc. Trans. U.K.* **17**:184–185.

5. Carter, M. J., 1972, *Biol. Rev.* **47**:465–513.
6. Carter, N. D., 1972, *Comp. Biochem. Physiol. B* **43**:743–747.
7. Chegwidden, W. R., Walz, D. A., Hewett-Emmett, D., and Tashian, R. E., 1988, *Biochem. Soc. Trans. U.K.* **16**:974–975.
8. Collins, F. S., and Weissman, S. M., 1984, *Prog. Nucleic Acid Res. Mol. Biol.* **31**:315–458.
9. Curtis, P. J., 1983, *J. Biol. Chem.* **258**:4459–4463.
10. Curtis, P. J., Withers, E., Demuth, D., Watt, R., Venta, P. J., and Tashian, R. E., 1983, *Gene* **25**:325–332.
11. Davis, M. B., West, L. F., Barlow, J. H., Butterworth, P. H. W., Lloyd, J. C., and Edwards, Y. H., 1987, *Somat. Cell Mol. Genet.* **13**:173–.
12. Debuire, B., Henry, C., Benaissa, M., Biserte, G., Claverie, J. M., Saule, S., Martin, P., and Stehelin, D., 1984, *Science* **224**:1456–1459.
13. DeSimone, J., Linde, M., and Tashian, R. E., 1973, *Nature (London) New Biol.* **242**:55–56.
14. Deutsch, H. F., 1984, *Ann. N.Y. Acad. Sci.* **429**:183–194.
15. Dodgson, S. J., Forster, R. E., Storey, B. T., and Mela, L., 1980, *Proc. Natl. Acad. Sci. USA* **77**:5562–5566.
16. Dodgson, S. J., Forster, R. E., and Storey, B. T., 1984, *Ann. N.Y. Acad. Sci.* **429**:516–527.
17. Dodgson, S. J., and Contino, L. C., 1988, *Arch. Biochem. Biophys.* **260**:334–341.
18. Edwards, Y. H., Lloyd, J., Parkar, M., and Povey, S., 1986, *Ann. Hum. Genet.* **50**:44–47.
19. Edwards, Y. H., Barlow, J. H., Konialis, C. P., Povey, S., and Butterworth, P. H. W., 1986, *Ann. Hum. Genet.* **50**:123–129.
20. Eicher, E. M., Stern, R. H., Womack, J. E., Davisson, M. T., Roderick, T. H., and Reynolds, S. C., 1976, *Biochem. Genet.* **14**:651–660.
21. Engberg, P., Millqvist, E., Pohl, G., and Lindskog, S., 1985, *Arch. Biochem. Biophys.* **241**: 628–638.
22. Feldstein, J. B., and Silverman, D. N., 1984, *J. Biol. Chem.* **259**:5447–5453.
23. Fernley, R. T., 1988, *Trends Biochem.* **13**:356–359.
24. Fernley, R. T., Congiu, M., Wright, R. D., and Coghlan, J. P., 1984, *Ann. N.Y. Acad. Sci.* **384**:212–213.
25. Fernley, R. T., Wright, R. D., and Coghlan, J. P., 1988, *Bichemistry* **27**:2815–2520.
26. Forsman, C., Behravan, G., Jonsson, B.-H., Liang, Z.-W., Lindskog, S., Ren, X., Sandström, J., and Wallgren, K., 1988, *FEBS Lett.* **229**:360–362.
27. Fraser, P., Cummings, P., and Curtis, P., 1989, *Mol. Cell. Biol.* **9**:3308–3313.
28. Fraser, P. J., and Curtis, P. J., 1986, *J. Mol. Evol.* **23**:294–299.
29. Graham, D., Reed, M. L., Patterson, B. D., and Hockley, D. G., 1984, *Ann. N.Y. Acad. Sci.* **384**:222–237.
30. Green, S., and Chambon, P., 1986, *Nature* (London) **324**:615–617.
31. Green, S., Walter, P., Kumar, V., Krust, A., Bornert, J.-M., Argos, P., and Chambon, P., 1986, *Nature* (London) **320**:134–139.
32. Hall, G. E., and Schraer, R., 1979, *Comp. Biochem. Physiol.* **63B**:561–567.
33. Henry, C., Coquillaud, M., Saule, S., Stehelin, D., and Debuire, B., 1985, *Virology* **140**:179–182.
34. Hewett-Emmett, D., Venta, P. J., and Tashian, R. E., 1982, in: *Macromolecular Sequences in Systematics and Evolutionary Biology* (M. Goodman, ed.), Plenum Press, New York, pp. 357–405.
35. Hewett-Emmett, D., Welty, R. J., and Tashian, R. E., 1983, *Genetics* **105**:409–420.
36. Hewett-Emmett, D., Hopkins, P. J., Tashian, R. E., and Czelusniak, J., 1984, *Ann. N.Y. Acad. Sci.* **429**:338–358.
37. Hewett-Emmett, D., Cook, R. G., and Dodgson, S. J., 1986, *Fed. Proc.* **45**:1661.
38. Jones, G. L., and Shaw, D. C., 1984, *Ann. N.Y. Acad. Sci.* **429**:249–261.
39. Kelly, C. D., Carter, N. D., Jeffrey, S., and Edwards, Y. H., 1988, *Biosci. Rep.* **8**:401–406.
40. Kendall, A. G., and Tashian, R. E., 1977, *Science* **197**:471–472.

41. Konialis, C. P., Barlow, J. H., and Butterworth, P. H. W., 1985, *Proc. Natl. Acad. Sci. USA* **82**:663–667.
42. Lee, B. L., Venta, P. J., and Tashian, R. E., 1985, *Hum. Genet. 69*:337–339.
43. Lewis, S. E., Erickson, R. P., Barnett, L. B., Venta, P. J., and Tashian, R. E., 1988, *Proc. Natl. Acad. Sci. USA* **85**:1962–1966.
44. Lloyd, J. C., Isenberg, H., Hopkinson, D. A., and Edwards, Y. H., 1985, *Ann. Hum. Genet.* **49**:241–251.
45. Lloyd, J., McMillan, S., Hopkinson, D., Edwards, Y. H., 1986, *Gene* **41**:233–239.
46. Lloyd, J., Brownson, S., Tweedie, S., Charlton, J., and Edwards, Y. H., 1987, *Genes Dev.* **1**: 594–602.
47. McKinley, D. N., and Whitney, P. L., 1976, *Biochim Biophys. Acta* **445**:780–790.
48. Montgomery, J. C., 1988, Ph.D. thesis, University of Michigan, Ann Arbor.
49. Montgomery, J. C., Shows, T. B., Venta, P. J., and Tashian, R. E., 1987, *Am. J. Hum. Genet.* **41**:A229.
50. Montgomery, J. C., Venta, P. J., Tashian, R. E., and Hewett-Emmett, D., 1987, *Nucleic Acids Res.* **15**:4687.
51. Murakumi, H., and Sly, W. S., 1987, *J. Biol. Chem.* **262**:1382–1388.
52. Murakami, H., Marelich, G. P., Grubb, J. H., Kyle, J. W., and Sly, W. S., 1987, *Genomics* **1**:159–166.
53. Nakai, H., Byers, M. G., Venta, P. J., Tashian, R. E., and Shows, T. B., 1987, *Cytogenet. Cell Genet.* **44**:234–235.
54. Niles, E. G., Condit, R. C., Caro,, P., Davidson, K., Matusick, L., and Seto, J. 1986, *Virology* **153**:96–112.
55. Niles, E. G., and Seto, J., 1988, *J. Virol.* **62**:3772–3778.
56. Parsons, D. S., 1982, *Philos. Trans. R. Soc. London* **B299**:369–381.
57. Rogers, J. H., 1987, *Eur. J. Biochem.* **162**:119–122.
58. Saitou, N., and Nei, M., 1987, *Mol. Biol. Evol.* **4**:406–425.
59. Sap, J., Munoz, A., Damm, D., Goldberg, Y., Ghysdael, J., Leutz, A., Beug, H., and Vennström, B., 1986, *Nature* (London) **324**:635–640.
60. Shapiro, L. H., Venta, P. J., and Tashian, R. E., 1987, *Mol. Cell. Biol.* **7**:4589–4593.
61. Sly, W. S., 1989, in: *The Metabolic Basis of Inherited Disease* 6th ed. (C. R. Scriver, A. L. Beaudet, W. S. Sly, and D. Valle, eds.), McGraw-Hill, New York, pp. 2857–2866.
62. Sly, W. S., Hewett-Emmett, D., Whyte, M. P., Yu, Y.-S. L., and Tashian, R. E., 1983, *Proc. Natl. Acad. Sci. USA* **80**:2752–2756.
63. Sly, W. S., Whyte, M. P., Sundaram, V., Tashian, R. E., Hewett-Emmett, D., Guibaud, P., Vainsel, M., Baluarte, H. J., Gruskin, A., Al-Mosawi, M., Sakati, N., and Ohlsson, A., 1985, *N. Engl. J. Med.* **313**:139–145.
64. Sutherland, G. R., Baker, E., Fernandez, K. E. W., Callen, D. F., Aldred, P., Coghlin, J. P., Wright, R. D., and Fernley, R. T., 1989, *Cytogenet. Cell Genet.* **50**:149–150.
65. Tashian, R. E., 1977, in: *Isozymes: Current Topics in Biological and Medical Research*, Volume 2 (M. C. Rattazzi, J. G., Scandalios, and G. S. Whitt, eds.), Alan R. Liss, New York, pp. 21–62.
66. Tashian, R. E., 1989, *Bioessays* **10**:186–192.
67. Tashian, R. E., and Carter, N. D., 1976, *Adv. Hum. Genet.* **7**:1–55.
68. Tashian, R. E., Hewett-Emmett, D., and Goodman, M., 1983, in: *Isozymes: Current Topics in Biological and Medical Research*, Vol. 7 (M. C. Ratazzi, J. G. Scandalios, and G. S. Whitt, eds.), Alan R. Liss, New York, pp. 79–100.
69. Tashian, R. E., Kendall, A. G., and Carter, N. D., 1980, *Hemoglobin* **4**:635–651.
70. Tu, C. K., Silverman, D. N., Forsman, C., Jonsson, B.-H., and Lindskog, S., 1989, *Biochemistry* **28**:7913–7918.
71. Tweedie, S., and Edwards, Y., 1989, *Biochem. Genet.* **27**:17–30.
72. Venta, P. J., Shows, T. B., Curtis, P. J., and Tashian, R. E., 1983, *Proc. Natl. Acad. Sci. USA* **80**:4437–4440.

73. Venta, P. J., Montgomery, J. C., Wiebauer, K., Hewett-Emmett, D., and Tashian, R. E., 1984, *Ann. N.Y. Acad. Sci.* **429**:309–323.
74. Venta, P. J., Montgomery, J. C., Hewett-Emmett, D., and Tashian, R. E., 1985, *Biochim. Biophys. Acta* **826**:195–201.
75. Venta, P. J., Montgomery, J. C., Hewett-Emmett, D., Wiebauer, K., and Tashian, R. E., 1985, *J. Biol. Chem.* **260**:12130–12135.
76. Venta, P. J., Montgomery, J. C., and Tashian, R. E., 1987, in: *Isozymes: Current Topics in Biological and Medical Research*, Volume 14 (M. C. Rattazzi, J. G. Scandalios, and G. S. Whitt, eds.), Alan R. Liss, New York, pp. 59–72.
77. Vincent, S. H., and Silverman, D. N., 1982, *J. Biol. Chem.* **257**:6850–6855.
78. Wade, R., Gunning, P., Eddy, R., Shows, T. B., and Kedes, L., 1986, *Proc. Natl. Acad. Sci. USA* **83**:9571–9575.
79. Weinberger, C., Thompson, C. C., Ong, E. S., Lebo, R., Gruol, D. J., and Evans, R. M., 1986, *Nature* (London) **324**:641–646.
80. Wendorff, K. M., Nishita, T., Jabusch, J. R., and Deutsch, H., 1985, *J. Biol. Chem.* **260**:6129–6132.
81. Whitney, P. L., and Briggle, T. V., 1982, *J. Biol. Chem.* **257**:12056–12059.
82. Wiborg, O., Pederson, M. S., Wind, A., Berglund, L. E., Marcker, K. E., and Vuust, J., 1985, *EMBO J* **4**:755–759.
83. Wistrand, P. J., 1984, *Ann. N.Y. Acad. Sci.* **429**:195–206.
84. Wistrand, P. J., and Knuuttila, K.-G., 1989, *Kidney Int.* **35**:851–859.
85. Yoshihara, C. M., Lee, J.-D., and Dodgson, J. B., 1987, *Nucleic Acids Res.* **15**:753–770.
86. Zhu, X. L., and Sly, W. S., 1990, *J. Biol. Chem.* **265**:8795–8801.
87. Amor-Gueret, M., and Levi-Strauss, M., 1990, *Nucleic Acids Res.* **18**:1646.
88. Lowe, N., Brady, H. J. M., Barlow, J. H., Sowden, J. C., Edwards, M., and Butterworth, P. H. W., 1990, *Gene* **93**:277–283.
89. Brady, H. J. M., Edwards, M., Linch, D. C., Knott, L., Barlow, J. H., and Butterworth, P. H. W., 1990, *Brit. J. Haemat.* **76**:135–142.
90. Fukuzawa, H., Fujiwara, S., Yamamoto, Y., Dionisio-Sese, M. L., and Miyachi, S., 1990, *Proc. Nat. Acad. Sci. USA* **87**:4383–4387.
91. Fukuzawa, H., Fujiwara, S., Tachiki, A., and Miyachi, S., 1990, *Nucleic Acids Res.* **18**:6441–6442.
92. Burnell, J. N., Gibbs, M. J., and Mason, J. G., 1989, *Plant Physiol.* **92**:37–40.
93. Fawcett, T. W., Browse, J. A., Nolokita, M., and Bartlett, S. C., 1990, *J. Biol. Chem.* **265**:5414–5417.
94. Roeske, C. A., and Ogren, W. L., 1990, *Nucleic Acids Res.* **18**:3413.
95. Kato, K., 1990, *FEBS Lett.* **271**:137–140.

CHAPTER 3

X-Ray Crystallographic Studies of Carbonic Anhydrase Isozymes I, II, and III

A. ELISABETH ERIKSSON and ANDERS LILJAS

The crystallographic structures of human carbonic anhydrases (CAs) (EC 4.2.1.1.) were determined in the early 1970s. For reviews of the early work, see references 1, 18, 19, 34, 48.

The refined structures of CA I,[18] II,[10] and III[9] are homologous with the root mean square deviation between their α-C positions of less than 1 Å (Fig. 1).

The CA isozymes have a 10-stranded β-sheet that halves the molecule. Most hydrophobic interactions are in the lower half (Table I); the active site and a smaller hydrophobic region are in the upper half.

1. The Active Site

The active site is a 15-Å wide, 15-Å deep cavity made up of residues with small crystallographic temperature factors and lacking major shifts when inhibitors are present. One half of the active site has only nonpolar residues[34] (Table II).

At the bottom of the cavity, zinc is ligated to the ϵ-N of His-94 and His-96, the ∂-N of His-119 and a water molecule.[10,18] The H bonds from the second N of the His ligands (Table III) stabilize the active-site structures and are conserved in the available amino acid sequences.[13] His-119 is H bonded to the unprotonated and buried Glu-117.

The zinc-bound water in the crystals of CA I and CA II is OH^-, since these

A. ELISABETH ERIKSSON • Institute of Molecular Biology, University of Oregon, Eugene, Oregon 97403-1229. *ANDERS LILJAS* • Department of Molecular Biophysics, Chemical Center, University of Lund, S-221 00, Sweden.

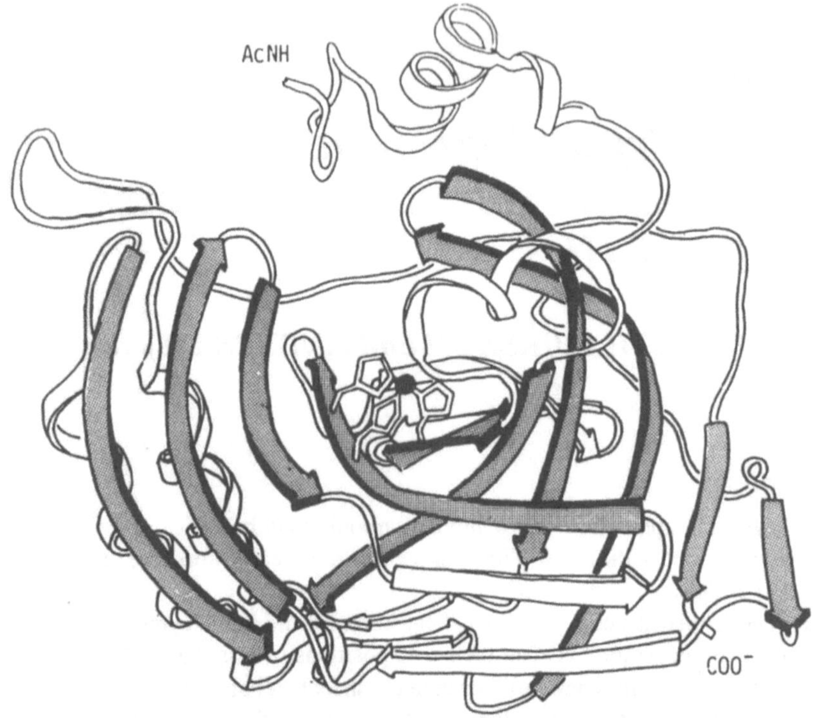

Figure 1. Polypeptide backbone of cytosolic CAs.

enzymes have pK_a values of about 7.0,[21,29,31,36,42] in contrast to a value of below 5.5 for CA III.[8,39,45] Locations of water molecules are illustrated in Fig. 2. Water molecule 338 is referred to as the "deep" water and forms H bonds to the zinc-bound water, 263, and to the amido group of Thr-199. The zinc-bound water is hydrogen bonded to the Oγ1 of the conserved residue Thr-199, which in turn is hydrogen bonded to the carboxyl group of Glu-106, a residue totally buried behind the zinc ion. This glutamate forms three H bonds (Table III) and is presumably negatively charged at high pH. A similar situation has been established in the case of trypsin[22] and is likely to be the case for many other enzymes.[37]

His-64 of CA II has been implicated in the mechanism of rapid proton transfer from the zinc-bound water molecule to buffer molecules outside the active-site region.[42] Its orientation and H bonding are thus of significant interest and have been inferred from the position of the surrounding water molecules. When the imidazole ring is oriented in the electron density with its ε-nitrogen in the direction toward the zinc ion, there are water molecules within hydrogen bonding distance from both nitrogens (Fig. 2). Both of these water molecules have high temperature factors, indicating a high mobility. However, in all crystal structures of inhibitor complexes with human CA II (HCA II), this orientation of His-64 remains the same.

TABLE I
Amino Acids of the Hydrophobic Core of HCA I, HCA II, and BCA III

Residue no.[a]	HCA I	HCA II	BCA III	Residue no.[a]	HCA I	HCA II	BCA III
47	Ile	Leu	Trp	148	Met	Leu	Leu
49	Val	Val	Ala	157	Leu	Leu	Phe
51	Tyr	Tyr	Tyr	160	Val	Val	Leu
59	Ile	Ile	Ile	161	Leu	Val	Leu
* 66	Phe	Phe	Cys	164	Leu	Leu	Leu
68	Val	Val	Val	167	Ile	Ile	Ile
* 70	Phe	Phe	Phe	*176	Phe	Phe	Phe
79	Leu	Leu	Leu	*179	Phe	Phe	Phe
90	Leu	Leu	Leu	184	Leu	Leu	Leu
* 93	Phe	Phe	Phe	185	Leu	Leu	Phe
* 95	Phe	Phe	Leu	210	Ile	Ile	Leu
* 97	Trp	Trp	Trp	212	Cys	Leu	Leu
118	Leu	Leu	Leu	216	Ile	Ile	Ile
120	Val	Leu	Leu	218	Val	Val	Val
122	His	His	His	223	Leu	Val	Ile
144	Ile	Leu	Val	*226	Phe	Phe	Leu
146	Val	Ile	Val	229	Leu	Leu	Leu

[a]Residues in the aromatic cluster in HCA I and II are indicated with asterisks.

CA I has three histidyl residues in the active site: His-64, His-67, and His-200; in contrast, CA II has His-64, Asn-67, and Thr-200. Thus, the arrangement of water molecules must be different for the CA I and CA II isozymes.[10,18]

CA III has three basic residues within its active site: Lys-64, Arg-67, and Arg-91. The locations of these residues in the three-dimensional structure (Fig. 3) show that only Arg-67 seems to be of relevance for the properties of the active

TABLE II
Amino Acids Constituting the Active Sites of HCA I, HCA II, and BCA III

Residue	Hydrophilic side			Residue	Hydrophobic side		
no.	HCA I	HCA II	BCA III	no.	HCA I	HCA II	BCA III
7	Tyr	Tyr	Tyr	91	Phe	Ile	Arg
62	Val	Asn	Asn	121	Ala	Val	Val
64	His	His	Lys	131	Leu	Phe	Tyr
67	His	Asn	Arg	135	Ala	Val	Leu
92	Gln	Gln	Gln	143	Val	Val	Val
106	Glu	Glu	Glu	198	Leu	Leu	Phe
199	Thr	Thr	Thr	207	Val	Val	Ile
200	His	Thr	Thr	209	Trp	Trp	Trp

TABLE III
Hydrogen Bonding Distances in the Active Site of HCA II

Bonding site		Distance (Å)
263 OHH	— 318 OHH	2.8
263 OHH	— 338 OHH	2.7
263 OHH	— Oγ1 Thr-199	2.8
264 OHH	— 292 OHH	3.0
264 OHH	— Oη Tyr-7	2.7
264 OHH	— O His-64	3.0
265 OHH	— Oη Tyr-7	2.4
265 OHH	— Oε2 Glu-106	2.8
265 OHH	— O Thr-199	3.0
292 OHH	— 318 OHH	2.8
292 OHH	— 369 OHH	2.7
318 OHH	— Oγ1 Thr-200	2.9
332 OHH	— Nδ1 His-64	3.1
369 OHH	— Nδ2 Asn-62	2.8
369 OHH	— Nδ2 Asn-67	2.8
385 OHH	— 393 OHH	2.5
385 OHH	— Nε2 His-64	2.6
Oδ1 Asn-62	— N His-64	3.0
Oε1 Gln-92	— Nδ1 His-94	2.7
Nδ1 His-96	— O Asn-244	2.7
Oε1 Glu-106	— Oγ1 Thr-199	2.4
Oε2 Glu-106	— N Ile-246	2.9
Oε1 Glu-117	— N Glu-106	2.9
Oε1 Glu-117	— N His-107	2.8
Oε2 Glu-117	— Nδ1 His-107	2.9
Oε2 Glu-117	— Nε2 His-119	2.6

site.[9] Arg-67 may be involved in the phosphatase activity that is unique to the CA III isozymes.[24]

The main crystallographic method used to explore the structural details of the catalytic mechanism is analysis of the binding of inhibitors, since the binding of substrates can be studied only in exceptional cases.

For HCA II, three different types of inhibitors have been studied by X-ray crystallography: anions, sulfonamides, and Hg^{2+}. All provide information about the function of the active site.[11] A reexamination of sulfonamide binding to HCA I has been done by using the phase angles from the refined native structure.[3] The structures of the inhibitor complexes display very small changes in the protein coordinates.[11]

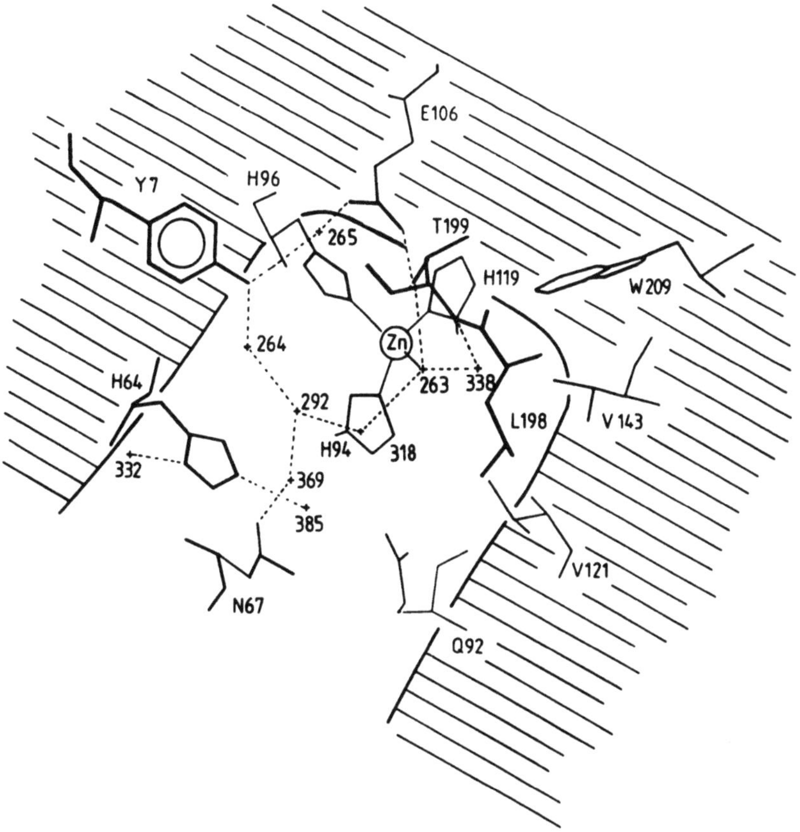

Figure 2. Schematic illustration of the active-site cavity of HCA II.[10] Nine water molecules have been identified. Hydrogen bonds are indicated with dashed lines. The zinc-bound water molecule (263) forms three hydrogen bonds, to the Oγl atom of Thr-199 and to water molecules 318 and 338. The three water molecules 264, 292, and 369 are situated 3.4–3.5 Å from the side chain of His-64.

1.1. Hg^{2+}

The metal ions Cu^{2+} and Hg^{2+} are inhibitors of CA II,[32,46] presumably by binding to His-64 and thus preventing the rapid transfer of protons from the zinc-bound water molecule to the external buffer.[46] Crystallographic investigation was able to confirm this hypothesis.[11] Both nitrogens of His-64 were found to bind mercury at about half occupancy. (Fig. 4). The most probable interpretation is that the two nitrogens bind mercury with about equal affinity, thereby eliminating the possibility for His-64 to participate in proton transfer.

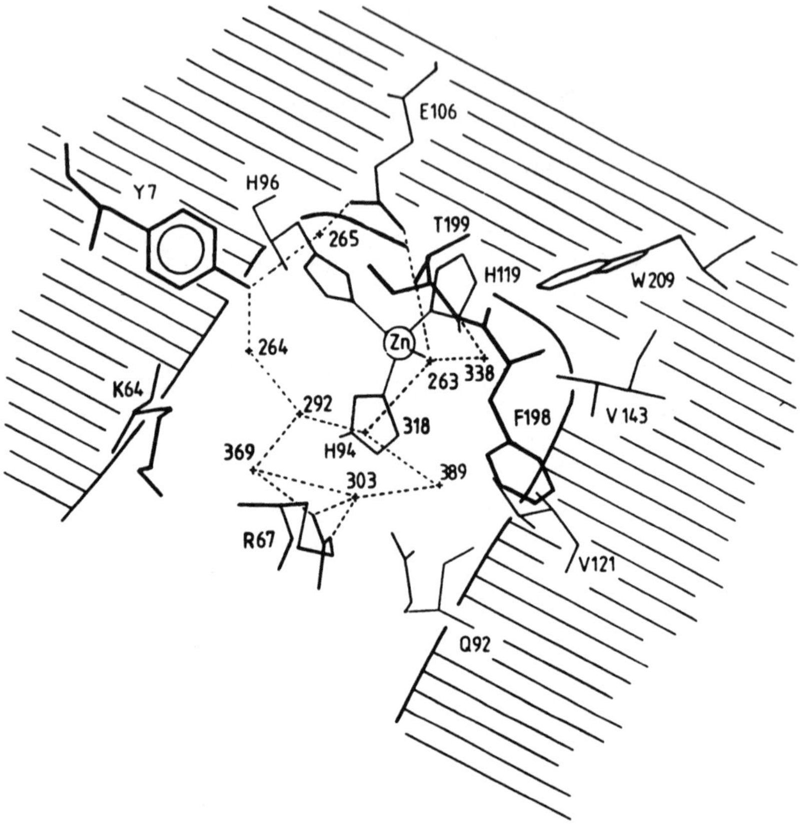

Figure 3. Schematic illustration of the active-site cavity of BCA III.[9] So far, eight water molecules have been identified during refinement. The positions of water molecules 263, 318, and 338 associated with the zinc ion, have not been accurately defined. Three water molecules seem to connect the zinc ion with the guanidinium group of Arg-67, which is situated 9 Å from the zinc. Lys-64 and Arg-91 are both directed out of the active-site cavity, forming salt bridges to Glu-4 and Asp-72, respectively.

1.2. Sulfonamide

Sulfonamides are a class of strong CA inhibitors (for reviews see references 30 and 33). A few sulfonamides bound to CAs have been investigated crystallographically. The results presented here are derived from the refined complex of HCA II and 3-acetoxymercury-4-aminobenzenesulfonamide (AMS) at 2-Å resolution.[11] HCA II was also studied in complex with acetazolamide (Diamox) at 1.9-Å resolution.[47] Furthermore, HCA I has been investigated in complex with AMS and Diamox.[3] A general description of the sulfonamide-binding HCA II follows.

The sulfonamide group is bound close to the zinc ion, presumably with its negatively charged NH^- group liganded to the zinc.[4,15] The NH^- group has the

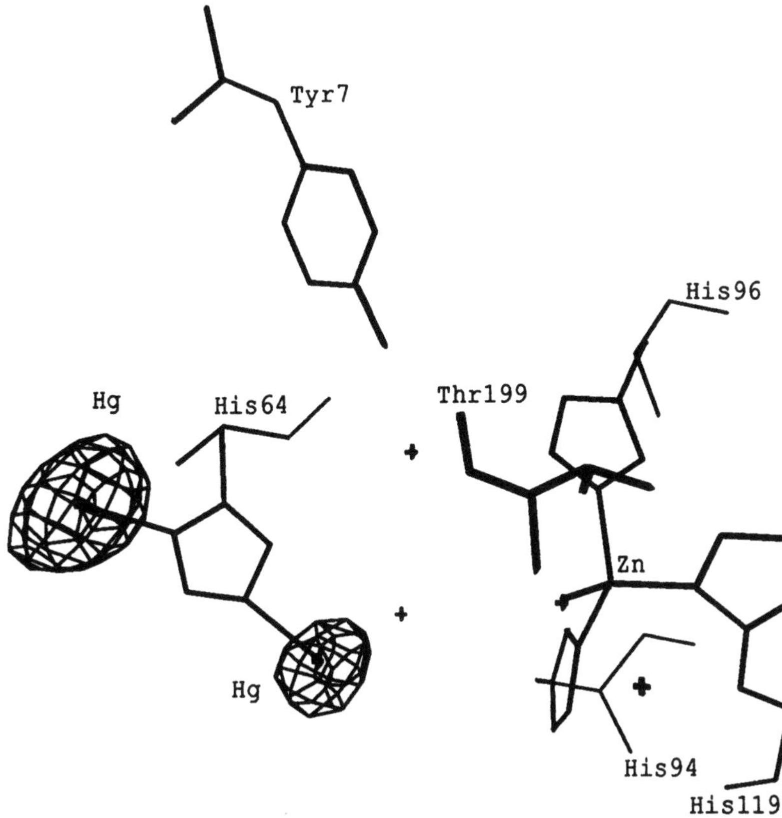

Figure 4. The $|F_{obs}(complex)|-|F_{calc}(native)|$ electron density for Hg^{2+} bound to HCA II. (From reference 11.)

same position as the zinc-bound hydroxyl ion in the native structure (Fig. 5); thus, the zinc coordination remains tetrahedral upon sulfonamide binding. In addition, the NH^- group forms a hydrogen bond to Oγ1 of Thr-199, and one of the sulfonamide group oxygens displaces the deep water molecule and makes a hydrogen bond to the peptide amide of Thr-199. The second sulfonamide oxygen has no direct contact with the protein and is 3.1 Å away from the zinc, consistent with the distance found for the unrefined complex with HCA I.[3] The interactions with the rest of the inhibitor molecule depend on which substituents are situated at various locations on the aromatic ring of the inhibitor. The mercury of the AMS molecule binds to His-64 in HCA II, and the benzene ring is held against the hydrophobic wall in the active site at van der Waals distances from Gln-92, Val-121, Leu-198, and Thr 200. The ring structure of Diamox is also bound in the same area.[47] The oxyen of the acetamido group forms a H bond to the side chain nitrogen

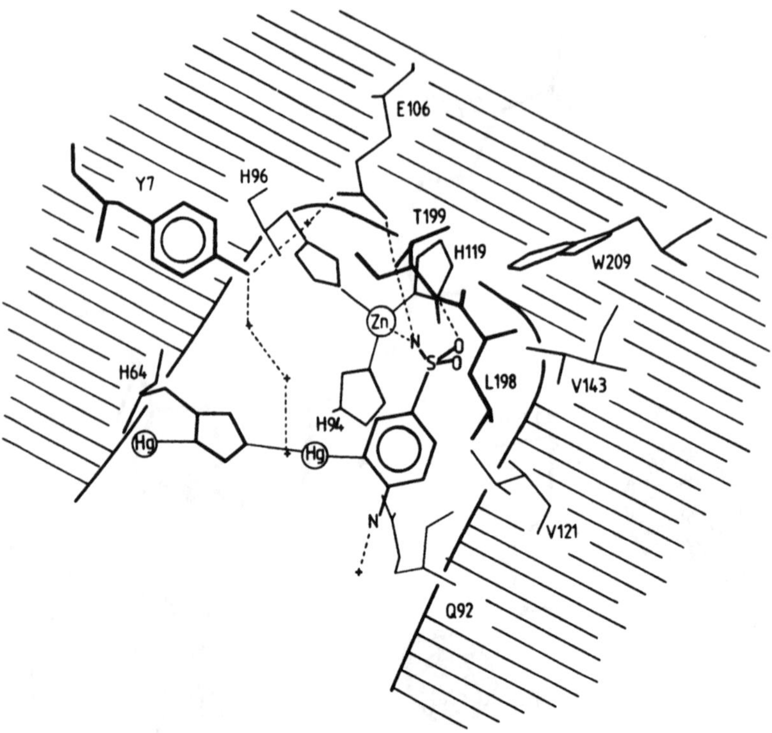

Figure 5. Schematic drawing of the AMS binding to HCA II. (From reference 11.)

of Gln-92. These interactions are similar for HCA I except that Thr-200 is replaced by a histidyl residue.[3] The difficulty that most sulfonamides have binding to CA III is probably because Phe, not Leu, is at position 198.[9]

1.3. Anions

The results from spectroscopic investigations of CA in which the zinc ion has been replaced by cobalt[5,29] are in good agreement with the crystallographic finding that the native enzyme at high pH and the sulfonamide-inhibited CAs have a tetrahedrally coordinated metal.[2] However, a number of other inhibitors display a distinctly different type of spectrum which can be interpreted as penta-coordinated. One of these inhibitors is thiocyanate (SCN^-).[2] The crystallographic investigation of the HCA II–SCN^- complex at high pH confirms this finding (Fig. 6). The anion is bound at the zinc ion where the sulfur moiety is 4.1 Å from the amide nitrogen of Thr-199, displacing the deep water molecule.[11] The SCN^- nitrogen binds to the zinc ion at a distance of 1.9 Å but 1.3 Å away from the position of the hydroxyl ion in the native enzyme. The penta-coordination of the metal ion is

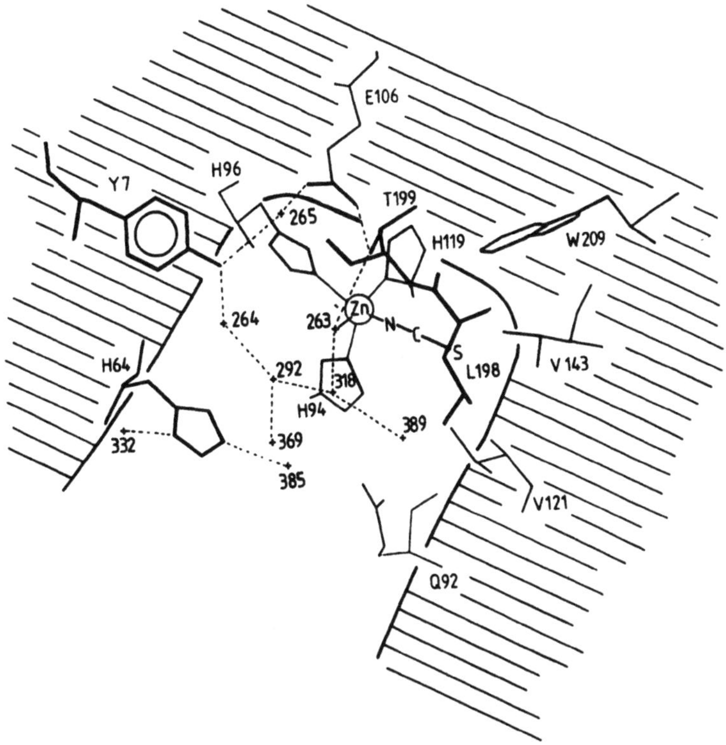

Figure 6. Schematic drawing of the SCN^- binding to HCA II. (From reference 11.)

observed as an additional electron density at the zinc. This density, interpreted as a water molecule, is situated 2.2 Å from the zinc, 1.2 Å away from the original OH^-. The distance between these two zinc ligands is 2.4 Å. The rearrangement of the zinc coordination is probably due to the position and environment of Thr-199. As has been described above, the OH- group of Thr-199 is a hydrogen bond donor to the charged residue Glu-106. Thus the Thr-199 Oγ1 atom can only function as an acceptor for further H bonds, as in the case of the zinc-bound hydroxyl or the nitrogen of the sulfonamide group. This hypothesis is consistent with the observed distances between the Oγ1 of Thr-199 and the thiocyanate nitrogen and between the Oγ1 atom and the penta-coordinated water, 3.6 and 2.7 Å, respectively. The HCA II–SCN^- complex gives crystallographic evidence for penta-coordination and provides structural details for the interactions around the active site. It is worth reemphasizing that there are no significant alterations of the atomic positions of the residues around the active site. The short hydrogen bond between Thr-199 and Glu-106 remains short in the thiocyanate complex (2.5 Å). Furthermore the SCN^- ion is in van der Waals contact with residues Val-143, Leu-198, and Trp-209.[11]

1.4. Imidazole

Of the few known competitive inhibitors of the hydration of CO_2, only imidazole (which inhibits CA I[21]) has been crystallographically investigated. The crystallographic analysis of the HCA I–imidazole complex at pH 8.7 at 2.0-Å resolution shows the inhibitor at a distance no closer than 2.8 Å from the zinc ion.[19,20] The metal-bound OH^- has not been displaced. In the spectroscopic analysis, this is observed as an intermediate between tetra-coordinated and penta-coordinated metal.[2] The general position of the imidazole is in the hydrophobic part of the active site close to residue Ala-121 but also in the vicinity of His-94. In general, the imidazole binds in the same area as the thiocyanate and the sulfonamide group, where probably also CO_2 must bind during catalysis.

2. Structural Aspects of the Catalytic Mechanism

The catalytic mechanism of CA is best described as several separate reactions[31,41]:

$$E{\cdot}OH^- + CO_2 + H_2O = E{\cdot}H_2O + HCO_3^-$$
$$E{\cdot}H_2O = H^+{\cdot}E{\cdot}OH^-$$
$$H^+{\cdot}E{\cdot}OH^- + B = E{\cdot}OH^- + BH^+$$

where E is the enzyme, $H^+{\cdot}E$ is the protonated form of the enzyme, and B and BH^+ are different protonation states of the buffer. Each reaction is discussed below, along with one catalytic mechanism that no structural evidence contradicts.[11,20,31]

2.1. Hydration of Carbon Dioxide

It is generally accepted that the hydration of CO_2 takes place at the zinc ion and that its bound OH^- reacts with the substrate.[41] Imidazole binds near the zinc in the hydrophobic part of the active-site cavity.[20] Thiocyanate and sulfonamide inhibitors bind to approximately the same area and displace the deep water molecule[11] and one oxygen of carbon dioxide; the second one is possibly located as the second oxygen of the sulfonamide group at a distance around 3 Å from the zinc ion. Thus, CO_2 and OH^- are close to the nitrogen and oxygens of the sulfonamide group. The binding of the substrate in this position corresponds to step 1 of the catalytic mechanism as shown in Fig. 7.[11] A nucleophilic attack by the hydroxyl ion on CO_2 causes a charge delocalization from the hydroxyl oxygen to the carbon dioxide oxygen in the proximity of zinc (step 2). This leads to a change in position of the bicarbonate so that it binds to the zinc ion with two of its oxygens (step 3) and gives the metal a penta-coordinated stage like for the thiocyanate complex of HCA II.[25]

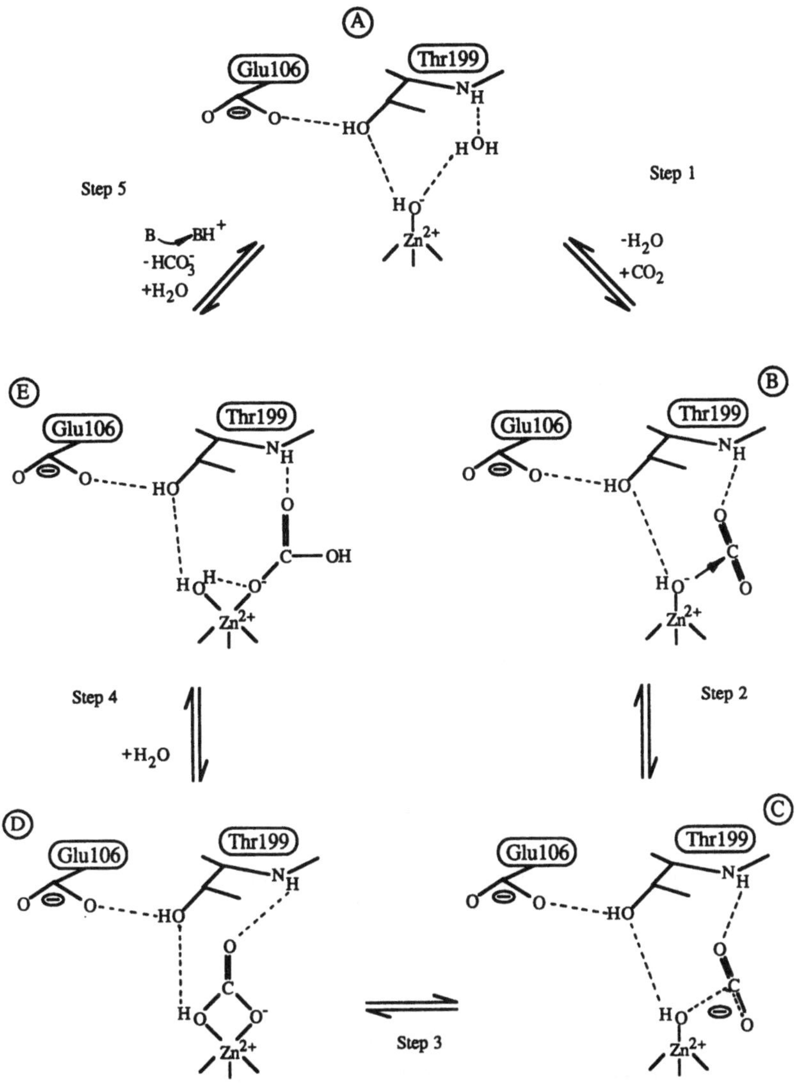

Figure 7. The catalytic mechanism of CA. (From reference 11.)

Model-building experiments have shown that bicarbonate can be fitted into the active site in the manner shown in stage D, with two H bonds to Thr-199, one to the Oγ1, and one to the amide nitrogen. The bond between the zinc ion and the OH moiety of the bicarbonate breaks, and a water molecule moves into its place (step 4). Finally, step 5 includes the release of bicarbonate from the active site, addition of a water molecule at the deep water site, and transfer of the proton from the zinc-bound water to buffer, possibly mediated by active-site residues like His-64 in CA II, as will be discussed in the next section. This makes the enzyme return to stage A, ready to bind another carbon dioxide molecule. This last step could proceed according to several plausible alternatives. Most kinetic data[41] suggest a HCO_3^- release followed by the proton transfer to His-64. However, bicarbonate with a negatively charged oxygen ligated to the zinc ion does not seem to be a good leaving group. Thus, it seems more likely that the proton transfer from the zinc-bound water molecules to the buffer molecules is what enables the release of the bicarbonate ion.

The unique binding of bicarbonate has been explored.[35] The authors investigated a range of alkyl carbonates for their possible function as substrates of CA and found that these compounds are efficient inhibitors of the enzyme, with binding constants comparable to those of the alkyl carboxylates. In the CA structures, the $CH_3OCO_2^-$ ion cannot bind at the site near Thr-199 and donate a hydrogen bond to its Oγ1, which seems to be required for a substrate. Presumably, alkyl carbonates bind in a manner similar to SCN^- at van der Waals distance from the threonine.

A proposed catalytic mechanism which involves an internal proton transfer within HCO_3^- during catalysis came from analysis of a number of models for the mode of binding and dissociation of bicarbonate from the active site.[26] The authors point out that there is possibly a requirement for the bicarbonate proton to be transferred from the metal-binding oxygen to one of the other oxygens after the formation of the bicarbonate if there is to be microscopic reversibility without formation of a bidentate intermediate of stage D. In the opposite direction of the reaction, bicarbonate would initially bind with the negatively charged oxygen. It is proposed that either Thr-199 or a suitably located water molecule participates in this facilitated proton transfer. The authors do not favor the mechanism that is the basis for our proposal,[31] mainly on the grounds that it cannot explain the inability of CA to catalyze the hydration of alkyl carbonate esters, $ROCO_2^-$.[35]

2.2. Protolysis of Water and Proton Transfer

The last step in the catalytic mechanism of CA is the protolysis of water and the transfer of a proton from the zinc environment to bulk solvent. This transfer in CA I and CA II depends on buffer molecules as proton acceptors, since the concentration of hydroxyl ions in water is too low to account for the catalytic rate of the enzyme.[6] This finding has been experimentally confirmed (reviewed in reference 41). Some step in this reaction is rate- limiting for catalysis by CA. In HCA II, a number of studies indicate that this rate-limiting step is an intramolecular proton transfer from the zinc-bound water molecule to His-64.[41,42] Inhibition studies with Cu^{2+} and Hg^{2+} ions[46] combined with the crystallographic localization of Hg^{2+} at His-64[11] have verified the suggestions that this residue is essential. The subsequent exchange of His-64 for other residues by protein engineering has further confirmed this conclusion (B.-H. Jonsson, S. Lindskog, C. K. Tu, and D. N. Silverman, private communication).

While the transfer of a proton from His-64 to buffer is very rapid and essentially diffusion limited,[38] the transfer of a proton from the zinc-bound water molecule to His-64 is slower than expected. An ice-like structure of water molecules bridging the two groups by H bonds would lead to a transfer of protons many orders of magnitude faster than the observed rate of 10^6 sec^{-1}.[6,41] The possible reason for this less than maximal rate has been discussed in structural terms. An ice-like structure of water molecules was observed for HCA II in the initial high-resolution map[27] and was refined crystallographically.[10] The refined positions of this chain of water molecules do not directly connect the zinc hydroxyl to His-64 by hydrogen bonds. Distances of 3.2 Å or less have been accepted as H bonds with an average error for atomic positions of around 0.2 Å. The zinc-bound hydroxyl ion has only one connection toward the opening of the active site, through water molecule 318 (Fig. 2). This water has H bonds in two directions, to Oγ1 of Thr-200 and to water 292. From here, the H-bond chain continues to water 369 and branches to Asn-62 and Asn-67. Thus, the refined structure does not support a direct involvement of His-64 with this H-bond network. The latter residue is H bonded to other water molecules through both of its nitrogens. For the proton transfer to take place, some local rearrangements need to occur: either the imidazole side chain is mobile enough to temporarily become properly connected to the water molecules or the water molecules may be capable of suitable rearrangements. The crystallographic temperature factors of these water molecules and His-64 are higher than those of other water molecules or residues in the active site. Furthermore, the line width of one ^1H nuclear magnetic resonance for His-64 also indicates that this residue is relatively mobile.[41] Thus, even though there is a discontinuity in the H-bond chain, there is sufficient mobility for the groups involved that a rearrangement can take place to permit the proton transfer. This view is consistent with an analysis of the temperature dependence of the catalytic parameters for CA II.[12]

3. Catalytic Differences of the Isoenzymes Related to Their Structures

CA I, CA II, and CA III differ greatly in the histidine content of their active sites. CA III has no histidyl residue but does have three basic residues in the proximity of the active site, with only one of these residues, Arg-67, in the active site.[9] This residue may be the reason for the low pK_a of the zinc-bound water. There is also a rate-limiting proton transfer from the zinc water molecule. There seems to be no buffer effect, nor is there any evidence for an intramolecular proton transfer. The proton is probably transferred at a rate of about 3×10^3 sec^{-1} directly to water.[41]

For CA II, the protons are transferred from the zinc-bound water to His-64 through a discontinuous chain of H bonds which becomes the rate-limiting step in the catalytic mechanism of this isoenzyme. His-64 subsequently transfers the protons to buffer molecules in the surrounding solvent. Thus, the enzyme carries a covalently bound buffer molecule at the entrance of the active site.

CA I has three histidyl residues (His-64, His-67, and His-200) in the active site, which could influence the catalytic rate of the isoenzyme favorably. However, in this case the turnover number if five times less than for CA II.[21] Also, here there seems to be a rate-limiting proton transfer, even though it is less well established than for CA I and CA III.[31] The histidyl residues have less than ideal pK_a values for the proton transfer: 4.7, 6.0, and 6.1, respectively.[31] The proximity between the histidyl residues may lower their pK_a values. His-200 is ideally located to receive a proton, a short distance from the zinc, but the other two histidyl residues might shield the further transport of the proton to buffer molecules. The locations of the water molecules in the active site of HCA I are not available for analysis of the hydrogen bond contacts between the zinc water and the histidyl residues.

In addition to the differences in histidyl and basic residues in the active sites of the CA isozymes, there are variations in the general volumes of the inner part of the active sites which may very well affect their catalytic behavior and inhibition. Calculation of the surface accessibilities in the active-site region shows that HCA II has the largest volume available for inhibitors and substrate molecules. The cavity size of HCA I is reduced by the introduction of His-67 and His-200, whereas the cavity of bovine CA III (BCA III) is reduced by Arg-67 and Phe-198. The latter residue is likely to influence the activity of CA III, since it is situated close to the binding site of thiocyanate and sulfonamide inhibitors. It seems probable that Phe-198 prevents the binding of many sulfonamide inhibitors to CA III.

It is obvious that a significant number of experiments of site-directed mutagenesis on CA will be rewarding in the continuing analysis of this family of enzymes. These experiments should be complemented by further crystallographic studies on the whole range of isozymes, on mutant enzymes, and on complexes of the isoenzymes with inhibitors and substrate analogs to obtain a detailed understanding of the molecular roles of different features of enzyme.

ACKNOWLEDGMENTS. We are grateful to Professors T. A. Jones and S. Lindskog for support and collaboration.

References

1. Bergstén, P.-C., Waara, I., Lövgren, S., Liljas, A., Kannan, K. K., and Bengtsson, U., 1971, in: *Oxygen Affinity of Hemoglobin and red cell acid-base status* (M. Rörth, and P. Astrup, eds.), Munksgaard, Copenhagen, Academic Press, New York, pp. 363–383.
2. Bertini, I., and Luchinat, C., 1983, *Acc. Chem. Res.* **16**:272–279.
3. Chakravarty, S., Yadava, V. S., Kumar, V., and Kannan, K. K., 1985, *Proc. Int. Symp. Biomol. Struct. Interact. Suppl. J. Biosci.* **8**:491–498.

4. Chen, R. F., and Kernohan, J. C., 1967, *J. Biol. Chem.* **242**:5813–5823.
5. Coleman, J. E., 1967, *J. Biol. Chem.* **242**:5212–5219.
6. Eigen, M., and Hammes, G. G., 1963, *Adv. Enzymol.* **25**:1–38.
7. Engberg, P., and Lindskog, S., 1984, *FEBS Lett.* **170**:326–330.
8. Engberg, P., Millqvist, E., Pohl, G., and Lindskog, S., 1985, *Arch. Biochem. Biophys.* **241**: 628–638.
9. Eriksson, A. E., 1988, *Acta Univ. Ups. Compr. Summ. Upps. Dissert. Fac. Sci.* **164**:1–36.
10. Eriksson, A. E., Jones, T. A., and Liljas, A., 1988, *Proteins, Structure, Function and Genetics* **4**:274–282.
11. Eriksson, A. E., Kylsten, P. M., Jones, T. A., and Liljas, A., 1988, *Proteins, Structure, Function and Genetics* **4**:283–293.
12. Ghannam, A. F., Tsen, W., and Rowlett, R. S., 1986, *J. Biol. Chem.* **261**:1164–1169.
13. Hewett-Emmett, D., Hopkins, P. J., Tashian, R. E., and Czelusniak, J., 1984, *Ann. N.Y. Acad. Sci.* **429**:338–358.
14. Holmes, R. S., 1977, *Eur. J. Biochem.* **78**:511–520.
15. Kanamori, K., and Roberts, J. D., 1983, *biochemistry* **22**:2658–2664.
16. Kannan, K. K., 1980, in: *Biophysics and Physiology of Carbon Dioxide* (C. Bauer, G. Gros, and H. Bartels, eds.), Springer-Verlag, Berlin, pp. 184–205.
17. Kannan, K. K., Notstrand, B., Fridborg, K., Lövgren, S., Ohlsson, A., and Petef, M., 1975, *Proc. Natl. Acad. Sci. USA* **72**:51–55.
18. Kannan, K. K., Ramanadham, M., and Jones, T. A., 1984, *Ann. N.Y Acad. Sci.* **429**:49–60.
19. Kannan, K. K., Vaara, I., Notstrand, B., Lövgren, S., Borell, A., Fridborg, K., and Petef, M., 1977, in: *Drug Action at the Molecular Level* (G. C. K. Roberts, ed.), The Macmillan Press, pp. 73–93.
20. Kannan, K. K., Petef, M., Fridborg, K., Cid-Dresdner, H., and Lövgren, S., 1977, *FEBS Lett.* **73**:115–119.
21. Khalifah, R. G., 1971, *J. Biol. Chem.* **246**:2561–2573.
22. Kossiakoff, A. A., and Spencer, S. A., 1981, *Biochemistry* **20**:6462–6474.
23. Koester, M. K., Register, A. M., and Noltmann, E. A., 1977, *Biochem. Biophys. Res. Commun.* **76**:196–204.
24. Koester, M. K., Pullan, L. M., and Noltmann, E. A., 1981, *Arch. Biochem. Biophys.* **211**: 632–642.
25. Led, J. J., and Neesgaard, E., 1987, *Biochemistry* **26**:183–192.
26. Liang, J.-Y., and Lipscomb, W. N., 1987, *Biochemistry* **26**:5293–5301.
27. Lijas, A., Kannan, K. K., Bergstén, P.-C., Waara, I., Fridborg, K., Strandberg, B., Carlbom, U., Järup, L., Lövgren, S., and Petef, M., 1972, *Nature (London) New Biol.* **235**:131–137.
28. Lindskog, S., 1966, *Biochemistry* **5**:264–2646.
29. Lindskog, S., and Coleman, J. E., 1973, *Proc. Natl. Acad. Sci. USA* **70**:2505–2508.
30. Lindskog, S., and Wistrand, P. J., 1987, in: *Design of Enzyme Inhibitors as Drugs* (H. Sandler and H. J. Smith, eds.), Oxford University Press, New York, pp. 698–723.
31. Lindskog, S., Engberg, P., Forsman, C., Ibrahim, S. A., Jonsson, B.-H., Simonsson, I., and Tibell, L., 1984, *Ann. N.Y. Acad. Sci.* **429**:61–75.
32. Magid, E., 1967, Scand. J. Haematal. **4**:257–270.
33. Maren, T. H., 1987, *Drug Devl. Res.* **10**:255–276.
34. Notstrand, B., Vaara, I., and Kannan, K. K., 1975, in: *Isozymes I. Molecular structure* (C. L. Markert, ed.), Academic Press, Inc., New York, pp. 575–599.
35. Pocker, Y., and Deits, T. L., 1983, *J. Am. Chem. Soc.* **105**:980–986.
36. Pocker, Y., and Sarkanen, S., 1978, *Adv. Enzymol.* **47**:149–274.
37. Quiocho, F. A., Sack, J. S., and Vyas, N. K., 1987, *Nature* **329**:561–564.
38. Rowlett, R. S., and Silverman, D. N., 1982, *J. Am. Chem. Soc.* **104**:6737–6741.
39. Sanyal, G., 1984, *Arch. Biochem. Biophys.* **234**:576–579.

40. Sanyal, G., Swenson, E. R., Pessah, N. I., and Maren, T. H., 1982, *Mol. Pharmacol.* **22**: 211–220.
41. Silverman, D. N., and Lindskog, S., 1988, *Acc. Chem. Res.* **21**:30–36.
42. Steiner, H., Jonsson, B.-H., and Lindskog, S., 1975, *Eur. J. Biochem.* **59**:253–259.
43. Tashian, R. E., 1977, *Curr. Top. Biol. Med. Res.* **2**:21–62.
44. Tashian, R. E., Hewett-Emmett, D., Stroup, S. K., Goodman, M., Yu, Y.-S. L., 1980, in: *Biophysics and Physiology of Carbon Dioxide* (C. Bauer, G. Gros, and H. Bartels, eds.), Springer-Verlag, Berlin, pp. 165–176.
45. Tu, C. K., Sanyal, G., Wynns, G. C., and Silverman, D. N., 1983, *J. Biol. Chem.* **258**:8867–8871.
46. Tu, C., Wynns, G. C., and Silverman, D. N., 1981, *J. Biol. Chem.* **256**:9466–9470.
47. Vidgren, J., Liljas, A., and Walker, N. P. C., 1990, *Int. J. Biol. Macromol.*, in press.
48. Waara, I., Lövgren, S., Liljas, A., Kannan, K. K., and Bergstén, P. C., 1972, in: *Hemoglobin and Red Cell Structure and Function* (G. J. Brewer, ed.), Plenum Press, New York, pp. 169–184.

CHAPTER 4

Carbonic Anhydrase Kinetics and Molecular Function

RAJA G. KHALIFAH and DAVID N. SILVERMAN

Detailed kinetic and mechanistic studies of the zinc metalloenzyme carbonic anhydrase (CA) became possible only after the discovery of its existence in the form of two distinct though homologous erythrocyte isozymes, CA I and CA II. For a comprehensive early review of the structure and function of CA, see reference 72. Studies in the last decade have led to the discovery of the low-specific-activity CA III in skeletal muscle that has now been kinetically investigated. Other CA isozymes (IV–VII) have now also been identified[102] but have not been sufficiently characterized by kinetics to cover here.

Comparative analysis of the catalytic mechanism of CO_2 hydration by the isozymes has long been the goal of CA enzymatic investigations in order to help us understand the variety of physiological functions that CA might play in the organism.[73] Although other catalytic activities, such as aromatic ester hydrolysis and keto hydration, have been associated with CA,[66,67,81] the physiological role of such activities has yet to be demonstrated. All physiological functions of CA known are believed to be associated with the rapid and reversible production of an anion (bicarbonate) and a proton from the reversible hydration of CO_2. From a molecular viewpoint, a comparison of isozymes with closely related yet distinct active-site structures and specific activities should lead to a detailed molecular interpretation of the catalysis by this model zinc metalloenzyme. The fast and efficient kinetics of CA has made it the object of diverse and extensive biophysical studies.[6,7,67,68]

Despite their differences, CA I, CA II, and CA III share basic catalytic

RAJA G. KHALIFAH • Biochemistry Department, Kansas University Medical School, and Research Service, Veterans Administration Medical Center, Kansas City, Missouri 64128. ***DAVID N. SILVERMAN*** • Department of Pharmacology, University of Florida College of Medicine, Gainesville, Florida 32610

features of hydration–dehydration of CO_2. Important developments have recently occurred on two fronts. First, crystallographic advances[25] have yielded the first three-dimensional structure for the very low specific activity muscle CA III.[24] Moreover, crystallographic refinements have finally been carried out on the earlier structures of CA I[42] and CA II[25,27] and some of their inhibitor complexes. This refinement makes it possible to confidently examine the roles of active-site residues at the atomic detail needed to reach mechanistic conclusions, and it provides accurate starting coordinates for carrying out theoretical computations on complexes of CA with substrates and inhibitors.[62–64,111,112] Second, a major breakthrough is rapidly unfolding in the successful application of molecular cloning techniques to produce site-specific mutants that are being kinetically characterized. Dramatic new advances in our understanding of CA are thus anticipated from these exciting approaches.

1. Kinetic Properties and Inhibition of the Isozymes

1.1. Kinetic Properties of the Native Isozymes

We emphasize here the catalysis of the hydration of CO_2 and the dehydration of HCO_3^-. The other reactions catalyzed by CA, such as the hydration of aldehydes and the hydrolysis of esters,[81] will be noted only briefly, since it is becoming evident that they may display striking mechanistic differences from the CO_2 hydration catalysis.[77,80,108] Methods for studying the CO_2 hydration kinetics are covered elsewhere[32]; it is noted that the kinetic information comes from two major approaches. The first involves the initial-rate, stop-flow method that yields classical Michaelis–Menten parameters.[44] The other approach involves measuring the catalysis at chemical equilibrium in which the interconversion of substrate and product is followed by ^{13}C nuclear magnetic resonance (NMR)[58,97] or by mass spectrometry of ^{18}O-labeled substrates.[91] The synthesis of results obtained by these two complementary approaches has provided insights into the mechanism of CA that could not have been obtained from either approach alone.

Table I presents maximal values for the CO_2 Michaelis–Menten parameters for CA I, CA II, and CA III at 25°C. CA II has the largest turnover number (10^6 sec^{-1}) and is among the most rapid enzyme-catalyzed reactions known. The maximal value of k_{cat}/K_m for CA II is at or near the diffusion-controlled limit for the encounter rate of enzyme and substrate. This fact suggests that catalyzed hydration of CO_2 occurs as fast as this substrate can diffuse into the active site. In this respect, CA II fits the description of a "perfectly evolved enzyme."[56] For comparison, the bimolecular rate constant for the uncatalyzed reaction of the hydroxide ion with CO_2 is 8.5×10^3 M^{-1} sec^{-1}. The Michaelis–Menten parameters of Table I are smaller for CA I and CA III than for CA II, and a major emphasis of recent work with site-directed mutants of CA has been to identify the residues responsible for these differences.[5,31,69,109]

TABLE I
Maximal Steady-State Constants for CO_2 Hydration Catalyzed by Isozymes of CA[a]

Isozyme	k_{cat} (sec^{-1})	k_{cat}/K_m(M^{-1} sec^{-1})
CA I (human)	2×10^5	5×10^7
CA II (human)	1.4×10^6	1.5×10^8
CA III (feline)	1×10^4	3×10^5

[a]Data are at 25°C. Human CA I and II values are from reference 44; those for CA III are from reference 43.

Besides differences in the magnitudes of these constants, there are remarkable differences in pH profiles (Fig. 1). Whereas the K_m values are pH independent, the turnover numbers k_{cat} for CA I and CA II appear to depend primarily on the ionization of a major catalytic group, the pK_a of which is near 7, with maximal hydration activity occurring at high pH.[44,70] There is much experimental data to assign this pK_a to the protolysis of a zinc-bound water.[70,92] The precise appearance of the pH profile indicates that there are other modifying ionizing groups with smaller effects on the activity that are different, or have different pK_a values, in CA I and CA II.[10,44,96] In contrast, K_{cat} for CA III does not vary much with pH in this range.[107] This has been interpreted as being due to the low pK_a of the controlling ionization, which in this case must be below 6 to account for the data. A transient observation of the low-pH form of Co(II)-CA III has now been achieved by rapid-scanning stop-flow spectrophotometry after production of this species during substrate turnover[86,106] (Fig. 2).

Little kinetic information is known for the remaining isozymes of mammalian CA identified to date. Wistrand and Knuuttila[119] have purified the membrane-

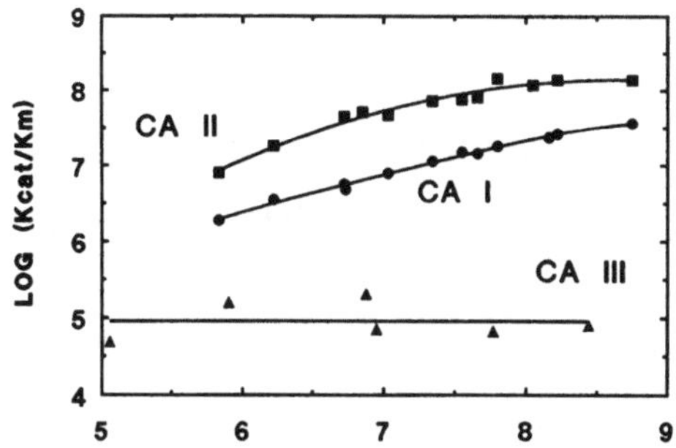

Figure 1. pH dependence of log(k_{cat}/K_m) for CO_2 hydration by human CA I and II at 25°C[44] and feline CA III at 10°C.[107] Since the K_m values are pH independent, the trends shown also apply for the variation of k_{cat} with pH.

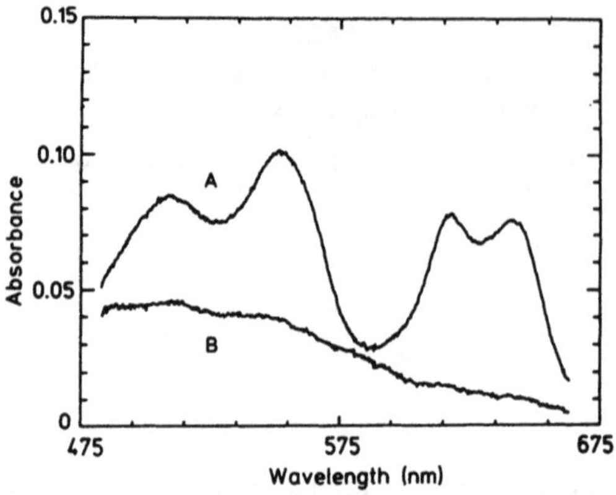

Figure 2. Rapid-scan stop-flow spectrum of activated 0.56 mM Co(II)-CA III during CO_2 hydration. (A) No CO_2; (B) spectrum recorded 6–7 msec after mixing with Na_2SO_4 saturated with CO_2. (From reference 86 with permission.)

bound CA IV from human kidneys. It resembles CA II in catalytic activity toward CO_2 hydration at pH 7.0. A similar result was reported for experiments at pH 7.4 in suspensions containing the membrane-bound CA IV from the brush border of the dog kidney, in which the Michaelis–Menten parameters were found to be nearly the same as with cytosolic CA II.[114] A CA isolated from the mitochondria of guinea pig liver has been identified as a unique CA V, mostly on the basis of a partial amino acid sequence.[35] This is a sulfonamide-sensitive enzyme like the high-activity CA I and CA II.[20] CA secreted into the saliva of the sheep has been fully sequenced and identified as CA VI.[30] The activity of this enzyme in the hydration of CO_2 is comparable to that of sheep red cell CA II. Moreover, the purification of rat salivary CA VI yields an enzyme equivalent in Michaelis–Menten parameters to the rat red cell CA II.[29] CA VI has a carboxyl-terminal tail of 45 residues not present in CA I, II, and III, and it also differs in containing carbohydrate attachments.[29,30]

1.2. Site-Specifically Modified Isozymes

1.2.1. Chemical Modifications and Metal Ion Replacements

Information on CA specifically modified at selected active-site residues or groups complements kinetic studies on the naturally occurring isozymes. Replacement of the intrinsic Zn(II) of CA by the closely related Co(II) was found to confer substantial catalytic activity toward CO_2 hydration, thus anchoring a central catalytic role for the metal ion.[72] Detailed kinetic studies on Co(II)-substituted CA II reveal subtle differences that can be interpreted in terms of essential mechanistic similarities to the zinc enzyme.[59,65] In contrast, the activity of Cd(II)-CA II is appreciable only at high pH, apparently as a result of the shift of the pH inflection to near 9.1.[3,103] A smaller alkaline shift of the esterase activity has been seen with Mn(II)-CA II[60] thus permitting the study of the CO_2 activity of Mn(II)-CA I at chemical equilibrium by NMR techniques.[62]

Histidines not involved in metal ion binding have been major targets of chemical modifications. Specific carboxymethylation of the highly nucleophilic His-200 of CA I,[117] which is absent in CA II and III, leads to a severalfold reduction in k_{cat} and complex changes in the pH dependence of k_{cat} and K_m.[47,48] The esterase activity displays a major shift in the controlling pK_a from about 7.5 to 9.2.[115] His-64 of CA II can be similarly alkylated with bromopyruvate,[33] and its high residual CO_2 hydration[47,48] and esterase[52] activities have been studied. However, the modified enzyme has never been completely purified and characterized.

Acrolein reacts with CA II, presumably by a (nonspecific) reaction with histidine side chains, such that k_{cat} for CO_2 hydration decreases by nearly 2 orders of magnitude and K_m decreases by 1 order of magnitude.[80] In contrast, the esterase and aldehyde hydration activities are actually slightly increased. A site-specific mutant of human CA II in which His-64 is replaced with alanine did not undergo activity changes when reacted with this reagent.[110]

Not all modifications produce decreases in catalytic activity. It has been reported that the catalysis of CO_2 hydration by the cysteine-rich bovine CA III can be activated severalfold by using certain thiol reagents that modify partially buried Cys-66, Cys-203, and Cys-206.[23,85] It has not yet been possible to identify the specific reactive residue(s) responsible for this activation. In contrast, thiol modification of human CA I at Cys-212[50] and CAII at Cys-204[15] produces a slight activity decrease only in the case of CA II. None of the cysteines actually protrude into the active-site cavity, although some are quite near. Activation of human and bovine CA III esterase and bicarbonate dehydration activities have also been reported with the arginine reagent 2,3-butanedione,[16] but it is not known whether cysteine modification was monitored in that study. Such activation with this reagent was not observed by Engberg and Lindskog,[23] using a CO_2 hydration assay. The reaction of porcine CA III with the arginine reagent phenylglyoxal leads only to a selective decrease in the weak *p*-nitrophenyl phosphatase activity of this CA without affecting the esterase or hydratase activity.[84]

1.2.2. Site-Specific Mutants

The successful cloning and expression of CA genes has culminated in the production of the first site-specific mutants of human CA II.[31] Some of the kinetic studies on such mutants are still preliminary[5,31,109] but already useful. To assess the catalytic influence of His-200, which is unique to CA I, the T200H mutant of CA II was produced and studied.[5] As shown in Table II, this simple substitution not only reproduces the lower k_{cat} and K_m for CO_2 seen in CA I as compared with CA II, but it also mimics the higher pK_a for the activity and the higher anion affinity of CA I.

TABLE II
Properties of CA I, CA II, CA II Mutant T200H, and CkeHis-64 CA II[a]

Parameter	CA II	T200H	CA I	CkeHis-64 CA II
k_{cat} (sec^{-1})	1.0×10^6	0.23×10^6	0.14×10^6	0.28×10^6
K_m (M)	8×10^{-3}	3.5×10^{-3}	4.8×10^{-3}	8×10^{-3}
k_{cat}/K_m (M^{-1} sec^{-1})	1.2×10^8	0.66×10^8	0.30×10^8	0.35×10^8
pK_a	6.9[b]	7.8[b]	7.1[b]	7.9[c]
$K_i^{SCN^-}$ (M)[d]	0.9×10^{-3}	0.04×10^{-3}	0.09×10^{-3}	

[a]The CO_2 hydration parameters k_{cat}, K_m, and k_{cat}/K_m refer to pH 8.7–8.8 and 25°C in 50 mM 1,2-dimethylimidazole–H_2SO_4 buffers. Data are from reference 5 except those for ckeHis-64 CA II, which are from reference 47.
[b]Obtained from catalysis of *p*-nitrophenyl acetate hydrolysis at 25°C.
[c]Obtained from k_{cat}/K_m.
[d]The K_i values refer to inhibition of esterase activity by NaSCN at pH 6.8.

However, this is not the only modification of CA II that produces a mimic of CA I. Similar results were earlier obtained (Table II) when His-64 of CA II was modified with bromopyruvate[47] (ckeHis-64 CA II), although effects on anion affinity were not studied.

Much attention has been focused on replacing His-64 of CA II by other amino acids in view of the proposed role of this side chain as a proton transfer group[31,109] (see Section 2.2). The H64E replacement, in which a glutamic residue is introduced, produces a mutant with a biphasic pH dependence of the esterase activity.[31] This is reminiscent of the results with the His-64 carboxyketoethylated CA II.[52] The other substitutions, H64K, H64Q, and H64A, did not produce this pH perturbation.[31]

Mutants involving the conserved residues Thr-199, which hydrogen bonds to the water ligand of the zinc,[25,40,41] and Glu-106, which is buried and hydrogen bonds to Thr-199,[41] have also been studied. Replacement of Thr-199 by alanine produces an inactive CA II mutant with no sulfonamide affinity, whereas replacement of Glu-106 by glutamine produces a 600-fold reduction in k_{cat} and a 13-fold reduction in K_m.[5]

1.3. Inhibition of Kinetics

CA inhibitors include the pharmacologically important sulfonamides and metal ion poisons as well as monoanions like nitrate and perchlorate, neutral compounds like imidazole and phenol, and dipositive metal ions like Cu(II) and Hg(II). The subject has been extensively reviewed[6,55,66,67,72–74]; we are concerned here mostly with implications regarding the mechanism of catalysis resulting from the mode of inhibition. In this regard, for example, the slowness of the dissociation of sulfonamides from the enzyme compared with the turnover by the substrates[103] limits their utilization.

Early studies on the anion inhibition of CO_2 hydration of CA indicated that the anions behave noncompetitively with regard to substrate CO_2 and competitively with regard to the anionic product bicarbonate.[72] More recent investigations have revealed that the pattern of inhibition of CO_2 hydration activity of CA II by anions is actually complex and pH dependent. Typically, uncompetitive inhibition patterns are observed at high pH near 9, while noncompetitive patterns are observed near neutral pH.[79,104] These results can be rationalized by kinetic models and simulations in which account is taken of slow proton transfer steps in the mechanism.[60,89,104] Very recently, a kinetic study of the inhibition of Co(II)-substituted CA III by azide was reported.[85] A remarkable pH-dependent difference was observed between the K_i for inhibition by kinetics and by equilibrium that increased with increasing pH, strongly implicating slow proton transfer steps.

The lack of competition by anions in CO_2 hydration kinetics and the discovery of imidazole as a unique competitive inhibitor of CA I[44] may be of much relevance to the question of CO_2 binding mode. This possibility has led to many further spectroscopic studies[1,4,54,120] as well as to the crystallographic determination of the structure of the CA I–imidazole complex.[41] In view of some of the issues raised by these studies (see Section 2.3), it is important to emphasize that the competitive behavior of imidazole was reported only at a single pH of 7.0. A unique CO_2 competitive inhibitor of CA II has also been discovered, this being phenol.[99] Comparison of its binding with that of imidazole is of great interest, and further studies on the interaction of phenol with CA II have been initiated.[49]

2. Molecular Basis for Catalysis and Inhibition

2.1. Role of the Metal Ion

Both k_{cat} and k_{cat}/K_m for CO_2 hydration and dehydration vary with pH in a manner that indicates major control by an ionizing group of pK_a near 7 for CA I and CA II.[70] This pH dependence is proposed to be a reflection of the protolysis of the zinc-bound water, with the zinc-bound hydroxide active in the hydration direction (Eq. 1)[18]

$$CO_2 + H_2O + EZnOH^- \rightleftharpoons HCO_3^- + EZnH_2O \qquad (1)$$

The possibility that one of the titratable histidines is this major catalytic group can be eliminated on several grounds. NMR experiments identified proton resonances from each of these residues and determined that none of them had the properties of the catalytic group.[13,14] The three histidine ligands of the metal, residues 94, 96, and 119, are not likely as the catalytic group, since the pK_a expected for an imidazole ring coordinated to a zinc is near 13[75]; there are exceptions, however.[54] Furthermore, they are not good candidates on the basis of mechanistic studies with model complexes.[12,34] In any case, the crystal structure of the active site indicates that these imidazole rings are H bonded to other residues in the protein and are not available for a catalytic function.[40]

The chemical modification studies on the active-site His-200 of CA I and His-64 of CA II also do not support their assignment as critical catalytic groups, since the modified enzymes retain substantial residual activity. Indeed, the studies of carboxymethylated CA I provide support for the assignment of a metal-bound water as the catalytic ionizing group. The pK_a of the residual esterase activity is shifted from about 7.5 to 9.2, in parallel with the shift of the pK_a of the visible spectrum of carboxymethylated Co(II)-CA I.[115] The latter spectral change is assignable to the ionization of the water ligand.[6–8] Furthermore, the alterations in catalytic properties accompanying carboxymethylation of His-200 of CA I have been shown by ^{13}C NMR to be due to perturbations at the metal ion. The introduced carboxymethyl carboxylate was shown to coordinate to the metal ion[51,52] in an "intramolecular reversible inhibition" mode first proposed by Coleman[17] (Fig. 3).

Another possibility is the conserved Glu-106. This buried side chain is H bonded to Thr-199, which in turn is H bonded with the water ligand of the zinc.[40,41] It is difficult to evaluate its role, but the shift in the apparent pK_a of the catalytic group with change in the metal, such as that seen with Cd(II), is not consistent with this residue being the activity-controlling group. Natural abundance ^{13}C NMR studies on the glutamates of CA I have also not yet revealed any such residue titrating in the neutral pH range.[53] Preliminary studies on the E106Q mutant of CA II,[5] in which it is replaced by glutamine, reveal that k_{cat}/K_m is diminished to about 2% of the value for native CA II. Studies on the pH dependence of this residual activity and the visible spectrum of its Co(II) derivative should be most useful.

Metal ion substitutions on CA I and CA II[36,71] as well as CA III[22] support the hypothesis that a metal-bound water ionizing to hydroxide is the main activity-controlling ionizing group. For example, the alkaline shift of the apparent pK_a for *p*-nitrophenyl acetate hydrolysis by CA II from about 7 in the Zn(II) enzyme to above 9 for the Cd(II)-substituted enzyme is in agreement with the pK_a values from pH titrations of ^{113}Cd NMR chemical shifts.[39] Co(II)-substituted CA shows excellent correlations between the visible absorption spectra and the pH rate profiles in a wide body of experiments.[66,67] For example, anion inhibitors, such as I^-, that are known to bind to the metal in CA II induce a shift to a higher pK_a in the visible absorption spectra that are parallel with the shift found for the catalytic activity.[9,82,83] These spectra also show that anions either may replace a metal-bound water to form four-coordinate complexes or may bind as an additional ligand to form a five-coordinate complex.[6–8] X-ray diffraction studies show in detail the position of SCN^- binding to the zinc of CA II as a fifth ligand.[24,25] These studies also confirm that the high-pH form of the active site has tetrahedral coordination about the zinc. Crystal structures of the low-pH forms of CA I and CA II, where visible spectra and electron spin correlation times indicate higher coordination number and lower symmetry,[6–8] remain to be investigated.

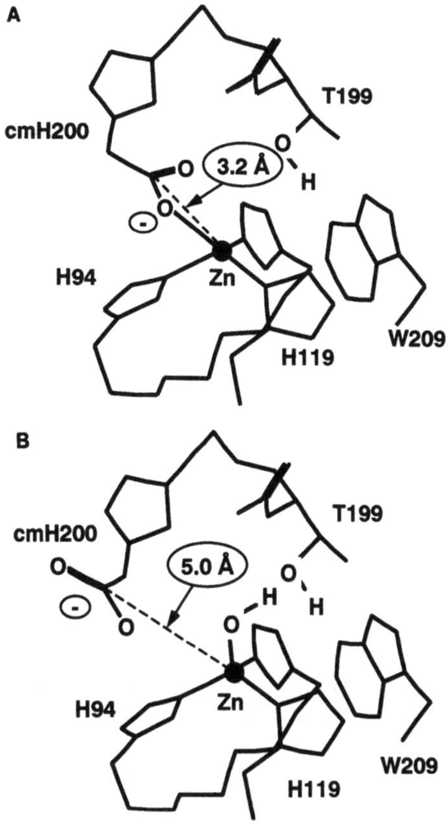

Figure 3. Distances between the metal ion and the carboxymethyl carboxyl carbon of human CA I alkylated at His-200, as determined by paramagnetic ^{13}C NMR measurements.[51,52] The carboxylate coordinates to the metal ion at pH 8 or below (B) but is displaced by a solvent hydroxyl at pH 10.4 (A) or by external inhibitors (not shown). Structures are drawn for illustrative purposes, and it is not known whether a water ligand is simultaneously present at low pH.

The assignment of the zinc-bound OH^- as the catalytic group for the CA-catalyzed hydration of CO_2 has its problems. The pK_a of about 7 assumed for the zinc-bound water is appreciably lower than expected from the pK_a of $Zn(H_2O)_6^{2+}$, which is near 10. However, the low pK_a in the enzyme could be due to the lowered coordination number about the metal in CA.[121] This low pK_a does not preclude the zinc-bound OH^- from acting in catalysis as a group of considerable nucleophilicity despite low basicity.[12,76] It is striking that the pK_a of the zinc-bound water could be as low as 5 for the case that His-64 is protonated[96] and may be lower for CA III.[22,107]

The catalytic hydration of CO_2 by the zinc-bound OH^- could follow a pathway involving a direct nucleophilic attack of the zinc-bound hydroxide on CO_2 or one in which the zinc-bound hydroxide acts as a general base extracting a proton from an intervening water molecule with attack of the incipient hydroxide on CO_2. This hypothesis is not certain; the arguments put forward so far are based on the solvent hydrogen isotope effects on the catalysis in steady-state and equilibrium experiments. There is no change in the ratio k_{cat}/K_m for hydration catalyzed by CA II when solvent is changed from H_2O to D_2O.[78,100] This ratio contains rate constants for steps up to and including the first irreversible step in the catalysis, which in initial velocity experiments is the departure of HCO_3^-. It is concluded that the actual steps involved in the interconversion of CO_2 and HCO_3^-, the steps of Eq. 1, do not contain a rate-contributing proton transfer. This finding supports but does not confirm the direct nucleophilic attack of the zinc-bound hydroxide on CO_2.

Further support comes from ^{13}C NMR and ^{18}O studies of exchange at equilibrium, both of which also reveal no effect of the change of solvent to D_2O on the steps of Eq. 1.[95,97] These results do not preclude internal proton transfer within the zinc-bound HCO_3^-[62] that may not be rate contributing. In fact, this internal transfer was calculated to have a low energy barrier when a water molecule is included for proton relay. Molecular orbital calculations have led to the conclusion that the reaction between CO_2 and OH^- in the gas phase proceeds with no energy barrier,[38] suggesting that the energy barrier for the uncatalyzed reaction in aqueous solution is due to solvation effects. Perhaps the hydrophobic nature of the active-site cavity is important in the CA catalytic effect.

An important aspect of the function of the metal is in the rate of protolysis of the zinc-bound water, which process must occur at least as fast as the turnover rate for the CA II-catalyzed CO_2 hydration of 10^6 sec^{-1}. Here is another important role of the zinc in the proton transfer that regenerates the zinc-bound hydroxide from the zinc-bound water. Molecular orbital calculations show that the energy barrier for this proton transfer is significantly lowered when the water, as proton donor, is coordinated to a Zn(II) ion.[63] The zinc ion lowers the pK_a of the bound water molecule and also repels the proton electrostatically. Moreover, these calculations show that the intramolecular proton transfer by which the zinc-bound hydroxide is formed is energetically more favorable for the four-coordinated than for the five-coordinated zinc at the active site.

2.2. Proton Transfer in the Catalysis

2.2.1. Intermolecular Proton Transfer

There are two classes of proton transfer to consider in the catalytic mechanism of CA, intermolecular and intramolecular.[94] The issue of intermolecular proton transfer was first described by DeVoe and Kistiakowsky,[19] who recognized that the very large turnover rate of CA II presented problems in terms of the transfer of a proton from enzyme to solution. It was known[21] that the fastest that an ionizable group with a pK_a near 7 could transfer a proton to water was about 10^3 sec^{-1}. In CA II such a proton transfer must occur as fast as the turnover rate of 10^6 sec^{-1}, since the catalyzed reaction releases a proton along with the bicarbonate product (Eq. 1). Alternatively, the product of the reaction could be carbonic acid, and no proton transfer would be involved. This is an equally difficult option, since the encounter rate of carbonic acid and enzyme needed to explain the dehydration reaction also exceeds the limits of diffusion control. A suggested solution was that the proton transfer in the hydration direction occurs directly from the enzyme to buffer in solution rather than to solvent water,[45,70,83] a process first discussed by Alberty.[2] Instead of H_2O, a poor acceptor of protons, or OH^-, which is not abundant at physiological pH, the proton acceptor was suggested to be buffers that in kinetic experiments are usually in the millimolar range.

Verification of the buffer hypothesis first came from experiments carried out at chemical equilibrium where buffers are not necessary. The exchange of ^{18}O between CO_2 and water catalyzed by CA II was enhanced by buffers of the imidazole–piperazine–pyridine type in the millimolar range.[93] Subsequently, this result was found in stop-flow experiments in which the initial velocities of the hydration of CO_2 were enhanced with similar buffers in the millimolar range.[37] The magnitude of the enhancement of catalytic rate by buffers was two- to fivefold and occurred with buffer concentrations of up to 10 mM in most cases. This effect is accounted for by the nature of the experimental techniques and does not detract from the support of the hypothesis of intermolecular proton transfer. Thus, the buffer hypothesis represented in Eqs. 1 and 2 was verified, and the suggestion that carbonic acid is the product of the catalytic reaction[57] was discarded. The implication of Eqs. 1 and 2 is that the catalytic pathway consists of two separate and distinct sequences, referred to as a ping-pong mechanism.

$$EZnH_2O + \text{buffer} \overset{k_4}{\rightleftharpoons} EZnOH^- + {}^+H\text{-buffer} \qquad (2)$$

The dependence of the catalyzed velocity of the hydration of CO_2 on the acidity of the acceptor buffer was also investigated. The rate constants k_4 in Eq. 2 for the transfer of a proton from the enzyme to buffer are consistent with the Bronsted law; i.e., k_4 depends on the difference in pK_a between the enzyme as proton donor with pK_a about 7 and the buffer as acceptor.[82,90] The maximal k_4 value of 1×10^9 sec^{-1} was consistent with a diffusion-controlled proton transfer for the case in which the pK_a of the buffer as acceptor is greater than that of the enzyme as donor. The behavior of this proton transfer is thus very similar to proton transfer between small molecules. No trend was detected in the structure of buffers that could transfer protons with CA II, which seemed uncharacteristic of the specificity associated with active-site processes. It is now believed (see next section) that the proton donor group on the enzyme with a pK_a near 7 is not the zinc-bound hydroxide 15 Å deep in the active-site cavity but the imidazole side chain of His-64, which is located much closer to the surface of the protein, about 7.5 Å from the metal.[24–27]

The slower CA I and CA III have not been studied as thoroughly for buffer effects. There is a report that CO_2 hydration catalyzed by CA I can be enhanced by buffers.[105] There are no general buffer effects as described by the Bronsted curve found with CA III.[43] No buffer effect is anticipated with the slower catalyses of the esterase function of the enzyme. There has been no careful consideration to date of the cytoplasmic buffers that are likely to contribute to the maximal catalytic activity of CA II inside the cell. Such a rate enhancement that is dependent on the supply of protons delivered by buffers in solution has not been well studied in other enzymes, and in this respect CA is unique. Similar processes are likely in other fast enzymes that produce or consume protons, such as acetylcholinesterase and superoxide dismutase.

2.2.2. Intramolecular Proton Transfer

The buffer studies established that the catalytic activity of CA II could be limited by intermolecular proton transfer when buffer concentration is less than about 10 mM. What limits the maximal catalytic turnover for CA II in the presence of abundant buffers as expected in a cytoplasmic environment? An answer to this question was first suggested from the studies of the kinetics of human and bovine CA II measured by using D_2O as solvent.[78,100] For the catalyzed hydration of CO_2, the change in solvent from H_2O to D_2O had no effect on the ratio k_{cat}/K_m, but the value of k_{cat} was greater by a factor of 3.8 in H_2O compared with D_2O. Very similar results were obtained in the dehydration direction. A solvent hydrogen isotope effect as great as 3.8 is consistent with a primary proton transfer in the catalytic pathway. This proton transfer was interpreted as being intramolecular between the zinc-bound water and another group near the active site in the steps that regenerate the zinc-bound hydroxide form of CA.[100] In the active-site cavity there is only one residue that qualifies for this role. This is His-64, whose imidazole side chain was the appropriate pK_a of 7.1, determined by NMR,[13,14] to support proton transfer as rapid as k_{cat} in both the hydration and dehydration directions. On the basis of the solvent hydrogen isotope effects on k_{cat}, it was suggested that the proton transfer involves intervening water molecules.[113] This proton transfer in CA II is represented in Eq. 3, where the proposed proton transfer group is identified as His-64:

$$\text{His64-EZnH}_2\text{O} \rightleftharpoons {}^{+}\text{H-His64-EZnOH}^{-} \overset{\text{buffer}}{\rightleftharpoons} \text{His64-EZnOH}^{-} + {}^{+}\text{H-buffer} \quad (3)$$

There is much evidence for this intramolecular proton transfer and for His-64 being a proton shuttle residue.[67,93,94] Major support has come from inhibition experiments, such as the inhibition by HCO_3^- of the hydration of CO_2 catalyzed by CA II.[101] The initial velocity of this catalysis is consistent with a term $[CO_2][HCO_3^-]$

in the denominator of the rate expression that is characteristic of a rate-limiting isomerization step such as in Eq. 3. Anion inhibition of the catalysis of CO_2 hydration is noncompetitive[79,104] with respect to CO_2 at pH 7 or lower and uncompetitive at pH near 9. This is consistent with intramolecular proton transfer due to the accumulation, at pH 9 and $[CO_2] > K_m$, of the species of enzyme occurring before the rate-limiting step. This is the zinc-bound water form of the active site to which anions bind most tightly.[104] These aspects of the catalytic pathway of CA II and its inhibition by anions have been confirmed by computer calculations of the catalysis which simulate many features described here, including the pH dependence of the Michaelis–Menten constants and the mode of anion inhibition.[68,89] Measurements of the catalysis at chemical equilibrium also confirm the hypothesis that a step outside the actual interconversion of CO_2 and HCO_3^- in Eq. 1 is rate limiting. Thus, the maximal rate at chemical equilibrium of Eq. 1, which does not include the intramolecular proton transfer, is greater than the maximal rate at steady state that does include that transfer.[95,97]

There is substantial evidence pointing to the involvement of His-64 of CA II in this process. Cupric and mercuric ions at concentrations in the micromolar range inhibit ^{18}O exchange catalyzed by CA II in a manner consistent with rate-limiting proton transfer. These metal ions had no effect on the rate at chemical equilibrium of the steps of Eq. 1, but they did inhibit the release from the enzyme of ^{18}O-labeled water. This finding was interpreted to indicate inhibition of Eqs. 3 and 4 by the coordination of these metal ions to the imidazole ring of His-64.[94]

$$^{+}\text{H-His64-EZn}^{18}\text{OH}^{-} \rightleftharpoons \text{His64-EZn}^{18}\text{OH}_2 \overset{\text{H}_2\text{O}}{\rightleftharpoons} \text{His64-EZnH}_2\text{O} + \text{H}_2{}^{18}\text{O} \qquad (4)$$

This suggestion is supported by the crystal structure of CA II inhibited by mercuric ions.[26] A mixture of two complexes was found, with Hg^{2+} bound to either of the two nitrogens of the imidazole ring of His-64.

In another approach, the nonspecific chemical modification of CA II by acrolein decreased by as much as 95 to 98% the steady-state turnover rate k_{cat} for the hydration of CO_2, an effect suggested to be due primarily to the formylethylation of His-64, even though other residues are also modified.[80] This assignment has been supported by ^{18}O exchange kinetics that showed that acrolein modification had an effect on the catalysis similar to the effect of inhibition by cupric ions. The rate of Eq. 1 was not affected (Fig. 4), but the rate of release of ^{18}O-labeled water from the enzyme, and hence the intramolecular proton transfer (Eqs. 3 and 4), was blocked (Fig. 5).[110] In addition, the ^{18}O exchange pattern of a mutant CA II in which the His-64 was replaced with alanine (H64A CA II) was similar to that of the acrolein-modified CA II. As anticipated from the above, the acrolein reaction of H64A CA II did not significantly alter its ^{18}O exchange kinetics.[110] These results support the hypothesis that His-64 is a proton transfer group in the catalysis.

Direct information on the role of His-64 in CA II can be obtained with site-

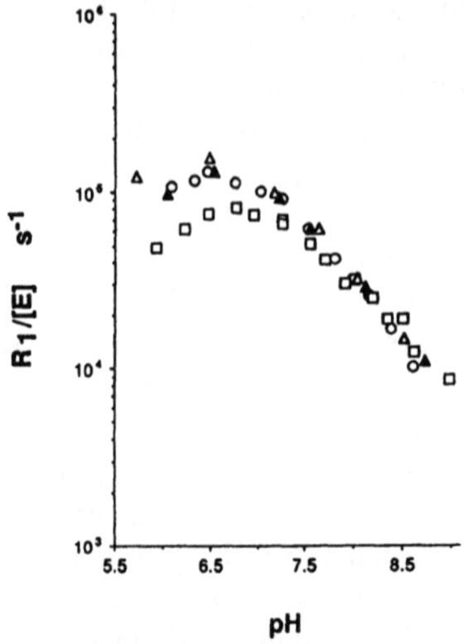

Figure 4. Dependence of the CO_2–HCO_3^- equilibrium exchange rate R_1/E on pH for human CA II (○), acrolein-modified CA II (Δ), H64A CA II mutant (□), and acrolein-modified H64A CA II (▲). The rate was determined by following the ^{18}O distribution between CO_2 species and water. (From reference 110 with permission.)

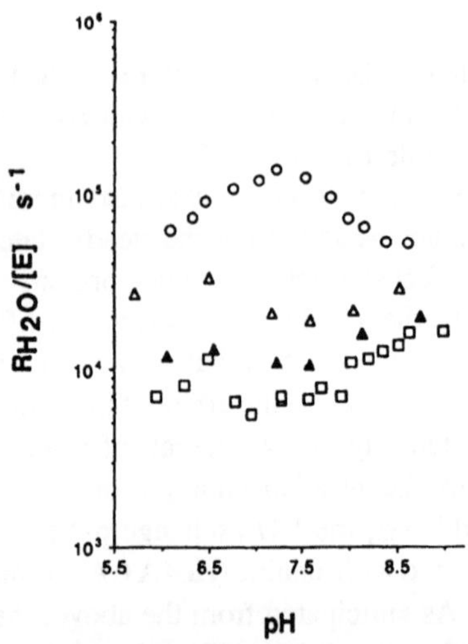

Figure 5. Dependence of the rate of ^{18}O-labeled water release from enzyme, R_{H_2O}/E, on pH for human CA II (○), acrolein-modified CA II (Δ), H64A CA II mutant (□), and acrolein-modified H64A CA II (▲). (From reference 110 with permission.)

specific mutants in which this residue is replaced by others with no proton transfer capability. Initial reports that His-64 is not necessary for high CO_2 hydration activity were based on studies at pH 8.8 in the presence of the buffer 1,2-dimethylimidazole.[31] In the absence of buffers or with buffers of larger size, both ^{18}O exchange and stop-flow kinetic data indicate that His-64 functions as a proton transfer group.[109] The ^{18}O-exchange pattern of the mutant H64A CA II is similar to that of both the cupric-inhibited and the acrolein-modified enzymes in that the rate of the actual interconversion of CO_2 and HCO_3^- in Eq. 1 is not significantly altered (Fig. 4), but the proton transfer-dependent release of ^{18}O-labeled water from the enzyme in Eq. 4 is decreased about 10-fold (Fig. 5).[109] Measured at pH 7.5 at steady state by using the buffer HEPES, the turnover number k_{cat} for the hydration of CO_2 was also decreased by 10-fold for the lysine mutant H64K CA II compared with the wild-type enzyme. The lower catalytic activity for the position 64 mutants is overcome by increasing the concentration of buffers of small size, such as imidazole and 1-methylimidazole. It is likely that buffer molecules of small size are able to enter the active-site cavity and effect proton transfer with the zinc-bound OH^-. More bulky buffers such as HEPES and MOPS had no effect on the kinetics, presumably because they could not fit into the active-site cavity.[109]

In the slower CA I, the function of His-64, if any, is unknown. This residue in CA I does not protrude into the active-site cavity as in CA II because of its shielding by His-67 and His-200.[40] Possibly His-200 in CA I functions as a proton shuttle group, as it is close enough to the metal to affect anion binding.[46,116] Rates of interconversion of CO_2 and HCO_3^- determined by ^{13}C NMR are as much as 50 times slower for CA I than for CA II,[98] indicating that the overall slower catalysis by CA I is not due to the proton transfer rate but to the slower rate of Eq. 1. In CA III, the residue at position 64 is a lysine pointing away from the zinc.[24,25] There is no evidence for intramolecular proton transfer in the catalysis of CO_2 hydration by CA III, but there is evidence for rate-limiting intermolecular proton transfer between the active site from studies of solvent hydrogen isotope effects, anion inhibition, and kinetics of chemically modified CA III.[43,85,86] These studies confirm, as in the case for CA II, that the hydration of CO_2 catalyzed by CA III is limited by the protolysis of the zinc-bound water. In the case of CA III there is no general buffer effect, however, and it is possible that the acceptor of protons from the zinc-bound water is another water molecule in the active-site cavity. Such an explanation is consistent with the turnover rate of 10^3 to 10^4 sec^{-1} for CO_2 hydration catalyzed by CA III.

2.3. Binding Sites of Substrates and Competitive Inhibitors

The small size of the neutral CO_2 substrate, its weak affinity for CA, and its limited solubility give little promise for determining its mode of binding by crystallographic techniques. The prospects for determining the structure of the CA complex with the reverse substrate HCO_3^- are better, in principle, especially with

CA I, for which HCO_3^- has higher affinity. However, no such crystallographic studies have been reported. Current information on the binding of these substrates thus comes from consideration of competitive inhibitor binding modes, from the application of physical techniques in solution, and from molecular modeling or similar approaches.

Since monoanions inhibit CA and compete with HCO_3^- for binding at the essential zinc,[6–8,25,65–67] it is probable that this substrate directly coordinates to the metal ion. The low extinction of the visible spectrum of its Co(II)-CA complex[6–9] suggests five-coordinate or a very distorted four-coordinate complex. Paramagnetic ^{13}C NMR measurements on the binding of HCO_3^- to Co(II)-CA I have revealed that the distance between the metal ion and the carbon is 3.2 Å.[118] Assuming penta-coordination, these results are not sufficient to distinguish whether HCO_3^- binds through two of its oxygens after displacing the water ligand or whether it binds through a single oxygen without displacing the water. Molecular mechanics studies[112] seem to only exclude a five-coordinate binding mode whereby the two coordinating oxygens of the substrate are equidistant from the metal, since the carbon-to-metal distance would be far shorter than observed by NMR. It seems reasonable that the negatively charged oxygen would bind closest to metal ion, with the other non-protonated oxygen interacting with the metal at a longer distance, as seen in sulfonamide binding.[24,25] Unfortunately, this study only considered other structures in which the protonated oxygen is closest to the metal ion. All of the above mentioned studies are consistent with a mechanism whereby a coordinated OH^- attacks a nearby CO_2 to produce a coordinated HCO_3^-.

Little progress has been achieved by direct and indirect approaches to understanding how CO_2 binds to the active site of CA. The affinity of this substrate for CA is entirely unknown. Initial speculations[48] that the K_m for hydration (Table I) might reflect binding affinity were based on an enzyme kinetic scheme that did not accommodate subsequent results on proton transfer involvement.[101] Computer simulations of kinetic mechanisms for CA II indicate a possible affinity for CO_2 that is at least 2 orders of magnitude weaker than the K_m,[68,89] thus greatly diminishing prospects for direct observation of CO_2 complexes of CA. Early attempts to observe such complexes by infrared spectroscopy[87] were considered not to reflect substrate binding,[44] but the method is intrinsically sound and may still be fruitful now that competitive inhibitors are known. Paramagnetic ^{13}C NMR studies on Co(II)-CA[118] and the inactive Cu(II)-CA[11] have not succeeded in shedding light on the binding of this substrate and its distance from the metal.

In view of the above, elucidation of the mode of binding of the competitive inhibitors imidazole[44] and phenol[99] is of interest. A crystal structure had been reported for the imidazole complex of CA I at a nominal pH of 8.7[41] and had been interpreted in terms of neutral imidazole binding as a distant fifth ligand of the zinc without displacing the water ligand (Fig. 6). This structure appeared to be supported by the finding that imidazole affinity for CA I was independent of the catalytic group ionization on the enzyme.[120] However, visible spectroscopic studies of imidazole bound to Co(II)-CA I revealed a striking pH-dependent

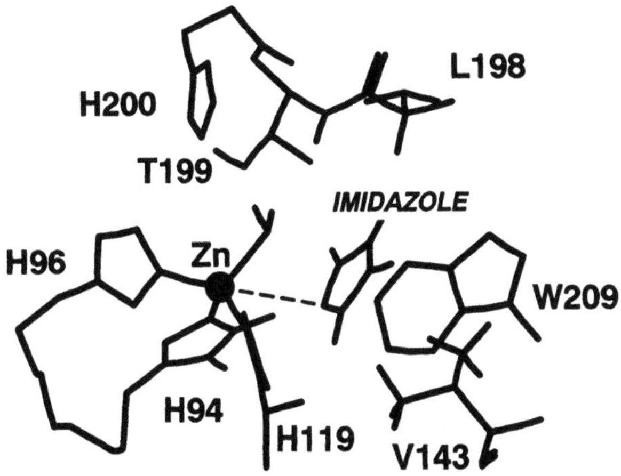

Figure 6. Approximate placement for the CO_2-competitive inhibitor imidazole in the active site of human CA I, based on information reported by Kannan *et al.*[41] for the unrefined crystal structure of the complex at a nominal pH of 8.7. Note that imidazole is shown not to displace the solvent water ligand. The distance from the nitrogen to the metal ion (dashed line) is reported to be about 2.7 Å, i.e., outside normal coordination, and the metal ion is not in the plane of the imidazole ring.

intensification of the visible spectrum of the complex that occurs in going from low pH to high pH, with a pK_a near 8.[1,4,54] Since the intense spectrum at high pH, but not that at low pH, is characteristic of tetrahedral complexes, it has been suggested[6] that the spectral changes reflect an equilibrium between two CA–imidazole complexes, with water being necessarily displaced at high pH by an imidazolate anion (normal pK_a of 13–14 in solution) (Fig. 7). A more detailed investigation of the pH dependence of binding has supported this suggestion.[54] Furthermore, it has been shown by ^{15}N NMR that amide inhibitors and substrates, which are deprotonated with a similar pK_a above 13, do bind as anions to the metal ion of CA.[77,88] Only the low-pH, five-coordinate imidazole complex (Fig. 6) competes for certain with CO_2. Further crystallographic studies at higher resolution and at different pH would help verify these conclusions, especially since the reported crystal structure[41] may be at a pH where a mixture of the two complexes are in equilibrium. Studies to elucidate the mode of phenol binding to CA II and to compare it to imidazole interaction with CA I have also been initiated using ^{13}C NMR[49] and have revealed a binding mode in which phenol does not displace the water ligand of the zinc. We cannot distinguish at present whether CO_2 interacts weakly at a distance with the zinc or whether it binds loosely in a hydrophobic cavity nearby.

2.4. Active-Site Residues and Catalysis

Our understanding of the role of specific active-site groups and residues in the mechanism of CA is limited. The only essential participatory group that has been

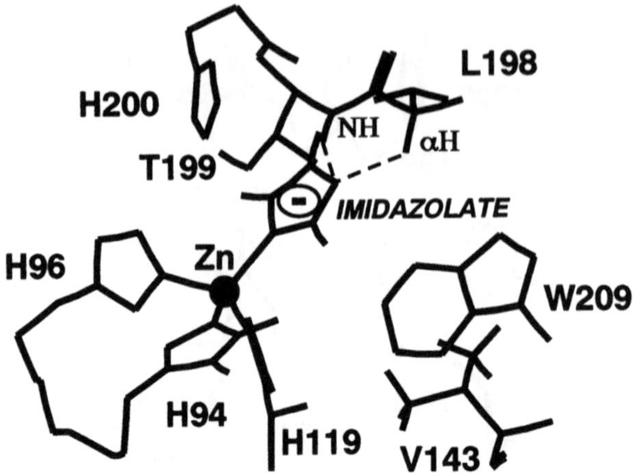

Figure 7. Presumed tetrahedral mode of binding of the imidazolate anion to human CA I at high pH deduced by model building using molecular graphics. Note the stabilizing interaction of the peptide NH of T199 with the deprotonated "pyrrole" nitrogen of the imidazolate anion (dashed line). A hydrogen at this pyrrole nitrogen would encounter severe steric hindrance with this peptide NH, as well as with the alpha-carbon proton of L198. This may possibly account for the inability of neutral imidazole to tetrahedrally coordinate in CA I (Fig. 6).

established is the metal ion and its water–hydroxide ligand. By its nature, the substrate CO_2 does not lend itself to easily studied interactions such as H bonding or coordination. The active site of CA may have evolved to complement this substrate by providing a nonpolar active-site environment, and yet CA can discriminate between two nearly identical neutral molecules such as CO_2 and N_2O.[44] The side chain of the conserved Thr-199 H bonds to the water-hydroxide ligand of the zinc and could also be a critical residue, especially since the T199A mutant is nearly inactive.[5] Its role could be to orient the reactants or stabilize transition states through H bonding. The extent to which its interaction with the buried Glu-106 influences the catalysis[41] remains to be established, but the results with the E106Q mutant[5] suggest this role is significant.

The nonconserved residues in the active-site appear to "fine tune" the catalysis and must be responsible for the wide isozymic differences in catalytic efficiency.[69] The only proven mechanistic role is that of His-64 in CA II, where it functions as a remarkable proton transfer group.[92] His-200 is likely to interact with nearby ligands of the zinc, such as anions and the substrate HCO_3^-.[46,69] A large part of its catalytic influence may be in shielding His-64 and preventing participation of the latter (see Table II). Despite early speculation, the unique presence of Arg-67 and Lys-64 residues in the active site of CA III has not been linked to the diminished catalytic activity of this isozyme. Crystallographic studies have suggested that the presence of the bulky Phe-198 in CA III may be very influential.[24,25]

Site-directed mutants on CA I and CA III are eagerly awaited to resolve such questions.

Aside from the above, there are other moieties in the active site that deserve attention. The most prominent of these is the peptide NH of Thr-199 that can H bond and interact with substrates and ligands of the metal ion.[24–27,69] It will be a worthwhile challenge to evaluate the role of this group which is not amenable to study by site-directed mutagenesis.

ACKNOWLEDGMENTS. Research support was provided by the Medical Research Service of the Veterans Administration (to R.G.K.) and the National Institutes of Health (grant GM 25154 to D.N.S.). It is a pleasure to express our gratitude to Professor Sven Lindskog for his collaboration, to our many colleagues whose contributions have been cited, and to Drs. S. Lindskog, A. Liljas, and A. E. Eriksson for providing helpful information prior to publication. This account is dedicated to Professors Thomas H. Maren and John T. Edsall, who first introduced us to carbonic anhydrase and who continued to stimulate our interest in it.

References

1. Alberti, G., Bertini, I., Luchinat, C., and Scozzafava, A., 1981, *Biochim. Biophys. Acta* **668**: 16–26.
2. Alberty, R. A., 1962, *Brookhaven Symp. Biol.* **15**:18–31.
3. Bauer, R., Limkilde, P., and Johansen, J. T., 1976, *Biochemistry* **15**:334–342.
4. Bauer, R., Limkilde, P., and Johansen, J. T., 1977, *Carlsberg. Res. Commun.* **42**:325–339.
5. Behravan, G., Forsman, C., Jonsson, B.-H., Liang, Z.-W., and Lindskog, S., 1989, *Highlights Mod. Biochem.* **1**:291–297.
6. Bertini, I., and Luchinat, C., 1983, *Acc. Chem. Res.* **16**:272–279.
7. Bertini, I., and Luchinat, C., 1984, *Adv. Inorg. Biochem.* **6**:71–111.
8. Bertini, I., Canti, G., Luchinat, C., and Scozzafava, A., 1978, *J. Am. Chem. Soc.* **100**:4873–4877.
9. Bertini, I., Luchinat, C., and Scozzafava, A., 1982, *Struc. Bonding* **48**:45–92.
10. Bertini, I., Dei, A., Luchinat, C., and Monnanni, R., 1985, *Inorg. Chem.* **24**:301–303.
11. Bertini, I., Luchinat, C., Monnanni, R., Roelens, S., and Moratal, J. M., 1987, *J. Am. Chem. Soc.* **109**:7855–7856.
12. Buckingham, D. A., 1977, *Biological Aspects of Inorganic Chemistry* (A. W. Addison, W. R. Cullen, D. Dolphin, and B. R. James, eds.), John Wiley & Sons, New York, pp. 141–196.
13. Campbell, I. D., Lindskog, S., and White, A. I., 1974, *J. Mol. Biol.* **90**:469–489.
14. Campbell, I. D., Lindskog, S., and White, A. I., 1975, *J. Mol. Biol.* **98**:597–614.
15. Carlsson, U., Aasa, R., Henderson, L. E., Jonsson, B.-H., and Lindskog, S., 1975, *Eur. J. Biochem.* **52**:25–36.
16. Chegwidden, W. R., Hewett-Emmett, D., and Tashian, R. E., 1984, *Ann. N.Y. Acad. Sci.* **429**:179–181.
17. Coleman, J. E., 1975, *Annu. Rev. Pharmacol.* **15**:221–242.
18. Davis, R. P., 1959, *J. Am. Chem. Soc.* **81**:5674–5678.
19. DeVoe, H., and Kistiakowsky, G. B., 1961, *J. Am. Chem. Soc.* **83**:274–280.

20. Dodgson, S. J., Forster, R. E., Storey, B. T., and Mela, L., 1980, *Proc. Natl. Acad. Sci. USA* **77**:5562–5566.
21. Eigen, M., and Hammes, G. G., 1964, *Adv. Enzymol.* **25**:1–37.
22. Engberg, P., and Lindskog, S., 1984, *FEBS Lett.* **170**:326–330.
23. Engberg, P., and Lindskog, S., 1986, *Eur. J. Biochem.* **156**:407–412.
24. Eriksson, A. E., 1988, Doctoral dissertation, Uppsala University, Uppsala, Sweden.
25. Eriksson, A. E., and Liljas, A., this volume, Chapter 3.
26. Eriksson, A. E., Jones, T. A., and Liljas, A., 1986, in: *Zinc Enzymes* (I. Bertini, C. Luchinat, W. Maret, and M. Zeppezauer, eds.), Birkhauser, Boston, pp. 317–328.
27. Eriksson, A. E., Jones, T. A., and Liljas, A., 1988, *Proteins, Structure, Function, and Genetics,* **4**:274–282.
28. Eriksson, A. E., Kylsten, P. M., Jones, T. A., and Liljas, A., 1988, *Proteins, Structure, Function, and Genetics* **4**:283–293.
29. Fedlstein, J. B., and Silverman, D. N., 1984, *J. Biol. Chem.* **259**:5447–5453.
30. Fernley, R. T., Wright, R. D., and Coghlan, J. P., 1988, *Biochemistry* **27**:2815–2820.
31. Forsman, C., Behraven, G., Jonsson, B.-H., Liang, Z. W., Lindskog, S., Ren, X., Sandstrom, J., and Wallgren, K., 1988, *FEBS Lett.* **229**:360–362.
32. Forster, R. E., II, this volume, Chapter 6.
33. Gothe, P. O., and Nyman, P. O., 1972, *FEBS Lett.* **21**:159–164.
34. Harrowfield, J. M., Norris, V., and Sargeson, A. M., 1976, *J. Am. Chem. Soc.* **98**:7282–7289.
35. Hewett-Emmett, D., Cook, R. G., and Dodgson, S. J., 1987, *Isozyme Bull.* **19**:13.
36. Hunt, J. B., Rhee, M. J., and Storm, C. B., 1977, *Anal. Biochem.* **79**:614–617.
37. Jonsson, B.-H., Steiner, H., and Lindskog, S., 1976, *FEBS Lett.* **64**:310–314.
38. Jönsson, B., Karlström, G., and Wennerström, H., 1978, *J. Am. Chem. Soc.* **100**:1658–1661.
39. Jonsson, B.-H., Tibell, L., Evelhoch, J. L., Bell, S. J., and Sudmeier, J. L., 1980, *Proc. Natl. Acad. Sci. USA* **77**:3269–3272.
40. Kannan, K. K., Notstrand, B., Fridborg, K., Lovgren, S.,Ohlsson, A., and Petef, M., 1975, *Proc. Natl. Acad. Sci. USA* **72**:51–55.
41. Kannan, K. K., Petef, M., Fridborg, K., Cid-Dresdner, H., and Lovgren, S., 1977, *FEBS Lett.* **73**:115–119.
42. Kannan, K. K., Ramanadham, M., and Jones, T. A., 1984, *Ann. N.Y. Acad. Sci.* **429**:49–60.
43. Kararli, T., and Silverman, D. N., 1985, *J. Biol. Chem.* **260**:3484–3489.
44. Khalifah, R. G., 1971, *J. Biol. Chem.* **246**:2561–2573.
45. Khalifah, R. G., 1973, *Proc. Natl. Acad. Sci. USA* **70**:1986–1989.
46. Khalifah, R. G., 1977, *Biochemistry* **16**:2236–2240.
47. Khalifah, R. G., 1980, in: *Biophysics and Physiology of Carbon Dioxide* (C. Bauer, G. Gros, and H. Bartels, eds.), Springer-Verlag, Berlin, pp. 206–215.
48. Khalifah, R. G., and Edsall, J. T., 1972, *Proc. Natl. Acad. Sci. USA* **69**:172–176.
49. Khalifah, R. G., and Rogers, J. I., 1988, *FASEB J.* **2**:A1118.
50. Khalifah, R. G., Sanyal, G., Strader, D. J., and Sutherland, W. M., 1979, *J. Biol. Chem.* **254**:602–604.
51. Khalifah, R. G., Rogers, J. I., Harmon, P., Morley, P. J., and Carroll, S. B., 1984 *Biochemistry* **23**:3129–3136.
52. Khalifah, R. G., Rogers, J. I., Mukherjee, J., and Morley, P. J., 1984, *Ann. N.Y. Acad. Sci.* **429**:114–128.
53. Khalifah, R. G., Rogers, J. I., and Mukherjee, J., 1986, in: *Zinc Enzymes* (I. Bertini, C. Luchinat, W. Maret, and M. Zeppezauer, eds.), Birkhauser, Boston, pp. 357–370.
54. Khalifah, R. G., Rogers, J. I., and Mukherjee, J., 1987, *Biochemistry* **22**:7057–7063.
55. King, R. W., and Burgen, A. S. V., 1976, *Proc. R. Soc. London B Ser.* **193**:107–125.
56. Knowles, J. R., and Albery, W. J., 1977, *Acc. Chem. Res.* **10**:105–111.
57. Koenig, S. H., and Brown, R. D., 1972, *Proc. Natl. Acad. Sci. USA* **69**:2422–2425.

58. Koenig, S. H., Brown, R. D., Needham, T. E., and Matwiyoff, N. A., 1973, *Biochem. Biophys. Res. Commun.* **53**:624–630.
59. Kogut, K. A., and Rowlett, R. S., 1987, *J. Biol. Chem.* **262**:16417–16424.
60. Lanir, A., Gradstajn, S., and Novan, G., 1975, *Biochemistry* **14**:242–248.
61. Led, J. J., Neesgard, E., and Johansen, J. T., 1982, *FEBS Lett.* **147**:74–80.
62. Liang, J.-Y., and Lipscomb, W. N., 1987, *Biochemistry* **26**:5293–5301.
63. Liang, J. Y., and Lipscomb, W. N., 1988, *Biochemistry* **27**:8676–8682.
64. Liang, J.-Y., and Lipscomb, W. N., 1989, *Biochemistry* **28**:9124–9133.
65. Lindskog, S., 1966, *Biochemistry* **5**:2641–2646.
66. Lindskog, S., 1982, *Adv. Inorg. Biochem.* **4**:115–170.
67. Lindskog, S., 1983, in: *Zinc Enzymes* (T. G. Spiro, ed.), John Wiley & Sons, New York, pp. 78–121.
68. Lindskog, S., 1984, *J. Mol. Catal.* **23**:357–368.
69. Lindskog, S., 1986, in: *Zinc Enzymes* (I. Bertini, C. Luchinat, W. Maret, and M. Zeppezauer, ed.), Birkhauser, Boston, pp. 307–316.
70. Lindskog, S., and Coleman, J. E., 1973, *Proc. Natl. Acad. Sci. USA 70*:2505–2508.
71. Lindskog, S., and Malmström, B. G., 1962, *J. Biol. Chem.* **237**:1129–1137.
72. Lindskog, S., Henderson, L. E., Kannan, K. K., Liljas, A., Nyman, P. O., and Strandberg, B., 1971, in: *The Enzymes*, 3rd ed., Volume 5 (P. D. Boyer, ed.), Academic Press, New York, pp. 587–665.
73. Maren, T. H., 1967, *Physiol. Rev.* **47**:595–781.
74. Maren, T. H., and Sanyal, G., 1983, *Annu. Rev. Pharmacol. Toxicol.* **23**:439–459.
75. Martin, R. B., 1974, *Proc. Natl. Acad. Sci. USA* **71**:4346–4347.
76. Martin, R. B., 1976, *J. Inorg. Nucl. Chem.* **38**:511–513.
77. Mukherjee, J., Rogers, J. I., and Khalifah, R. G., 1987, *J. Am. Chem. Soc.* **109**:7232–7233.
78. Pocker, Y., and Bjorkquist, D. W., 1977, *Biochemistry* **16**:5698–5707.
79. Pocker, Y., and Deits, T. L., 1982, *J. Am. Chem. Soc.* **104**:2424–2434.
80. Pocker, Y., and Janjic, N., 1988, *J. Biol. Chem.* **263**:6169–6176.
81. Pocker, Y., and Sarkanen, S., 1978, *Adv. Enzymol.* **47**:149–274.
82. Pocker, Y., Janjic, N., Miao, C. H., 1986, in: *Zinc Enzymes* (I. Bertini, C. Luchinat, W. Maret, and M. Zeppezauer, eds.), Birkhauser, Boston, pp. 341–356.
83. Prince, R. H., and Woolley, P. R., 1973, *Bioorg. Chem.* **2**:337–344.
84. Pullan, L. M., and Noltmann, E. A., 1985, *Biochemistry* **24**:635–640.
85. Ren, X., Jonsson, B. H., Millqvist, E., and Lindskog, S., 1988 *Biochim. Biophys. Acta* **953**: 79–85.
86. Ren, S., Sandstrom, A., and Lindskog, S., 1988, *Eur. J. Biochem.* **173**:73–78.
87. Riepe, M. E., and Wang, J. H., 1967, *J. Biol. Chem.* **243**:2779–2787.
88. Rogers, J. I., Mukherjee, J., and Kahlifah, R. G., 1987, *Biochemistry* 26:5672–5679.
89. Rowlett, R. S., 1984, *J. Protein Chem.* **3**:369–393.
90. Rowlett, R. S., and Silverman, D. N., 1982, *J. Am. Chem. Soc.* **104**:6737–6741.
91. Silverman, D. N., 1982, *Methods Enzymol.* **87**:732–752.
92. Silverman, D. N., and Lindskog, S., 1988, *Acc. Chem. Res.* **21**:30–36.
93. Silverman, D. N., and Tu, C. K., 1975, *J. Am. Chem. Soc.* **97**:2263–2269.
94. Silverman, D. N. and Vincent, 1983, *Crit. Rev. Biochem.* **14**:207–255.
95. Silverman, D. N., Tu, C. K., Lindskog, S., and Wynns, G. C., 1979, *J. Am. Chem. Soc.* **101**:6734–6740.
96. Simonsson, I., and Lindskog, S., 1982, *Eur. J. Biochem.* **123**:29–36.
97. Simonsson, I., Jonsson, B.-H., and Lindskog, S., 1979, *Eur. J. Biochem.* **93**:409–417.
98. Simonsson, I., Jonsson, B.-H., and Lindskog, S., 1982, *Eur. J. Biochem.* **129**:163–169.
99. Simonsson, I., Jonsson, B.-H., and Lindskog, S., 1982, *Biochem. Biophys. Res. Commun.* **108**:1406–1412.

100. Steiner, H., Jonsson, B.-H., and Lindskog, S., 1975, *Eur. J. Biochem.* **59**:253–259.
101. Steiner, H., Jonsson, B.-H., and Lindskog, S., 1976, *FEBS Lett.* **62**:16–20.
102. Tashian, R. E., 1989, *BioEssays* **10**:186–192.
103. Thorslund, A., and Lindskog, S., 1968, *Eur. J. Biochem.* **3**:453–460.
104. Tibell, L., Forsman, C., Simonsson, I., and Lindskog, S., 1984, *Biochem. Biophys. Acta* **789**:302–310.
105. Tu, C. K., and Silverman, D. N., 1977, *J. Biol. Chem.* **252**:3332–3337.
106. Tu, C. K., and Silverman, D. N., 1986, *J. Am. Chem. Soc.* **108**:6065–6066.
107. Tu, C. K., Sanyal, G., Wynns, G. C., and Silverman, D. N., 1983, *J. Biol. Chem.* **258**:8867–8871.
108. Tu, C. K., Thomas, H. G., Wynns, G. C., and Silverman, D. N., 1986, *J. Biol. Chem.* **261**:10100–10103.
109. Tu, C. K., Silverman, D. N., Forsman, C., Jonsson, B.-H., and Lindskog, S., 1989, *Biochemistry* **28**:7913–7918.
110. Tu, C. K., Wynns, G. C., and Silverman, D. N., 1989, *J. Biol. Chem.* **264**:12389–12393.
111. Vedani, A., and Dunitz, J. D., 1985, *J. Am. Chem. Soc.* **107**:7653–7658.
112. Vedani, A., Huhta, D. W., and Jacober, S. P., 1989, *J. Am. Chem. Soc.* **111**:4075–4081.
113. Venkatasubban, K. S., and Silverman, D. N., 1980, *Biochemistry* **19**:4984–4989.
114. Vincent, S. H., and Silverman, D. N., 1980, *Biochem. Biophys.* **205**:51–56.
115. Whitney, P. L., 1970, *Eur. J. Biochem.* **16**:126–135.
116. Whitney, P. L., and Brandt, H. J., 1976, *J. Biol. Chem.* **251**:3862–3867.
117. Whitney, P. L., Nyman, P. O., and Malmstrom, B. G., 1967 *J. Biol. Chem.* **242**:4212–4220.
118. Williams, T. J., and Henkens, R. W., 1985, *Biochemistry* **24**:2459–2462.
119. Wistrand, P. J., and Knuuttila, K. G., 1989, *Kidney Int.* **35**:851–859.
120. Wolpert, H. R., Strader, C. D., and Khalifah, R. G., 1977, *Biochemistry* **16**:5717–5721.
121. Woolley, P. O., 1975, *Nature* (London) **258**:677–682.

CHAPTER 5

Carbonic Anhydrases in Mammalian Cell Culture and Tumors

PATRICK J. VENTA

This chapter describes some of the work that has been performed with tissue culture cells to understand how carbonic anhydrase (CA) is regulated in various cell types. The advantages of studying genes and gene products in tissue culture are that the treatments of the cells can be completely defined and, in many cases, homogeneous populations of cells can be used without the confounding effects of other cell types. The chapter is not all-inclusive, focusing primarily on the analysis of a number of leukemic cell lines and tissue types that have been studied most frequently or that may be most useful in dissecting the regulation of the CA isozymes and their genes at the molecular level. Many excellent studies using other cell types have not been included.

1. Carbonic Anhydrase in Primary Tissue Culture

1.1. Bone Culture

The importance of CA II in bone metabolism is shown by the osteopetrosis component of the human CA II deficiency syndrome.[40] The regulation of bone remodeling, both during development and in repair, is extremely complex,[11,28] and study of the regulation of the CA II gene may help to resolve this complexity. High concentrations of CA in bone are found almost exclusively in the osteoclasts and chondrocytes. To examine the regulation of CA II in bone, culture systems of bone have been developed in which regulating agents can be controlled.

PATRICK J. VENTA • Department of Human Genetics, University of Michigan Medical School, Ann Arbor, Michigan 48109-0618.

1.1.1. Induction by Vitamin D_3

The effect of 1,25-dihydroxyvitamin D_3 (vitamin D_3) on CA activity and bone resorption in cultured neonatal mouse calvaria has been investigated.[19] It was found that CA activity increased 2-fold at 48 hr and 2.5-fold at 96 hr of treatment with physiological (10 nM) concentrations of vitamin D_3, with a nearly 50% increase in culture medium calcium released from the bone which could be inhibited by 1×10^{-4} M acetazolamide. These findings are consistent with the hypothesis that CA is regulated by vitamin D_3 to increase bone resorption. Similar results were found in another study using cultured fetal rat long bones.[33] Incubation of the bones with vitamin D_3 increased bone resorption, as monitored by calcium released into the medium.

1.1.2. Induction by Parathyroid Hormone

The effect of parathyroid hormone (PTH) on CA in mouse calvaria bone culture in relation to the regulation of acidity in osteoclasts was investigated.[3] Changes in acidity were monitored by use of a technique in which acridine orange becomes concentrated in acid-containing subcellular compartments and which fluoresces more intensely with increased acid concentrations. The variation in light intensity was determined with an exposure meter attached to a microscope. A maximal fourfold rise over untreated bone osteoclasts in light intensity was seen after 24 hr of treatment with PTH. This PTH-stimulated increase in acidity was suppressed by acetazolamide.

The effect of PTH and acetazolamide on calcium released from bone was studied.[20] Acetazolamide (10^{-1} M) completely inhibited PTH-stimulated calcium release from neonatal mouse calvaria. This result in tissue culture supports the hypothesis that bone resorption can be mediated by a PTH effect on CA. Whether the effect is caused by a change in amount, location, or specific activity of CA is unknown. It is also unknown whether the effect of PTH on CA in osteoclasts is direct or mediated by other cell types; nevertheless, these bone culture methods should be useful for determining the nature of hormonal influences on CA in bone.[39] The results obtained in these studies are consistent with results of earlier studies that have examined the relationships of CA to bone resorption.[22,26,45]

Other hormones known to affect bone metabolism include calcitonin, calcitonin gene-related peptide, osteoclast-activating factor, transforming growth factors, and prostaglandin E_2.[28] The bone culture systems should be very useful in determining how each of these hormones affect CA expression, as well as the role of CA in bone resorption.

1.2. Glial Cell Culture

CA II is the main soluble CA isozyme present in brain and is found primarily in oligodendroglia, where it may be important for myelin synthesis. Small amounts of CA II may also be present in astroglia, although these findings are controversial.[35] Several groups have shown that CA is expressed in primary glial cultures.[6,18,23] The developmental expression of CA in rat glial cultures has been shown to parallel that seen in the whole animal.[16]

Basic fibroblast growth factor has been found to increase the amount of CA in primary rat glial cultures derived from neonatal cerebral hemispheres.[9] The cells were grown in a chemically defined medium after the primary culture was started in medium containing 10% fetal calf serum. The astrocytes and oligodendrocytes appear to differentiate from a common precursor cell (presumably similar to the O-2A progenitor cell found in developing optic nerve,[32] and the rate of differentiation slowed in the defined medium. The amount of CA per oligodendrocyte increased upon treatment with bovine fibroblast growth factor from 3- to 13-fold after 11 and 14 days of treatment, respectively. It would be of interest to observe the effects of platelet-derived growth factor and ciliary neurotrophic factor on CA expression in this system because of their known differentiation-inducing capacity in O-2A cells.[32]

In addition to these protein concentration and activity studies, it is now also possible to study the regulation at the nucleic acid level. For example, both CA II and myelin basic protein can be detected in oligodendrocytes by using double-labeling *in situ* hybridization.[15] The glial cell culture system should also be useful for studying the events that turn on CA II gene transcription in the brain.

2. Carbonic Anhydrase in Leukemia Cells

2.1. Mouse Erythroleukemia Cells

The bulk of the work on CA in erythroleukemia cells has been in mouse erythroleukemia (MEL or Friend) cells. These cells are arrested in their development at the proerythroblast stage after infection by a mouse erythroleukemia virus.[24] MEL cells can be induced to differentiate along an erythroid pathway by dimethyl sulfoxide (DMSO).[14] Many studies have been carried out with these cells to examine the regulation of the globins, which increase markedly upon induction of differentiation by DMSO or other chemicals.[24] Because CA I and CA II are the most abundant proteins in the red cell after the globins, several investigators have turned their attention to studying the regulation of these CA isozymes. In the earliest study,[21] CA activity in control and DMSO-induced cells was examined and found to be three times higher in induced cells. However, CA protein concentrations were not determined.

In another study,[41] CA I and CA II concentrations in control and induced cultures were examined by radial immunodiffusion analysis. Increases in globin concentration were used to follow the course of the cell induction. CA I levels declined or remained unchanged in treated cultures relative to untreated controls, and the CA II concentration remained about the same relative to control cultures.

However, both CA I and CA II levels increased about two- to three-fold in the untreated control cultures during the 6 to 8 days of the experiments.

With the cloning of the mouse CA II cDNA,[8] it became possible to examine the effect of DMSO-induced differentiation at the RNA level. An apparent two-fold increase in the transcription of the CA II gene was detected, using total RNA as a control. However, increased stability of the transcript in the nucleus as opposed to increased transcription of the CA II gene could not be completely ruled out in this experiment. In studies using mouse CA I and CA II cDNA probes,[13] a clear decrease in CA I gene transcription and an initial increase in CA II gene transcription, followed by a fall in transcription during the time course of DMSO treatment, were found. The maximum increase for the CA II transcription was six- to eight-fold at day 2. The CA I mRNA level fell sharply over this period, whereas the CA II mRNA level dropped slightly despite the increased transcription. It was concluded that the transcriptional activation–inactivation of the CA I gene precedes that of the globin genes, although the results for the CA II gene were more difficult to interpret.

2.2. Human Erythroleukemia Cells

Studies have been carried out on human as well as mouse erythroleukemia lines to study CA regulation. The effect of the chemical induction on the differentiation of the human HEL and mouse MEL cell lines has been examined.[12,47] CA I and CA II concentrations were observed to remain essentially constant during hemin induction of the human line and DMSO induction of the mouse line, as observed by the Western blotting technique. Globin synthesis was greatly increased in these cells during the induction.

In another study, CA I levels in HEL cells remained the same during DMSO induction, as seen by Western blot.[43] However, CA I protein decreased significantly during treatment with 12-*O*-tetradecanoyl phorbol-13 acetate (TPA). These cells are induced to a macrophage-like phenotype by TPA,[25,31] and CA I is not expressed in normal macrophages.[42] Interestingly, no CA I was found in a second erythroleukemia line, K562. CA II was found in K562 and HEL cells as well as KG-1 and HL-60 myeloid cell lines, although no differentiation studies were undertaken.

CA I has been used as a marker for erythroleukemias.[27,46] These studies indicate that CA I is probably one of the best markers for this leukemia type (M6 in the French-American-British classification[4]). The range of CA I-expressing erythroleukemia cells from nine patients was 25–91%.[44] The concentration of CA I protein was not investigated. Many of these cells do not express globin, and thus it appears that CA I is expressed earlier than the globins during erythroid differentiation.

2.3. Human Myeloid Cell Lines

CA II has been observed in the myeloid cell lines KG-1 and HL-60. These lines have been used extensively to study myeloid cell differentiation.[7] The HL-60 cell line undergoes differentiation along either a monocyte/macrophage pathway or a granulocyte pathway, depending on the inducing agent. The expression of CA I, CA II, CA III, and CA VII in the HL-60 cell line after induction along both pathways has been studied.[38] When induced with DMSO to undergo granulocytic differentiation, there is no induction of CA II protein or mRNA. However, when induced along the monocyte/macrophage pathway by 1,25-dihydroxyvitamin D_3, CA II increases at least 20-fold in mRNA and 10-fold in protein. With TPA, which also induces HL-60 cells along the monocyte/macrophage pathway, CA II mRNA increases seven-fold. No induction of CA I, CA III, or CA VII was observed with use of specific cDNA or genomic clones as probes on Northern blots.

2.4. The CA II Gene Promoter in Epithelial and Fibroblastic Cells

The function of the CA II gene promoter in human HeLa cells and mouse L cells, which are of epithelial and fibroblastic origin, respectively, has been examined.[37] Both of these cell lines allow the expression of many tissue-specific promoters that are not expressed in normal epithelial cells or fibroblasts, e.g., the α-globin gene.[48] Thus, they are useful for studying minimal promoter elements that may not in themselves confer tissue-specific regulation. The cloned human CA II promoter was fused with the bacterial chloramphenicol acetyltransferase (CAT) reporter gene,[17] and the DNA was transfected into the cell lines. Deletion analysis of this promoter region showed that maximal expression of the CAT reporter gene could be produced by retaining only 184 bp of DNA upstream of the transcription start site.

3. Carbonic Anhydrase in Tumors

3.1. Early Studies

Zinc concentrations in blood and tissues have been related to the presence or progression of cancer. Because CA contains zinc at its active site, the relationship between CA levels and cancer has been studied. An apparent decrease in zinc that correlated with the progression of various carcinomas suggested a relationship with CA.[1,2] Later studies have found both increases and decreases in zinc content in blood and tissues so that the predictive value of zinc, and thus total CA, for the presence or progression of cancer remains controversial.[10,36]

3.2. Neural Tumors

The examination of specific isozymes in tumors might be useful for determining the cell type from which a tumor originated. The presence of CA II in a variety of tumors from central and peripheral nervous systems has been examined.[29] CA II is expressed at detectable levels in only a subset of cells in nervous tissues,[34] and its presence or absence might be used as a marker for the origin of tumors from different cell types. The origins of the tumors in this study were first determined by independent criteria and then examined for CA II expression by using specific antibodies. It was found that CA II was not informative when used as a marker because it was expressed in inappropriate cell types in neural tumors. Thus, it appears that the regulatory mechanisms for CA II may not be restrictive enough to remain cell specific in the aberrant cellular environment of tumor cells, at least in those of neural origin.

3.3. Pancreatic Adenocarcinoma

The expression of CA II in a variety of cell lines derived from human pancreatic carcinoma has been investigated (M. Frazier and D. Hewett-Emmett, personal communication). Because CA II appears to be expressed exclusively in ductal cells of the pancreas, it might indicate from which cell type such tumors are derived. CA II mRNA as well as immunoreactive protein were found in five cell lines examined except that the protein appeared to be missing from one. An additional larger mRNA (4.4 kb) and protein (60 Da) were also found in some of the cell lines, which might represent an additional CA isozyme, although alternative interpretations of these results are also possible (D. Hewett-Emmett, personal communication). The presence of the CA II mRNA and protein in these cell lines supports the contention that adenocarcinomas of pancreas are of ductal cell origin.

4. Prospectives

Although the regulation of CA in cell cultures may not completely reflect that seen in normal cells, it is likely that some of the general features, particularly for the molecular analysis of the genes, will be useful. In addition, some of the isozymes may be indicative of the cell type from which various tumors are derived. Some prospective views on expression of the various CA isozymes in culture and tumors follow.

4.1. The CA I Gene

The primary erythroleukemia cells may be useful for studying the regulatory switch that turns on the CA I gene. Given that all of the primary erythroleukemias studied express CA I,[43] the K562 cell line, which does not express CA I, represents an exception. Although the HEL cell line does not produce a change in CA I concentrations upon treatment with DMSO[44] or hemin,[47] the line either may represent a later stage of erythroid development than most erythroleukemias or may require a different inducer such as granulocyte–macrophage colony-stimulating factor or erythropoietin to increase CA I expression. A variety of newer erythroleukemia lines are also becoming available that may be derived from earlier stages of erythroid development.[30]

4.2. The CA II Gene

Studies on mouse and human erythroleukemia cell lines indicate that the effect of differentiation on the CA II gene is relatively small. The CA II gene is already transcribing at a high rate, and studies of these small effects are variable and difficult to interpret. A better model system for the study of CA II gene regulation is the promyelocytic cell line HL-60. The effect of differentiation on the expression of the CA II gene is relatively large, giving readily interpretable results.

4.3. Other Carbonic Anhydrase Genes

Systems for studying the regulation other CA genes in immortal cell lines have, in large, not yet been developed. However, these genes are being studied in intact tissues. In addition, some CA genes are expressed in primary tissue culture. For example, CA III is expressed at a low levels in dissociated rat hepatocytes.[5] Development of these and other tissue culture systems should greatly increase our understanding of how the CAs are regulated in the intact organism and perhaps will lead to an understanding of why there are so many CA genes when it appears that one would be sufficient to carry out the primary function of the enzyme.

Note added in proof: The work by D. Hewett-Emmett and M. Frazier has been published (reference 49).

ACKNOWLEDGMENTS. I thank Dr. Richard E. Tashian for many helpful suggestions, Dr. David Hewett-Emmett and Dr. Martha Frazier for communication of their unpublished work, and Jayne Long for her careful preparation of the manuscript.

References

1. Addink, N. W. H., 1965, *Nature* (London) **207**:1271–1272.
2. Addink, N. W. H., and Bastings, L., 1967, *Nature* (London) **216**:72–73.
3. Anderson, R. E., W. S. S. Jee, and D. M. Woodbury, 1985, *Calcif. Tissue Int.* **37**:646–650.
4. Bennett, J. M., Catovsky, D., Daniel, M.-T., Flandrin, G., Galton, D. A. G., Gralnick, H. R., and Sultan, C., 1976, *Br. J. Haematol.* **33**:451–458.
5. Carter, N., Jeffery, S., Legg, R., Wistrand, P., and Lönnerholm, G., 1987, *Biochem. Soc. Trans.* **15**:667–668.
6. Church, G. A., Kimelberg, H. K., and Sapirstein, V. S., 1980, *J. Neurochem.* **34**:873–879.
7. Collins, S. J., 1987, *Blood* **70**:1233–1244.
8. Curtis, P. J., 1983, *J. Biol. Chem.* **258**:4459–4463.
9. Delaunoy, J. P., Langui, D., Ghandour, S., Labourdette, G., and Sensenbrenner, M., 1988, *Int. J. Dev. Neurosci.* **6**:129–136.
10. Diez, M., Cerdan, F. J., Arroyo, M., and Balibrea, J. L., 1989, *Cancer* **63**:726–730.
11. Evered, D., and Harnett, S. (eds.), 1988, *Ciba Found. Symp.* **136**:1–307.
12. Frankel, S. R., Walloch, J., Hirata, R. K., Boudurant, M. C., Villanueva, R., and Weil, S. C., 1985, *Proc. Natl. Acad. Sci. USA* **82**:5175–5179.
13. Fraser, P. J., and Curtis, P. J., 1987, *Genes Devel.* **1**:855–861.
14. Friend, C., Scher, W., Holland, J. G., and Sato, T., 1971, *Proc. Natl. Acad. Sci.* USA **68**:378–382.

15. Ghandour, M. S., and Skoff, R. P., 1988, *J. Cell Biol.* **107**:514a.
16. Ghandour, M. S., Vincendon, G., Gombos, G., Limozin, N., Filippi, D., Dalmasso, C., and Laurent, G., 1980, *Dev. Biol.* **77**:73–83.
17. Gorman, C. M., Moffat, L. F., and Howard, B. H., 1982, *Mol. Cell. Biol.* **2**:1044–1051.
18. Griot, C., and Vandevelde, M., 1988, *J. Neuroimmunol.* **18**:333–340.
19. Hall, G. E., and Kenny, A. D., 1985, *Calcif. Tissue Int.* **37**:134–142.
20. Hall, G. E., and Kenny, A. D., 1987, *Calcif. Tissue Int.* **40**:212–218.
21. Kabat, D., Sherton, C. C., Evans, L. H., Bigley, R., and Koler, R. D., 1975, *Cell* **5**:331–338.
22. Kenny, A. D., 1985, *Calcif. Tissue Int.* **37**:126–133.
23. Langui, D., Delaunoy, J. P., Ghandour, M. S., and Sensenbrenner, M., 1985, *Neurosci. Lett.* **60**:151–156.
24. Marks, P. A., and Rifkind, R. A., 1978, *Annu. Rev. Biochem.* **47**:419–448.
25. Martin, P., and Papayannopoulou, T., 1982, *Science* **216**:1233–1235.
26. Minkin, C., and Jennings, J. M., 1972, *Science* **176**:1031–1033.
27. Mitjavila, M. T., Villeval, J. L., Cramer, P., Henri, A., Gasson, J., Krystal, G., Tulliez, M., Berger, R., Breton-Gorius, J., and Vainchenker, W., 1987, *Blood* **70**:965–973.
28. Mundy, G. R., and Roodman, G. D., 1987, *Bone Mineral Res.* **5**:209–279.
29. Nakagawa, Y., Perentes, E., and Rubinstein, L. J., 1987, *J. Neuropathol. Exp. Neurol.* **46**:451–460.
30. Papayannopoulou, T., Nakamoto, B., Kurachi, S., Tweeddale, M., and Messner, H., 1988, *Blood* **72**:1029–1038.
31. Papayannopoulou, T., Nakamoto, B., Yokochi, T., Chait, A., and Kannagi, R., 1983, *Blood* **62**:832–845.
32. Raff, M. C., 1989, *Science* **243**:1450–1455.
33. Raisz, L. G., Simmons, H. A., Thompson, W. J., Shepard, K. L., Anderson, P. S., and Rodan, G. A., 1988, *Endocrinology* **122**:1083–1086.
34. Sapirstein, V. S., 1983, in: *Handbook of Neurochemistry*, 2nd ed. (A. Lajtha, ed.), Plenum Press, New York.
35. Sapirstein, V. S., Glynn, C., and Lees, M. B., 1980, *Brain Res.* **185**:373–383.
36. Schwartz, M. K., 1975, *Cancer Res.* **35**:3481–3487.
37. Shapiro, L. H., Venta, P. J., and Tashian, R. E., 1987, *Mol. Cell. Biol.* **7**:4589–4593.
38. Shapiro, L. H., Venta, P. J., and Tashian, R. E., 1989, *FEBS Lett.* **249**:307–310.
39. Silverton, S. F., Dodgson, S. J., Fallon, M D., and Forster, R. E., II, 1987, *Am. J. Physiol.* **253**:E670–E674.
40. Sly, W. S., Whyte, M. P., Sundaram, V., Tashian, R. E., Hewett-Emmett, D., Guibaud, P., Vainsel, M., Baluarte, H. J., Gruskin, A., Al-Mosawi, M., Sakati, N., and Ohlsson, A., 1985, *N. Engl. J. Med.* **313**:139–145.
41. Stern, R. H., Boyer, S. H., Conscience, J.-F., Friend, C., Margolet, L., Tashian, R. E., and Ruddle, F. H., 1977, *Proc. Soc. Exp. Biol. Med.* **156**:52–55.
42. Sundquist, K. T., Leppilampi, M., Järvelin, K., Kumpulainen, T., and Väänänen, H. K., 1987, *Bone* **8**:33–38.
43. Villeval, J. L., Testa, U., Vinci, G., Tonthat, H., Bettaieb, A., Titeux, M., Cramer, P., Edelman, L., Rochant, H., Breton-Gorius, J., and Vainchenker, W., 1985, *Blood* **66**:1162–1170.
44. Villeval, J. L., Cramer, P., Lemoine, F., Henri, A., Bettaieb, A., Bernaudin, F., Beuzard, Y., Berger, R., Flandrin, G., Breton-Gorius, J., and Vainchenker, W., 1986, *Blood* **68**:1167–1174.
45. Waite, L. C., Volkert, W. A., and Kenny, A. D., 1970, *Endocrinology* **87**:1129–1139.
46. Walloch, J., Frankel, S., Hrisinko, M. A., and Weil, S. C., 1986, *Blood* **68**:304–306.
47. Weil, S. C., Walloch, J., Frankel, S. R., and Hirata, R. K., 1984, *Ann. N.Y. Acad. Sci.* **429**:335.
48. Whitelaw, E., Hogben, P., Hanscombe, O., and Proudfoot, N. J., 1989, *Mol. Cell. Biol.* **9**: 241–251.
49. Frazier, M. L., Lilly, B. J., Wu, E. F., Ota, T., and Hewett-Emmett, D., 1990, *Pancreas* **5**: 507–514.

CHAPTER 6

Methods for the Measurement of Carbonic Anhydrase Activity

ROBERT E. FORSTER II

This chapter provides a summary of published methods to measure carbonic anhydrase (CA) catalysis of the reversible reactions of CO_2–HCO_3^-. For other reviews see references 12, 13, 31, 41, and 53.

1. General Comments on Measurement of Carbonic Anhydrase Activity

1.1. Mechanism of Carbonic Anhydrase Action

The physiological reaction catalyzed by CA is the reversible hydration–dehydration of CO_2 and H_2O:

$$CO_2 + HOH \leftrightharpoons H_2CO_3 \leftrightharpoons H^+ + HCO_3^- \qquad (1)$$

This reaction is exceptionally rapid in the absence of catalyst, with a half-time of about 3 sec under physiological conditions. The reversible dissociation of H_2CO_3 to give H^+ and HCO_3^- has a half-time of the order of 9×10^{-8} sec, instantaneous on a physiological time scale.[15]

In addition to its acceleration of the CO_2–HCO_3^- reactions, CA can act as an esterase and can catalyze the hydration of aldehydes, which reactions have been used to measure activity of the enzyme,[16,47] However, I will deal only with the reversible CO_2–HCO_3^- reactions.

To initiate the reaction, either acid is added to a bicarbonate solution, produc-

ROBERT E. FORSTER II • Department of Physiology, University of Pennsylvania School of Medicine, Philadelphia, Pennsylvania 19104-6085.

ing CO_2 and consuming H^+, or CO_2 is added to an alkaline solution, consuming CO_2 and producing H^+. Experimentally, one measures the rate of change of $[CO_2]$ or of $[HCO_3^-]$ and/or $[H^+]$ according to the schema*

$$d[CO_2]/dt = -k_u[CO_2] + k_v[H_2CO_3]$$
$$= -k_u[CO_2] + \frac{k_v[H+]}{k_{HA}}[HCO_3^-] \qquad (2)$$

where k_u is the hydration reaction velocity constant in seconds^{-1}, k_v is the dehydration reaction velocity constant in seconds^{-1}, and, K_{HA} is the ionization constant of $H_2CO_3 = [H+][HCO_3^-]/[H_2CO_3]$. Its value is 3.5×10^{-4} M.[54]

Solution of Eq. 2, describing the time course of $[CO_2]$, or $[H^+]$, for calculating k_u and k_v requires (1) knowing $[H^+]$ as a function of CO_2 produced or consumed (in other words, the effective buffer capacity of the system), and (2) maintaining total CO_2 content constant so that $[HCO_3^-]$ = total CO_2 content $-$ $[CO_2]$, and (3) the following equation:

$$K' = [H^+][HCO_3^-]/CO_2 = k_u K_{HA}/k_v \qquad (3)$$

With K', the Henderson–Hasselbalch constant, $= 10^{-6.1}$, an average value of $k_u/k_v = 1/440$. This ratio is approximately constant over the physiological range.

The complete solution of Eq. 2 may be complicated because of the nonlinear buffering of H^+ and possible exchanges across membranes and therefore is not often used. Instead, the initial value of $d[CO_2]/dt$ or $d[H^+]/dt$ (buffer capacity) is obtained as a tangent to a graph of $[CO_2]$ or $[H^+]$ against time and the initial values of $[CO_2]$ and/or $[H^+]$ and $[HCO_3^-]$ substituted in differential Eq. 2.

The catalysis of the reversible reaction of CO_2 and HCO_3^- by CA is generally considered to take place as follows[61]:

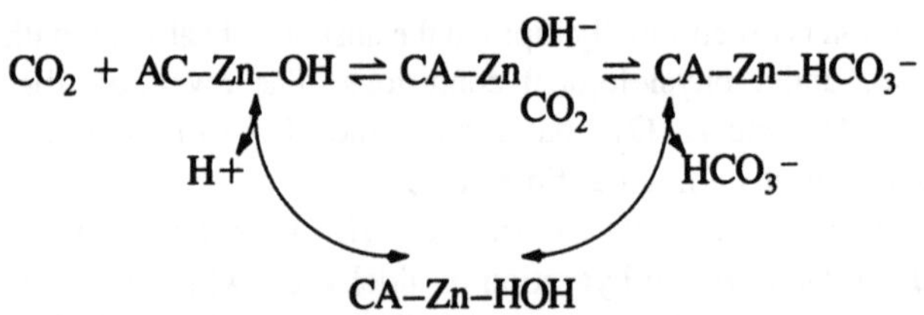

(4)

Zn represents the active site of the enzyme. The overall reaction appears to be rate limited by intermolecular and intramolecular transport.[56]

Intermolecular transport of H^+ produced or consumed from the active site to the ambient electrolyte solution about 15 Å, the depth of the cavity, is facilitated by

*I have used the original symbols of Roughton[54] for the hydration and dehydration reaction velocity constants because they are useful phenomenological constants as defined by Eq. 2 and may include several velocity constants and processes.

buffer, which should always be present (for example, 10 mM imidazole) so that intermolecular H^+ transport is not rate limiting. Intramolecular transport of H^+ takes place between the HOH or OH^- bound to the Zn atom and neighboring buffer groups of the CA, considered to be histidines; this is unaffected by buffers. Heavy water (DOD) instead of HOH slows the reaction with an isotope effect of about 3.4,[61] thus the normal rate-limiting molecular or ionic step must involve H^+ transport.

The velocity of exchange of ^{13}C between CO_2 and HCO_3^- through the catalytic enzyme path as measured by nuclear magnetic resonance (NMR) is about eight times faster than the velocity of net formation of HCO_3^- from CO_2 or the reverse.[27,62] Thus, the binding of CO_2 and HCO_3^- to CA and the reversible catalytic reaction, the classical Michaelis–Menten mechanism, are not rate limiting the net chemical reactions. The exchange of H^+ with the OH^- or HOH bound to the Zn^{2+} determines the rate of formation of HCO_3^- or CO_2.

When the rate of hydration of CO_2 is measured by an initial rate (a "steady-state") method, CA is mixed with a solution containing CO_2 but no HCO_3^-. Nevertheless, CA–Zn^{2+}–HCO_3^- is instantly formed and accumulates to a concentration nearly in equilibrium with CA–Zn^{2+}–CO_2, because of the slower release of HCO_3^- into the bulk solution. When the rate of dehydration is measured by adding acid to HCO_3^-, the CA–Zn^{2+}–CO_2 complex is formed instantly and accumulates to a concentration approximately in equilibrium with CA–Zn^{2+}–HCO_3^- for the same reason. When the reaction velocity is measured by ^{18}O exchange at chemical equilibrium, again the concentrations of the complexes of HCO_3^- and CO_2 with CA are almost precisely equilibrated, so that the same reaction rates will be obtained in all three cases.

1.2. Velocities of the CO_2–HCO_3^- Reactions and the Requirements They Place on the Method of Measurement

The reaction of Eq. 2 will proceed until $K_v[H^+][HCO_3^-]/k_u\ K_{HA})$ becomes equal to $[CO_2]$. If there is buffer present, $[H^+]$ varies less and the reaction half-time increases. At 37°C, the half-time of the uncatalyzed CO_2–HCO_3^- reaction under physiological conditions with sufficient buffer power in the reacting solution to hold pH relatively constant is <3.5 sec. If the buffer power is decreased, the half-time decreases, reaching several milliseconds as a limit. pK_{HA} for the dissociation of H_2CO_3 is 3.4 so that the buffer power of this acid is negligible at physiological pH. Buffer is added to increase the duration of the reaction and reduce dpH/dt. This buffer should neither accelerate the uncatalyzed reaction itself nor inhibit CA. On the other hand, at 0°C and a constant pH of 7.4, the uncatalyzed half-time is 350 sec, slow enough to be measured accurately by most techniques.

[CA] inside human red cells can accelerate the uncatalyzed hydration rate about 17,000-fold[30] at 37°C; thus, the half-time at a constant pH of 7.4 would be 0.0002 sec, too rapid for any initial rate method to follow.

Earlier methods of measuring CA activity analyze the reaction in a stirred vessel. The fastest mixing time of a mechanical stirrer is about 1 sec, setting a lower limit on the time resolution of the technique, which cannot be used to measure even the uncatalyzed rate at 37°C but can measure the rate at 0°C. Therefore, many methods are carried out at the unphysiological temperatures of 0–4°C.

At the start of a reaction that is too fast for the mixing rate or analytical system to follow, there is a delay in the pH record and then the slope slowly increases, which gives an incorrect estimate of the reaction velocity. Since the true pH curve is approximately exponential even though the overall response time of the system is too slow to record the initial changes in pH correctly, the recorded (dpH/dt)/pH, or the slope on a semilog graph, will eventually become equal to the true value, which can then be used to calculate reaction velocity. By this technique, G. Gros (personal communication) can determine reaction velocity and enzyme characteristics at 25°C for a [CA] that accelerates k_u sixfold, using a stirred chamber and glass pH electrode.

Rapid-mixing apparatus (either continuous-flow or stop-flow) is needed to measure CO_2 reactions with a half-time of less than several seconds.

In the continuous-flow method,[23] reactants are driven into a mixing chamber, mixed in about 1 msec, and thence driven into a glass observation tube about 0.2 cm in diameter. In the mixing chamber the reaction is initiated, and it continues in the outflowing solution. Flow conditions in the observation tube are maintained turbulent so that the reacting mixture flows as a square front and the duration of the reaction at any point along the tube equals the transit time of the fluid from the mixing chamber. The progress of the reaction is analyzed at several distances along the tube by measuring pH, pCO_2, optical density, or temperature to obtain a complete time course. $d[HCO_3^-]/dt$ at the start of the reaction, when the concentrations of reactants, including $[H^+]$, are well defined, is obtained from the slope of a graph of $[HCO_3^-]$ versus time. A 60- to 180-ml amount of reactants is required for each point on the time course, the apparatus is large and complicated and a completely defined curve requires up to 1 day's work. Provided there is sufficient volume of reactants to maintain flow until the measuring instrument has responded completely, the response time of the analytical instrument can be slow and still permit measurement of the chemical composition of the mixture at times as short as 1 msec.

In the stop-flow rapid-mixing instrument,[21,52] several tenths of a milliliter of reactants is impelled into the mixing chamber and out through a short observation tube that is connected to a third syringe. The flow of fluid through the system is suddenly (less than 0.001 sec) stopped when the plunger of the third syringe hits a barrier. The progress of the reaction is monitored spectrophotometrically, or by some other rapidly responding instrument, and either recorded on an oscilloscope or digitized and stored in an electronic memory. The stop-flow method produces a continuous record of the reaction from the first millisecond up to tens of seconds with 3 orders of magnitude less material than the continuous-flow method. It can be used with any rapidly responding analytical instrument, spectrophotometer, pH electrode, or even temperature recorder.[2]

1.3. Sensitivity of Measurements of CA Activity

Experimentally, the catalyzed reaction velocity is the measured reaction velocity less the uncatalyzed velocity. Thus, the minimal measurable enzyme concentration in all methods is equivalent to the error in measuring the uncatalyzed velocity. Absolute sensitivity, that is, the moles of CA detectable, is limited by the volume of the reacting mixture. Maren miniaturized a variant of the Philpot method, bubbling CO_2 through an 0.8-ml volume at 4°C, and could detect as little as 1×10^{-14} to 4×10^{-13} moles, of high-activity CA.[40] The ^{18}O exchange method can be miniaturized by placing 0.01 ml of reacting fluid on a Millipore filter disk and measuring the disappearance of labeled CO_2 species from a 1-ml gas phase.[3] This permits the detection of about 1⁄80 as much CA or 10^{-15} moles of human CA II (HCA II) at room temperature.

Decreasing the temperature from 37 to 0°C (Table I) reduces k_u to 1⁄86 but lowers the catalyzed (HCA II) velocity constant (k_{enz}) only to 1⁄16. This decreases k_u/normalized k_{enz} (fourth column in Table I) by a factor of 6. Lowering the temperature has the additional benefit of slowing the absolute reaction velocity so that the response times of the transducer and recorders and the mixing time of the reacting fluid are less important.

Most CA analytical methods cannot measure reaction velocities over a range greater than 10-fold. Whereas the lowest activity that can be detected is equal to the error in the uncatalyzed rate, the fastest is limited by response time of the method. Larger [CA] can be analyzed by diluting the original samples, which renders conditions different than *in vivo* and may alter enzyme function, although there is no evidence so far that CA activity is not proportional to [CA] up to the highest physiological concentrations, that in the red cells.[15,32,33]

TABLE I
Relative Changes in Uncatalyzed (k_u) and CA II-Catalyzed (k_{enz}) Hydration Reaction Velocity Constants with Temperature[a]

Temp. (°C)	k_u (sec^{-1})	Normalized k_{enz} (sec^{-1})	k_u/normalized k_{enz}
0	0.0022	0.068	0.032
4	0.0037	0.097	0.038
10	0.014	0.26	0.054
25	0.051	0.46	0.11
37	0.19	1.0	0.19

[a] k_u data are from S. J. Dodgson and R. E. Forster (unpublished), using ^{18}O exchange in 300 mosM NaCl at pH 7.4. The relative catalyzed velocity constant (k_{enz}) for CA II was calculated from data of Sanyal and Maren[55] for human CA II with ΔH = 8,150 cal.

1.4. Measurement of Reaction Inside Intact Cells and Organelles

If CA is inside a cell or other lipid membrane-bounded particle, the intracellular H^+ cannot exchange with the suspending solution rapidly enough to prevent large changes in intracellular $[H^+]$.[4] Thus, the total substrate consumed depends on the total buffer power inside the cells, and the duration of the reaction depends on the total intracellular enzyme activity in relation to intracellular buffer power. In most cells, the extent of the reaction is so small and its rate is so rapid it is difficult to determine reliably that CA is even present. Only in the ^{18}O exchange and ^{13}C NMR methods, where there is no net production or consumption of H^+ and the system is at chemical equilibrium, can intracellular CA activity be measured correctly. It is possible to estimate intracellular CA activity in two steps by methods depending on net chemical reaction. The activity is measured first in a suspension of intact cells and then again with the cells lysed. The difference represents the activity of the intracellular enzyme.

Any membrane surface has a stagnant layer of liquid upon it which presents a resistance to diffusing molecules and ions and which of itself will reduce the velocity of CO_2–HCO_3^- reactions inside this membrane. The continuous-flow rapid-mixing apparatus produces and maintains turbulent flow conditions in the mixture for the duration of the measured reaction, decreasing the thickness of the stagnant layer. Although the stop-flow instrument produces turbulent flow at the start of the reaction, stripping this stagnant layer, the layer rebuilds rapidly at stoppage. Measured values of the rate of CO uptake of erythrocytes in a stop-flow instrument are significantly slower that in a continuous-flow instrument, compatible with the expected behavior of the stagnant layer.[9,38]

2. Methods of Measuring the Velocity of the Reversible Reactions of CO_2–HCO_3^-

2.1. Methods Involving Net Chemical Reaction

2.1.1. Changing pH

2.1.1a. pH Indicator Dyes. H^+ is produced or consumed in the reactions of CO_2, and its rate of change provides an easily measured index of the reaction velocity. Including a pH indicator dye in the reaction mixture was the earliest and simplest method to determine the extent of the chemical process. The Philpots[46] described their method of this kind, actually for a classroom experiment. $[CO_2]$ is kept constant by bubbling the gas[7,46] through a solution containing a pH indicator that is acid. Alkaline solution is added, raising pH and changing the indicator color. The CO_2 hydrates, producing H^+, and the indicator becomes acid again. The time for this color change, measured by stopwatch, gives a relative measure of the speed of the hydration of CO_2. The equipment is simple and inexpensive, and the measurement is quick. Bruns and Gros describe a derivative method in this volume.[7] The reaction can be initiated by adding a volume of liquid equilibrated with CO_2 to a more alkaline solution.[25,64,67] The reactant mixture must be stirred mechanically, a function served in the Philpot technique by the CO_2 bubbles. The $[CO_2]$ now decreases during the reaction, but its starting value can be calculated from the volume and $[CO_2]$ of the liquid added, and conditions can be kept constant for comparative measurements.

A great improvement in the indicator dye method was the introduction of (1) spectrophotometric measurements of the ionization state of the dye and (2) rapid-mixing apparatus. The first technique makes it possible to measure pH instantaneously and sensitively (±0.001 pH), and the second initiates the reaction in less than 0.001 sec. The present standard of reference for this technique is the method of Khalifah.[34] The wave-length is chosen to fit the absorption curve of the indicator dye. The instrument records the absorbance (A) from the first 0.001 sec to the completion of the reaction, from which the pH curve can be computed by using the ionization constant of the indicator. The amount of HCO_3^- or H^+ produced at each point is calculated from the buffer dissociation curve. These two computations can be combined in a buffer factor and the rate of the reaction can be calculated as follows:

$$dx/dt = (\text{buffer factor}) \times dA/dt \qquad (5)$$

where x is the amount of H^+ or of HCO_3^- formed and buffer factor is the concentration of H^+ added to or removed from the solution divided by the change in absorbance (A), a function of the concentrations and ionization constants of both buffer and indicator and their initial state of ionization. Its value is obtained by titrating the indicator–buffer pair against A with acid or alkali. The buffer factor may vary with pH during the course of the reaction. Two modifications[34] eliminate this problem: (1) choosing a buffer with the same pK as the indicator so that the buffer factor becomes theoretically independent of pH, and (2) computing the total HCO_3^- or H^+ produced at final chemical equilibrium in each experiment from the concentration of associated and dissociated buffer, the initial $[CO_2]$, the ionization constant of the buffer, and the Henderson–Hasselbalch constant (see Appendix). Dividing this total change is $[HCO_3^-]$ by the total change in A gives an experimental determination of the buffer factor in each experiment.

2.1.1b. pH Electrode. pH electrodes were introduced in the 1930s and 1940s to avoid the inhibition of CA produced by some dyes.[8,64,67] Reliable, relatively inexpensive instrumentation is available off the shelf today and can provide stable, sensitive (0.001 pH units), and rapidly responding (down to milliseconds) recordings from the high impedance glass electrodes. The salt bridge and its exchanges with the sample electrolytes are the major source of instability and dragging response.

The reactions can be carried out as for the colorimetic methods in a small vessel stirred with a mechanical flea; the reaction is initiated by adding a volume of CO_2 solution.[25,42] The mixing time limits the response to over 1 sec. However, the method is convenient, rapid, and useful.

A glass pH electrode in a stirred vessel can be combined with a motor-driven syringe that delivers base or acid to the reaction mixture under electronic feedback control to keep pH constant (pH stat) during a reaction.[25] The amount of OH^- or H^+ delivered equals the amount of CO_2 that has reacted. The pH is constant, whereas it changes in other initial rate methods. The response time is limited by mixing time and response of the pH stat.

The pH electrode has been adapted to the continuous-flow mixing apparatus and has been used to measure acid and alkali exchange with red cells.[35,49]

The glass pH electrode has also been adapted to the stop-flow apparatus[7,11,63] by directing the outflow from the mixing chamber onto the end of a 3-mm-diameter glass electrode in order to reduce the thickness of the stagnant layer on its surface and increase the speed of response. The electrode itself has a very fast response time, provided buffer is present in the measured solution. pH-sensitive glass binds protons,[1,24] and the exchange of H^+ between the bulk solution and the glass through the stagnant layer that must exist on the surface can become rate limiting. The buffer acts to facilitate the flux of H^+ to the surface of the glass. This technique consumes as little as 1 ml of reacting fluid, is relatively inexpensive, is not affected by light absorption or scattering, and in the case of cell or organic suspensions measures extracellular pH as contrasted with cellular pH.

2.1.2. Changing CO_2 or pCO_2

2.1.2a. Gas Manometer. The reversible reaction of CO_2–HCO_3^- (Eq. 2) produces or absorbs CO_2, changing its concentration in the reaction mixture. These changes can be detected as changes in the volume or pressure of a closed gas phase in contact with the liquid. This was the technique developed by the Cambridge discoverers of CA.[43,51] They used a glass boat with two compartments in its bottom attached to a manometer. Placing HCO_3^- in one compartment and acid buffer in the other initiates the dehydration reaction on shaking and liberates CO_2 into the gas phase, which increases pressure in the manometer and permits the reaction to be followed continuously. It is impossible to maintain pCO_2 in the reacting liquid continuously in equilibrium with the gas in the manometer despite the agitation of the boat because of diffusion resistance. This slows the response time of the manometer to over 30 sec and requires empirical corrections.[43] The technique is sensitive but cumbersome and obsolete.

A modern version of the Cartesian diver with two small chambers containing different reactant solutions that are mixed to start the reaction has detected CA in single neurons. The production or absorption of CO_2 is measured by recording the magnetic field required to oppose the changes in buoyancy and to keep the diver, which contains a small magnet, at a constant position.[20]

2.1.2b. pCO_2 Electrode. The most convenient and rapid way to measure $[CO_2]$ in solution is with a CO_2 electrode. This instrument consists of a thin layer of HCO_3^- solution between a pH-sensitive glass electrode and a hydrophobic membrane (such as Teflon). CO_2 in the sample diffuses through the Teflon into the bicarbonate layer, where it equilibrates with H^+ and HCO_3^-, altering the pH. The $[HCO_3^-]$ is effectively constant because no cation can enter or leave the liquid layer, so log pCO_2 is a linear function of pH. The response time of this electrode is about 30 sec, but small and fragile versions have been made with response times approaching 10 sec[36]; it cannot be used to measure CO_2 reactions in a stirred vessel. It can be used in a continuous-flow rapid-reaction apparatus,[10,28,36] where the response time of the analytical instrument is not critical, to measure CO_2 reaction rates in solutions and in cell suspensions. Changes in $[CO_2]$ can be detected as soon as several milliseconds after initiation of a reaction.

CO_2 diffuses rapidly across lipid membranes, so the pCO_2 is in approximate equilibrium with the interior of any cells or particles as well as with the suspending fluid, in contrast to $[H^+]$ measured by pH electrode, which is only that in the extracellular fluid. Thus pCO_2 measurements reflect reactions inside as well as outside cells.

The equilibrated pCO_2, that is, $[H+][KCO_3^-]/(K' \times CO_2$ solubility), of a well-stirred HCO_3^- solution exposed to air will decrease as the gas escapes. This fall can be monitored with a pCO_2 electrode in the solution.[51] The escape of the gas is partially dependent on facilitated transport in the surface layers of the liquid and is therefore dependent on the CO_2/HCO_3^- reaction velocities and, under certain conditions, on [CA]. This method is a convenient technique for comparative data.

2.1.3. Changing Temperature

The dehydration of H_2CO_3 and the hydration of CO_2 have significant heats of reaction, and the resulting change in temperature of the reacting solution can be measured to follow the course of the reaction. Roughton and co-workers[33,50,52] applied this technique to CO_2 reactions in the continuous-flow rapid-mixing apparatus with thermocouples inside the observation tube to register the small temperature differences along the stream. For example, mixing 0.059 M HCl with 0.039 M $NaHCO_3^-$ at 18°C causes a rise of 0.023°C in less than 1 sec, followed by a slower increase of 0.031°C with a half-time of 0.05 sec. The rapid initial increment in temperature results from the heat released by the association of $H^+ + HCO_3^-$ to give H_2CO_3, whereas the slower increase in temperature is caused by the dehydration of H_2CO_3 to give CO_2. The increment in temperature for each molecule of CO_2 reacting is the same, so the time course of temperature parallels the time course of chemical reaction.

$$\text{Initial } dT/dt \times \Delta[CO_2]/\Delta T = k_v[H_2CO_3] \qquad (6)$$

where T is absolute temperature, $\Delta[CO_2]$ is the total concentration of CO_2 formed, and ΔT is the associated total change in temperature minus the rapid initial increment in temperature. This method can be used for any reaction with a heat of 1,000 to 10,000 cal/mole, with reactant concentrations of 0.0005 to 0.0005 M, for which one needs to be able to measure ±0.004°C to as low as ±0.00004°C, with a half-time of >1 msec. The temperature changes of the fluid include the contribution of chemical reactions inside any cells; they are unaffected by optical absorption or light scattering.[33] The amount of CO_2 reacted can be calculated and the associated change in temperature obtained from experimental data, permitting calculation of the heat of the reaction and other thermodynamic constants. The method has the disadvantages of the continuous-flow apparatus. There are several other chemical and mechanical processes which can produce or absorb heat, such as reaction of dilution, fluid friction, and conduction of heat to and from the measuring thermocouple. These effects are generally negligible, but if the temperature change of the reaction is small, corrections need to be made for them.

The CO_2–HCO_3^- reactions can also be followed thermally in a stop-flow rapid-mixing apparatus. The technical difficulty is in measuring the small temperature changes rapidly. Apparatus has been built that can measure changes of ±0.00002°C in 0.01 sec.[2]

2.1.4. Chemical Analysis

Chemical analysis cannot be accomplished fast enough to measure the velocity of reaction of CO_2–HCO_3^- in solution. However, chemical reactions plus cell exchanges in a suspension can be measured by filtering off very small samples for chemical analysis of suspending fluid at different times (distances) from the mixing chamber in a continuous-flow rapid-mixing apparatus.[36,65] The CO_2–HCO_3^- system is in chemical equilibrium within the sample, but exchanges between the intracellular and extracellular volumes can be calculated.

The apparatus may consist of a half a dozen lateral ports in the glass observation tube of a continuous-flow apparatus, each covered with a small Millipore filter disk. The pressure within the observation tube causes suspending fluid to flow through the Millipore and into a closed collecting vessel.[65] Alternatively, the outflow of the observation tube can be directed against the filter, and the separated fluid is collected on the other side.[36]

2.2. Methods Involving No Net Chemical Reaction

2.2.1. ^{18}O Exchange Between CO_2–HCO_3^- and HOH

Dry ^{18}O-labeled $NaHCO_3$ or $KHCO_3$ is added to a predetermined solution, and the disappearance of ^{18}O isotopes in CO_2 and/or HCO_3^- is monitored.[29,48,59] CO_2 reaches elemental chemical equilibrium with $[H+][HCO_3^-]$ with a half-time of several seconds under physiological conditions, approximately 2 orders of magnitude faster than the ^{18}O exchange (see Section 1). The ^{18}O exchanges with unlabeled oxygen in water; the reversible reactions of CO_2 and HCO_3^- are essentially irreversible because of the large dilution volume of 55 M. This reduces the concentration of labeled species at rates that are functions of the hydration reaction velocity constant, k_u or of the dehydration reaction velocity constant, k_v, and K'.

Early investigators collected and acidified aliquots of the reaction mixture and analyzed the gas produced as a batch in the mass spectrometer.[44,57] This method determines the abundance of the labeled isotopes in total ($CO_2 + HCO_3^-$) but does not distinguish between CO_2 and HCO_3^-. It is more convenient to measure continuously the concentrations of isotopic CO_2 in the reaction mixture by letting gases dissolved in it diffuse through a membrane in the bottom of a 3- to 10-ml glass reaction vessel directly into the ion source of a mass spectrometer, giving a response time of about 3 sec.[29,58]

Generally, the exchange is initiated in the absence of any catalyst to provide a control measure of k_u, and then the CA sample is added (Fig. 1). The exchange reactions are:

$$C^{18}O^{16}O + HOH \underset{\frac{2}{3}k_v}{\overset{k_u}{\rightleftharpoons}} H_2C^{18}O^{16}O_3, \qquad H_2C^{18}O^{16}O_3 \xrightarrow{\frac{1}{3}k_v} CO_2 + H^{18}OH \tag{7}$$

$$C^{18}O_2 + HOH \underset{\frac{1}{3}k_v}{\overset{k_u}{\rightleftharpoons}} H_2C^{18}O_2{}^{16}O, \qquad H_2C^{18}O_2{}^{16}O \xrightarrow{\frac{2}{3}k_v} C^{18}O^{16}O + H^{18}OH \tag{8}$$

There are two useful experimental variables that can be measured*:

1. α = the abundance of ^{18}O in CO_2,

$$\alpha = \frac{[C^{18}O_2] + [C^{18}O^{16}O]/2}{[C^{18}O_2] + [C^{18}O^{16}O] + [C^{16}O_2]} \tag{9}$$

2. The abundance of doubly labeled CO_2,

$$\frac{[C^{18}O_2]}{[C^{18}O_2] + [C^{18}O^{16}O] + [C^{16}O_2]} \tag{10}$$

The time course of disappearance of ^{18}O from CO_2 or from $C^{18}O_2$ is described by an equation of the form[44]

*Only the excess abundance of isotopes over the naturally occurring is of concern. It will be assumed that α, γ, and λ are excess abundance.

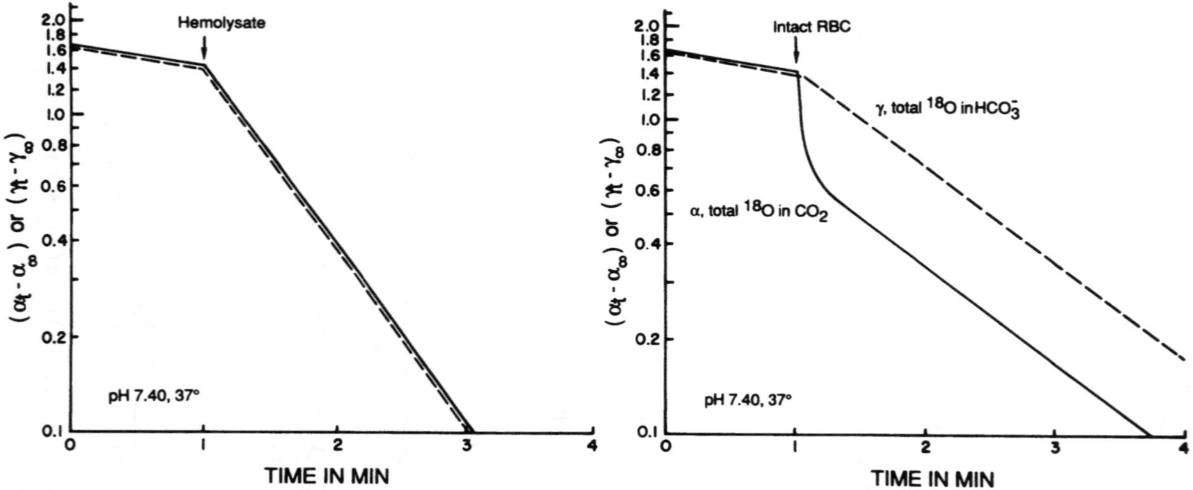

Figure 1. Semilogarithmic graphs of total (^{18}O abundance − ^{18}O abundance at final equilibrium) for CO_2, $\alpha_t - \alpha_\infty$ (—) and for HCO_3^-, $\gamma_t - \gamma_\infty$ (---) against time.[29] Twenty microliters of human red cell (RBC) suspension was added at the arrow in the right-hand panel and the same volume of hemolysate was added at the arrow in the left-hand panel, which gave a fractional cell water volume of 0.0006, or the equivalent resulting in a total [CA] of about 0.07 μM. The isotopic concentrations of CO_2 were measured by mass spectrometer, using a Teflon membrane inlet at the bottom of a 10ml stirred and temperature-regulated glass reaction chamber. The ^{18}O abundance in HCO_3^- was calculated by using the equations of Mills and Urey.[44] The first portion of the line up to the arrow represents the uncatalyzed reaction. The hemolysate accelerates the disappearance of both α and γ in an exponential manner. The intact cells accelerate the exchange of labeled CO_2 more than of labeled HCO_3^-, reducing α below γ.

$$\alpha = C_m e^{-mt} + C_n e^{-nt} \tag{11}$$

The solution for the exponents of Eq. 11 is

$$m, n = \left(\frac{k_u + k_{enz}}{2}\right)\left\{\left(1 + \frac{H^+}{K'}\right) \pm \sqrt{\left(1 + \frac{H^+}{K'}\right)^2 - 4\frac{[H^+]}{3K'}}\right\} \tag{12}$$

$[H^+]$ is adjusted rapidly at the start of the uncatalyzed exchange and remains constant. K′ is available from the literature, as is k_u, so k_{enz} can be calculated from the experimental value of n. The dehydration reaction velocity constant k_v is calculated from Eq. 3.

In homogeneous uncatalyzed CO_2–HCO_3^- solutions the first exponential term in Eq. 1 describes the establishment of a ratio of ^{18}O abundance in CO_2 to that in HCO_3^-, α/γ. This ratio is maintained during the remainder of the exponential disappearance of labeled CO_2 and HCO_3^-, as described by the second term, n. In uncatalyzed solutions and in solutions containing CA, the first term in Eq. 11 is small (at pH 7.4 it is only 1.6%) and rapid (half-time of 3.7 sec) and can be neglected.* Thus, the experimental time course is a single exponential with the exponent n. The half-time of the exchange at 37°C uncatalyzed is 320 sec, some 60 times slower than the initial rate measured by the production of H^+, which is 3.6 sec. This conveniently slower rate results from the facts that (1) only one-third of the molecules undergoing the dehydration reaction exchange ^{18}O with water and (2) the combined pool of HCO_3^- and CO_2, which must equilibrate with water, is 20 times greater than the pool of CO_2 alone, which determines the half-time for the production of H^+ or HCO_3^-. At 0°C, the half-time of the ^{18}O exchange increases 100-fold.

An equation analogous to Eq. 11 exists for the abundance of in HCO_3^-, γ, where γ is the abundance of 18O in HCO_3:

$$\gamma = \frac{2[HC^{18}O_2{}^{16}O^-] + [HC^{18}O^{16}O_2{}^-]}{3\{[HC^{18}O_2{}^{16}O^-] + [HC^{18}O^{16}O_2{}^-] + [HC^{16}O_2{}^-]\}} \tag{13}$$

The disappearance of mass 48, $C^{18}O_2$, is described by an equation analogous to Eqs. 11–13 except that the 4 within the radical sign in Eq. 13 becomes an 8. Thus, the decrease in the abundance of $C^{18}O_2$, that is, λ is almost exactly twice as fast as that of α, provided the exchange is rate limited by the chemical reactions, and provides another measure of $k_u + k_{enz}$. The relative concentrations of the three isotopes of CO_2, $C^{18}O_2$,$C^{18}O^{16}O$, and $C^{16}O_2$, during the exchange of the labeled oxygen with $H^{16}OH$ are given by the binomial rule, provided the exchange is rate limited by the chemical reactions. Thus, the abundance of $C^{18}O_2 = \alpha^2$, that of

* There is a small step in the disappearance of labeled CO_2 even in homogeneous solutions of CA, particularly for CA III.[55]

$C^{18}O^{16}O = 2\alpha(1 - \alpha)$, and that of $C^{16}O_2 = (1 - \alpha)^2$. This means that $\lambda/\alpha^2 = 1$ (symbolized as C/τ^2 in reference 60). If diffusion is partly limiting, the exchange λ/α^2 is > 1; this ratio is a convenient measure of the importance of diffusion in an ^{18}O exchange experiment.

In some experiments, the isotope exchange is partially rate limited by diffusion. The permeability through the red cell membrane of HCO_3^- compared with CO_2 is about 1/10,000 as great. When red cells are added to the solution, the labeled CO_2 disappears much faster than the labeled HCO_3^- (Fig. 1) until it establishes a ratio of α/γ which is much less than 1 rather than slightly greater than 1 as during an uncatalyzed run. This ratio is maintained during the remainder of the experiment. The exchange process (Appendix Eq. A2)[29] depends on two properties of the cell suspension, the total activity of CA per milliliter (A) and the permeability (centimeters per second) of the cell wall of $HCO_3^- \times$ total cell surface area per milliliter (P), needing two independent experimental data to be determined. The experimental data are not the exponents m and n in Eq. A2 because m is so large the duration of the first exponential term is brief and the mixing time of the reaction fluid is not fast enough for the system to follow the exchange correctly. Therefore, the magnitude, $C_m\ (n - m)$, of the first term is used instead. λ/α^2 is greater than 1 during ^{18}O exchange of intact red cells, confirming that the process is partially rate limited by diffusion.

Even in homogeneous CA solution, experimentally there is a significant sized first term of Eq. 11, or step, greatest for CA II[60] but also seen for CA II[29] and CA I (Y. Hayashi, L. Lin, and R. E. Forster, unpublished data). This lowering of α/γ is most likely explained on the grounds that the diffusion path from the bulk solution to the active Zn site has a higher resistance for HCO_3^- than for CO_2. The fact that the relative difference between α and γ is greater for CA III than for the other isozymes may be explained by the more restricted inlet to the pocket containing the active site.[19]

Michaelis–Menten enzyme characteristics, V_{max} and K_m, are derived from initial velocity measurements, in which there is no significant concentration of end product. The ^{18}O exchange method requires chemical equilibrium, so there are equivalent concentrations of end product and substrate. This should lower the concentration of free enzyme available to react with substrate at any instant, lowering the measured reaction velocity compared with the initial rate method, and give erroneous values of V_{max} and K_m. Experimentally, reaction velocity measured by ^{18}O is the same as that obtained with the standard initial rate techniques. This is because (see Section 1.1) the velocity of the catalyzed reactions is rate limited by H^+ exchange and not as much by the actual catalytic step.

The advantages of the ^{18}O exchange method are that it measures CA activity at constant pH, can measure it inside a cell or organelle, at physiological temperature and in the absence of buffer, and can measure as little as 10^{-15} moles of high-activity enzyme in a volume of 0.01 ml^3 and determine HCO_3^- permeability of the lipid membrane. It requires a mass spectrometer and dedicated inlet system, a large, expensive instrument needing constant expert attention, and is slow in that one measurement can take over 15 min.

To measure ^{18}O isotope abundance at times less than the mixing time of the reaction cell or the response time of the inlet system plus reaction cell, some 3 to 6 sec, one can use the continuous-flow rapid-mixing principle and drive solutions containing labeled HCO_3^-–CO_2 and CA into a mixing chamber, which may be only a T-tube, out of which the reacting mixture flows into a variable length of glass tubing and then to a 2-mm length of silastic tubing that is exposed directly to the high vacuum of the ion source.[14,66] The isotopes of CO_2 in the reacting mixture at that point diffuse through the plastic wall and are analyzed by the spectrometer. By varying the length of tubing from the mixing point to the silastic tubing, the CO_2 isotopes can be analyzed in the reacting mixture after as little as 2 sec.

In all of the applications of the ^{18}O exchange method described above, the processes take place in one liquid phase. It is also possible to measure CO_2–HCO_3^- reaction velocities in a thin liquid layer on a 50-μm-thick filter disk exposed to gas containing ^{18}O-labeled CO_2. The CO_2 isotopes in the gas maintain equilibrium with the CO_2 dissolved in the liquid. Although the volume of water is small, the number of molecules of oxygen in water is so great as to act as a semi-infinite sink. The disappearance of labeled CO_2 from the gas phase is measured with very small (<0.010 ml) gas samples introduced into a mass spectrometer. k_{enz} is obtained from the following equation, which is a slight modification of Eq. 12:

$$m, n = \left(\frac{k_u + k_{enz}}{2}\right)\left\{\left(\frac{1}{R} + \frac{\mathrm{H}^+}{K'}\right) \pm \sqrt{\left(\frac{1}{R} + \frac{\mathrm{H}^+}{K'}\right)^2 - 4\frac{[\mathrm{H}^+]}{3KR'}}\right\} \quad (14)$$

where R is a weighting factor = (total moles of CO_2 + HCO_3^- in liquid + gas phases)/moles of CO_2 in liquid. The advantage of this method is the small volume of HOH, which reduces the absolute velocity of the uncatalyzed reaction, increasing the absolute sensitivity to better than 10^{-15} moles of CA. The total mass of labeled CO_2 in gas is large in related to that of HOH, and the half-time for the uncatalyzed reaction at room temperature is over 100 min.

The same principles can be applied to measure the average CA activity in an isolated lung perfused with a fluid containing no CA by measuring the disappearance of labeled CO_2 from rebreathed gas.[45] The isotopes of CO_2 in alveolar gas are considered in constant equilibrium with CO_2 dissolved in the lung tissue.

2.2.2. ^{13}C Nuclear Magnetic Resonance

A solution containing ^{13}C-labeled CO_2 and HCO_3^- demonstrates two NMR peaks whose separation results from differences in the chemical shifts of the two molecules. When labeled ^{13}C exchanges chemically between CO_2 and HCO_3^-, it causes the line width (half-maximal peak height) to increase (broaden); the magnitude of this broadening is described by the relationship[56,62]

$$1/\tau = \pi\, \delta nu \tag{15}$$

where $1/\tau$ is the fraction of the molecular species reacting or exchanging, in seconds^{-1}, τ is the lifetime in seconds of the chemical species, and δnu is the increase in half-maximal peak height broadening caused by the chemical reaction exchange.

At equilibrium, the exchange between CO_2 and HCO_3^-, or the reaction velocity, must be the same in both directions.

$$1/\tau_{HCO_3^-} [HCO_3^-] = 1/\tau_{CO_2} [CO_2] \tag{16}$$

where subscripts refer to the two molecular species and $1/\tau_{HCO_3^-}$ and $1/\tau_{CO_2}$ are, respectively, k_v and k_u.

The apparent $1/\tau$, represented by the symbol $1/T_2$, obtained by experimental measurements, is usually greater than the actual rate of chemical exchange, $1/\tau$, according to the equation

$$1/T_2 = 1/T_{2'} + R + 1/\tau \tag{17}$$

where $1/T_2$ is the intrinsic rate of relaxation of the nuclear magnetic spin-spin and R is a rate component produced by inhomogeneities in the magnetic field.

The reaction velocity of CO_2–HCO_3^- in the presence of CA is nearly a magnitude faster than the rate of CO_2 or HCO_3^- production[61,62] because the exchange path does not include the exchange of H^+ or OH^- (see Section 1.1). Therefore, it is not clear that it provides physiological information. The method requires expensive and complicated instrumentation and may require minutes to hours because the spectra have to be averaged over hundreds to thousands of observations in order to reduce the noise/signal ration.

2.2.3. Facilitation of CO_2 Flux Through a Liquid Film

The transport of $^{13}CO_2$ through a thin layer of HCO_3^- solution between two gas phases at chemical equilibrium is facilitated by the diffusion flux of HCO_3^- formed on one surface and dehydrated on the other. When the film is thin enough, the chemical rates of the hydration–dehydration reactions become limiting. CA enhances this facilitated CO_2 transport.[39] $k_u + k_{enz}$ can be calculated from the measured facilitation, that is, (total CO_2 flux − CO_2 diffusion flux)/(CO_2 diffusion flux), and from film thickness, diffusion coefficient for CO_2 in the liquid, and $[H^+]/K'$.[15] $[H^+]$ is constant throughout the film because elemental chemical equilibrium exists between the two gas phases and the film in between. Total (elemental) CO_2 flux through a film is also facilitated by HCO_3^- diffusion and further augmented by CA, but because $[H^+]$ varies through the layer, mathematical equations cannot be solved conveniently to obtain $k_u + k_{enz}$.

The method can measure CA activity up to the physiological concentrations and theoretically down to the sensitivity of the ^{18}O exchange technique (10^{-15} moles[3]). It has the disadvantage of requiring analysis of labeled CO_2, with either stable or radio-isotopes.

2.2.4. Relaxation Methods

In relaxation methods, a HCO_3^-–CO_2 solution at equilibrium is subjected to a sudden (microsecond) change in physical conditions, such as electrical field, pressure, or temperature, which changes the chemical equilibrium state. The time course of the readjustment, or relaxation, of $[CO_2]$ and $[HCO_3^-]$ is monitored with an instantaneous analytical method. The great advantage of these methods is their ability to measure extremely rapid chemical reactions. Several methods that have been applied to the CO_2–HCO_3^- reactions are briefly presented as follows.

One method is the ionization of H_2CO_3[17] by electrical field pulse. This is accomplished by putting a transient pulse of 10–200 kV through a solution of CO_2–HCO_3^- and measuring the electrical conductivity as the field voltage varies. The high field causes H_2CO_3 to dissociate, increasing the electrical conductivity, which depends on the concentration of ions and their mobilities, and thus gives a measure of $[H^+]$ and $[HCO_3^-]$ produced, if there are no other ions in the solution. At 25°C and low ionic strength, the dissociation rate constant was $\approx 8 \times 10^6$ sec^{-1}, the recombination rate constant was 4.7 10^{10} M^{-1} sec^{-1}, and the half-time at physiological pH was 0.087 μsec, instantaneous on a physiological time scale. If there are other ions in the solution, they will increase the conductivity and interfere with the measurement of $[H^+]$ and $[HCO_3^-]$. It would be possible to use a pH indicator dye and record the changes in $[H^+]$ spectrophotometrically. Although this method can measure more rapid processes than many other techniques, it requires very special equipment and the speed is outside the physiological range.

In the pressure jump technique,[37] the HCO_3^-–CO_2 solution is exposed to a large step change in pressure, which alters the equilibrium constant proportionally to the increment in partial molal volume of the reaction. The solution starts at 100 atm and is returned to 1 atm in 50 μsec when a metal diaphragm ruptures. The electrical conductivity at 40 kHz is recorded as the system approaches equilibrium at the lower pressure. Temperature changes are compensated for by measuring the difference in conductivity in two identical cells subjected to the pressure changes, one containing the HCO_3^-–CO_2 solution and the other a solution whose conductivity is the equal but which does not alter with pressure. The method can follow reactions from <50 μsec to several minutes.

These methods require specialized instrumentation. The conductivity is affected by other ions, ruling out physiological solutions. To our knowledge, the methods have not been used for CA-catalyzed reactions.

Appendix

Calculation of $\Delta[HCO_3^-]$ in the Spectrophotometric Stop-Flow pH Method[32]

$$\Delta[HCO_3^-] = \frac{([CO_2]_0 + [B]_0 + [HB]_0 K_{HB}/K') \pm \{([CO_2]_0 + [B]_0 + [HB]_0 K_{HB}/K')^2 - 4[CO_2]_0[B]_0(1 - K_{HB}/K')\}^{1/2}}{2(1 - K_{HB}/K')} \quad \text{(A1)}$$

where:

$\Delta[HCO_3^-]$ is the total change in $[HCO_3^-]$ at final equilibrium.

$[CO_2]_0$, $[B]_0$, and $[HB]_0$ are, respectively, the initial concentrations of CO_2, unprotonated buffer, and protonated buffer.

K_{HB} is the dissociation equilibrium constant of the buffer.

Total Enrichment, α, of ^{18}O in CO_2 as a Function of Time in a Suspension of Cells or Liposomes Containing CA[29]

$$\alpha = [C_m/(n - m)]\,e^{-mt} + [C_n/(n - m)]\,e^{-nt} \quad \text{(A2)}$$

$$-m, -n = \frac{k_u}{1 + \theta}\left|(R + S) \pm \tfrac{1}{2}[(R + S)^2 - 8AS/3\,(1 - \theta)^2)]^{1/2}\right| \quad \text{(A3)}$$

$$C_m = -\alpha_{t0},\ m + \left(1 + \frac{A\theta}{1 + \theta}\right)\left(\gamma_{t0}k_u + \frac{\alpha_{t0}k_v}{1 + \frac{K_{HA}}{[H^+]}}\right) \quad \text{(A4)}$$

$$C_n = +\alpha_{t0},\ n - \left(1 + \frac{A\theta}{1 + \theta}\right)\left(\gamma_{t0}k_u + \frac{\alpha_{t0}k_v}{1 + \frac{K_{HA}}{[H^+]}}\right) \quad \text{(A5)}$$

where:

m and n are the exponential constants defined with + and − radical term, respectively.

C_m and C_n are the respective coefficients.

k_u is the uncatalyzed hydration velocity constant in seconds^{-1}.

k_v is the uncatalyzed dehydration velocity constant in seconds^{-1}.

$R = \tfrac{1}{2}\{2a/3-(A+1)(1+\theta)\}$.

A = [(effective intracellular hydration velocity constant)/(effective extracellular hydration velocity constant) intracellular water volume/milliliter]. The effective constants include uncatalyzed plus catalyzed reactions. However, for practical purposes, the intracellular reaction constant = catalyzed reaction velocity/[CO_2] and, the extracellular reaction velocity constant = uncatalyzed reaction velocity constant.

$S = -1/K'(1 + \theta + A\theta)$.

$\theta = PK_{HA}/Ak_v[H^+]$.

P is the permeability to HCO_3^- of the red cell membrane in centimeters × seconds^{-1} total membrane surface area in square centimeters per milliliter.

$K_{HA} = [H^+][HCO_3^-]/[H_2CO_3]$

$K' = [H^+][HCO_3^-]/H_2CO_3] = (1 + K_{HA}/[H^+]k_u/k_v$.

ACKNOWLEDGMENTS. The author wishes to thank Bayard T. Storey, Ph.D., Mortimer Civan, M.D., Lydia Lin, M.S., and of course Susanna Jane Dodgson, Ph.D., for advice and assistance in preparing this chapter.

References

1. Bates, R. G., 1954, *Determination of pH. Theory and Practice*. Wiley; New York, Chapter 10.
2. Berger, R. A., 1978, *Biophys. J.* **24**:2–20.
3. Bitterman, N., Lin, L., and Forster, R. E., 1988, *J. Appl. Physiol.* **65**:1902–1906.
4. Booth, V. H., 1938, *J. Physiol. (Lond.)* **93**:117–128.
5. Botré, C., Memoli, A., Mascini, M., and Mussini, E., 1983, *Anal. Lett.* **16**(B1):9–23.
6. Brinkman, R., and Margaria, R., 1931, *J. Physiol. (Lond.)* **72**:6P.
7. Bruns, W., and Gros, G., this volume, Chapter 9.
8. Clark, A. M., and Perrin, D. D., 1951, *Biochem. J.* **48**:495–503.
9. Coin, J. T., and Olson, J. S., 1979, *J. Biol. Chem.* **254**:1178–1190.
10. Constantine, H. P., Craw, M. R., and Forster, R. E., 1965, *Am. J. Physiol.* **208**:801–811.
11. Crandall, E. D., Klocke, R. A., and Forster, R. E., 1971, *J. Gen. Physiol.* **57**:664–683.
12. Davis, R. P., 1961, in: *The Enzymes*, (2nd ed., Volume 5 (P. D. Boyer, H. Lardy, and K. Myrback, eds.). Academic Press, New York, pp. 545–562.
13. Davis, R. P., 1963, *Methods Biochem. Anal.* **11**:307–328.
14. Degn, H., and Kristensen, B., 1986, *J. Biochem. Biophys. Methods* **12**:305–310.
15. Donaldson, T. L., and Quinn, J. A., 1974, *Proc. Natl. Acad. Sci. USA* **71**:4995–4999.
16. Edsall, J. T., 1968, *Harvey Lect.* **62**:191–230.
17. Eigen, M., Kustin, K., and Maas, G., 1961, *Z. Physik. Chem. NF* **30**:130–136.
18. Eigen, M., and De Maeyer, L., 1963, *Tech. Org. Chem.* **8**:977–1001.
19. Eriksson, A., 1988, *Acta Univ. Ups. Fac. Sci.* **164**.
20. Giacobini, E., 1962, *J. Neurochem.* **9**:169–177.
21. Gibson, Q. H., and Milnes, L., 1964, *Biochem. J.* **91**:161–171.
22. Hansen, P., and Magid, E., 1966, *Scand. J. Clin. Lab. Invest.* **18**:21–32.
23. Hartridge, H., and Roughton, F. J. W., 1923, *Proc. R. Soc. London Ser. B.* **94**:336–394.
24. Haugaard, G. N., 1937, *Nature (London)* **140**:66.
25. Henry, R. P., this volume, Chapter 8.
26. Ho, C., and Sturtevant, J. M., 1963, *J. Biol. Chem.* **238**:3499–3501.
27. Hoffman, and Henkens, O. 1987, *Biochim. Biophys. Acta* **143**:67–73.
28. Holland, R. A. B., and Forster, R. E., II, 1975, *J. Physiol. (Lond.)***228**:1589–1596.
29. Itada, N., and Forster, R. E., 1977, *J. Biol. Chem.* **252**:3881–3890.
30. Itada, N., and Peiffer, L., and Forster, R. E., 1978, *Frontiers of Biologial Energetics*. New York, Academic Press, pp. 715–724.

31. Kern, D. M., 1960, *J. Chem. Educat.* **37**:14–23.
32. Kernohan, J. C., Forrest, W. W., and Roughton, F. J. W., 1963, *Biochim. Biophys. Acta* **67**:31–41.
33. Kernohan, J. C., and Roughton, F. J. W., 1968, *J. Physiol. (London)* **197**:345–361.
34. Khalifah, R. G., 1971, J. Biol. Chem. **246**:2561–2573.
35. Klocke, R. A., Andersson, K. K., Rotman, H. H., and Forster, R. E., 1972, *Am. J. Physiol.* **222**:1004–1013.
36. Klocke, R. A., 1976, *J. Appl. Physiol.* 40:707–714.
37. Knoche, W., 1980, in: *Biophysics and Physiology of Carbon Dioxide* (C. Bauer, G. Gros, and H. Bartels, eds.), Springer-Verlag, Berlin, pp. 3–11.
38. Krawiecz, J. A., Forster, R. E., Gottliebsen, T. W., and Fish, D., 1983, *Fed. Proc.* **42**:993.
39. Longmuir, I. S., Forster, R. E., and Wool, C., 1966, *Nature (London)* **209**:393–394.
40. Maren, T. H., 1960, *J. Pharmacol. Exp. Ther.* **130**:26–29.
41. Maren, T. H., 1967, *Physiol. Rev.* **47**:595–781.
42. Maren, T. H., and Couto, E. O., 1979, *Arch. Biochem. Biophys.* **196**:501–510.
43. Meldrum, N. U., and Roughton, F. J. W., 1933, *J. Physiol. (London)* **80**:113–142.
44. Mills, G. A., and Urey, H. C., 1940, *J. Am. Chem. Soc.* **62**:1019–1026.
45. Nioka, S., Henry, R. P., and Forster, R. E., 1988, *J. Appl. Physiol.* **65**:2236–2244.
46. Philpot, F. J., and Philpot, J., 1936, *Biochem. J.* **30**:2191–2193.
47. Pocker, Y., and Meany, J. E., 1965, *Biochemistry* **4**:2535–2541.
48. Poulton, D. J, and Baldwin, H. W., 1967, *Can. J. Chem.* **45**:1045–1050.
49. Rossi-Bernardi, L., and Berger, R. L., 1968, *J. Biol. Chem.* **243**:1297–1302.
50. Roughton, F. J. W., 1941, *J. Am. Chem. Soc.* **63**:2930–2934.
51. Roughton, F. J. W., and Booth, V. H., 1946, *Biochem. J.* **40**:319–390.
52. Roughton, F. J. W., and Chance, B., 1963, *Tech. Org. Chem.* **8**:703–792.
53. Roughton, F. J. W., and Clark, A. M., 1951, in: *The Enzymes*, Volume 1, Part 2, Academic Press; New York, Chapter 43.
54. Roughton, F. J. W., 1964, in: *Handbook of Physiology–Respiration I* (W. O. Fenn and H. Rahn, eds.), American Physiological Society, Washington, D.C., pp. 767–825.
55. Sanyal, G., and Maren, T. H., 1981, *J. Biol. Chem.* *256*:608–612.
56. Shporer, M., Forster, R. E., and Civan, M. M., 1984, *Am. J. Physiol.* **246**:C231–C234.
57. Silverman, D. N., 1973, *Arch. Biochem. Biophys.* **155**:452–457.
58. Silverman, D. N., Tu, C., and Wynns, G. C., 1976, *J. Biol. Chem.* **251**:4428–4435.
59. Silverman, D. N., 1982, *Methods Enzymol.* **87**:732–752.
60. Silverman, D. N., and Tu, C. K., 1986, *Biochemistry* **25**:8402–8408.
61. Silverman, D. N., and Lindskog, S., 2988, *Acc. Chem. Res.* **21**:30–36.
62. Simonsson, I., Jonsson, B.-G., and Lindskog, S., 1979, *Eur. J. Biochem.* **93**:409–417.
63. Sirs, J. A., 1958, *Trans. Faraday Soc.* **54**:207–212.
64. Stadie, W. C., and O'Brien, H., 1933, *Biol. Chem.* **103**:521–529.
65. Tosteson, D. C., 1959, *Acta Physiol. Scand.* **46**:19–41.
66. Tu, C., Wynns, G. C., McMurray, R. E., and Silverman, D. N., 1978, *J. Biol. Chem.* **253**:8178–8184.
67. Wilbur, K. M., and Anderson, N. G., 1948, *J. Biol. Chem.* **176**:147–154.

PART II

Carbonic Anhydrase Analysis

CHAPTER 7

Purification of the Carbonic Anhydrases

W. RICHARD CHEGWIDDEN

1. Introduction and Historic Perspective

Carbonic anhydrase (CA) (EC 4.2.1.1.) is very widespread throughout the plant and animal kingdoms and is also present in certain bacteria. In mammals, it occurs in at least seven different isozyme forms, designated CA I to CA VII, six of which have been isolated in pure form.[109,117]

The enzyme was first isolated in 1933 from bovine erythrocytes by Meldrum and Roughton.[79] Keilin and Mann subsequently showed that these cells contain up to 2 g of enzyme per liter.[61]

Removal of the vast excess of hemoglobin is the first important step in the isolation procedure, and Keilin and Mann[61] achieved this ethanol–chloroform extraction. For so harsh a process, this appears to have little deleterious effect on the enzyme from most species and was much used subsequently in many large-scale isolation methods. By following this treatment with ammonium sulfate fractionation and alumina Cγ extraction, a relatively high degree of purification was achieved.[61] Some 20 years later, however, after the work of Nyman,[86] Rickli and Edsall,[95] and Laurent *et al.*,[67] who used more modern separation techniques of electrophoresis and chromatography also developed by Lindskog,[70] it became clear that in human erythrocytes CA is present principally in two distinct isozyme forms, now known as CA I and CA II.[110]

In the subsequent period, refinements in isolation procedures were concentrated on development of gentler methods for separation of hemoglobin and on separation of the isozymes. Edsall's laboratory achieved efficient recovery of human CA and separation from hemoglobin by gel filtration on Sephadex G-75,

W. RICHARD CHEGWIDDEN • Division of Biomedical Sciences, Sheffield City Polytechnic, Sheffield S1 1WB, United Kingdom.

followed by chromatography on hydroxyapatite to separate the isozymes.[96] Subsequently, they adopted a simpler and more rapid method of removing hemoglobin on Sephadex A-50 and, after precipitation with ammonium sulfate, separated the isozymes on DEAE–Sephadex.[4] Similar techniques were used by other workers to isolate erythrocyte CA from many other species, some requiring further purification by techniques such as isoelectric focusing or preparative electrophoresis.[5]

More recent times have seen significant progress in the development of highly specific affinity resins which permit both a considerable degree of purification and the resolution of isozymes in one step.[62,87,125] The resins are prepared by coupling a sulfonamide inhibitor of CA to a suitable inert, insoluble matrix. These resins specifically bind to the enzyme which, after washing off nonspecifically bound protein, is eluted by ionic strength gradient or, more commonly, a concentration gradient of a tightly binding ionic inhibitor to displace the sulfonamide.

Affinity chromatography employing such resins plays an important role in current isolation methods for all six mammalian CA isozymes hitherto isolated. Indeed, since virtually all known CAs in the animal kingdom, and those from bacteria and algae, are highly sulfonamide sensitive,[76] an even broader applicability of this technique is indicated.

Even plant CAs, which are much less sensitive to sulfonamide inhibition,[38,76] may well, at low ionic strength, bind to affinity resins used in the isolation of CA III, which is similarly less sensitive (see below).

Another important practical advance in recent years has been the development of a rapid ion exchange chromatography system, which was used to isolate porcine and bovine CAs.[8] After an initial ethanol–chloroform extraction to separate hemoglobin, the crude enzyme extract was applied to a DEAE–cellulose DE-23 (Whatman) column and eluted with a linear gradient of sodium chloride, under air pressure of up to 40 kPa/cm^2. The consequent high flow rate of up to 30 cm^3/min permitted the production of gram quantities of homogeneous enzyme preparation in a single step taking only 2–4 h.

2. General Methods Ancillary to Purification Procedures

2.1. Preparation of Affinity Resins

Early attempts to produce affinity resins specific for CAs[27,119] involved the attachment of sulfonamide inhibitors to Sepharose by an ester linkage with use of a cyanogen bromide method. Unfortunately, the effectiveness of these resins was diminished by the tendency for CA to hydrolyze the linkages.

This problem was overcome by Osborne and Tashian,[87] who coupled sulfonamides to CM Sephadex (Pharmacia) through a peptide bond. They were particularly successful with the inhibitor *p*-[(2,4-diaminophenyl)azo]benzenesulfonamide (Prontosil), which gave a stable, reusable resin of high binding capacity (20 mg of enzyme per cm^3 of gel).

2.1.1. Preparation of Prontosil

Prontosil is prepared by diazotization of sulfanilamide (*p*-aminobenzenesulfonamide) and coupling to *m*-phenylenediamine.

Sulfanilamide (10 g) is dissolved in 17.4 cm^3 of 30% (wt/vol) HCl and 120 cm^3 of H_2O. Keeping the solution cooled on ice (important), cold sodium nitrite solution (4.3 g of $NaNO_2$ in 30 cm^3 of H_2O) is added dropwise. Complete diazotization may be confirmed by the presence of excess nitrite producing a blue color with KI paper.

m-Phenylenediamine (6.2 g) dissolved in 10.2 cm^3 of 10% (wt/vol) HCl is added dropwise to this cooled, diazotized solution and allowed to stand for several hours on ice. The red precipitate of Prontosil is filtered off, washed with ice-cold water, and dried in air.

2.1.2. Coupling Prontosil to the Resin

The affinity resin is prepared by linking Prontosil onto CM Sephadex C50 by the method of Hoare and Koshland,[51] a procedure that is sensibly performed in a fume hood.

CM Sephadex C50 (10 g) is added to 5 g of Prontosil dissolved in 300 cm^3 of 50% acetone–water. The pH is maintained at 4.75 by dropwise addition of 1–2 M sulfuric acid with stirring, which is continued for 1 hr, after which 5 g of 1-(3-dimethylaminopropyl)-3-ethylcarbodiimide hydrochloride, freshly dissolved in 10 cm^3 of water, are added dropwise (pH maintained). After being stirred overnight at room temperature, the gel is washed successively with acetone, 50% acetone–water, and finally water adjusted to pH 10 with sodium carbonate on a large Büchner funnel (coarse or medium sinter) until all free Prontosil (red color) is removed. This process requires several liters of acetone. (N.B.: The prudent purifier will be equipped with a number of Büchner funnels, since their sinters will rapidly become blocked. They can be cleared with chromic acid, but this process takes a day or so.)

2.1.3. Other Affinity Resins

Subsequently, Khalifah *et al.*[62] modified this method by attaching the sulfonamide inhibitor *p*-aminomethylbenzene sulfonamide (PAMBS) to CM Bio-Gel A (Bio-Rad Laboratories).[62] The coupling technique is identical except that the inhibitor is added to preswollen gel washed in 50% acetone. This modification has been adopted by a number of laboratories, partly perhaps because PAMBS is readily available commercially and also because the bed volume of CM Bio-Gel A is less sensitive to variations in ionic strength than is CM Sephadex. Prontosil, however, possesses the advantages that its intense color permits visual assessment of the coupling to the gel and of the extent of the subsequent washing process. Also, perhaps because of its longer "arm," the Prontosil–CM Sephadex resin, but not the PAMBS–CM Bio-Gel A, will bind to certain weakly binding CA isozymes (e.g., bovine CA III).

2.2. Monitoring Progress of Purification

Ideally, the specific activity of the enzyme sample should be measured after each step in the purification procedure. The most commonly used routine assay, both in identification of active fractions and in monitoring purification, has been a colorimetric method of measuring the CO_2 hydration reaction.[96] Protein concentration is most simply monitored as absorbance at 280 nm (A_{280}), and specific activity is expressed as activity units divided by A_{280}.

2.3. Spot Test for Identifying CA Activity

The bromothymol blue (BTB) spot test developed in Tashian's laboratory[108] from the original method of Pihar,[88] is an extremely useful and sensitive method for the rapid identification of CA activity. It permits the qualitative identification of 100 or more fractions containing CA activity in a matter of minutes.

A suitably sized piece of Whatman 3MM chromatography paper is soaked with 0.15% BTB in 0.1 M Veronal buffer, pH 9.0. Excess buffer is dabbed off with paper towels and the indicator-soaked paper is placed in a small, shallow-sided tray, sitting on a bed of ice to slow down the reaction. Samples of 4 μl are applied to the paper in serried ranks and a transparent plate containing a small hole is placed over the tray. Carbon dioxide from a cylinder is then blown onto the samples for 3–4 sec through the hole, which is immediately sealed with a small piece of tape. Within a few seconds, the spots containing CA activity will turn yellow as the acid produced by the CA reaction changes the color of the BTB indicator. (N.B.: After about 30 sec, all spots will turn yellow and then slowly fade.) The test will detect less than 5 ng of the high-activity isozyme CA II in 5 mM Tris, pH 9 (J. Shelton and W. R. Chegwidden, unpublished data). Obviously, it is not effective for samples in acid solution.

The test can be modified to differentiate between CA isozymes by addition of an appropriate concentration of acetazolamide to the BTB solution; e.g., a concentration of 2 μM acetazolamide will completely inhibit CA I and CA II but leave CA III unaffected.

2.4. Assessment of Purity

The purified enzyme should be assessed for homogeneity by at least one method. Sodium dodecyl sulfate-polyacrylamide gel electrophoresis (SDS-PAGE) has been commonly used. Typically, this procedure may be performed as described previously,[65] using a slab gel apparatus.[94] Gel filtration, ultracentrifugation, and electrophoresis of the native (not SDS-denatured) protein have frequently served as additional methods. Comprehensive practical details of gel electrophoresis of proteins are available elsewhere.[42]

2.5. General Points

Concentration of sample fractions during isolation procedures is generally achieved by ultrafiltration. Many systems are readily available commercially. Vacuum dialysis, a very simple and inexpensive method that is described with the utmost clarity,[118] may be used with most CA isozymes. Alternative procedures, such as freeze drying, may be required for isozymes that are particularly labile. The membrane-bound form (CA IV) may well fall into this category, although vacuum dialysis has been used in purifying this isozyme from bovine lung.[120]

Information on storage conditions for the purified isozymes is not well documented in the literature. In general, the mammalian cytoplasmic isozymes are stable for many weeks stored at a concentration of at least 10 mg/cm^3 in buffer containing 1 mM dithiothreitol (DTT) at 4°C.

Although many columns can be run under gravity, it is best to use a suitable pump in order to achieve consistent, reproducible results. Where flow rates are not stated, the manufacturer's recommendations should be followed.

All practical procedures should be undertaken at 4°C unless otherwise stated, and glass-distilled water, or water of similar purity, should be used throughout.

3. Mammalian Carbonic Anhydrases

3.1. The Cytoplasmic Isozymes—CA I, CA II, and CA III

Next to hemoglobin, CA is the most abundant protein in the human erythrocyte. About 85% is present as CA I, and 15% is present as CA II, the higher-activity isozyme. CA II occurs in virtually all types of tissue and is present in the erythrocytes of all mammals investigated. CA I, on the other hand, is seemingly more limited in its tissue distribution, being generally abundant in the red cell and also present in a number of other cell types such as gastrointestinal mucosa and vascular endothelium.[109] It is, however, missing from the erythrocyte of all ruminants examined (e.g., ox, sheep, and goat), felids (e.g., cat, tiger, and lion), the wallaby, and the blue-white dolphin.

Both CA isozymes have been purified from human red cells as well as from those of, e.g., rhesus monkey,[24,44] various other primates,[114] pig,[107] rabbit,[32,72] horse,[35] guinea pig,[17] rat,[71] and deer mouse and house mouse.[114] CA II has also been purified from ox,[70] sheep,[106] deer and goat,[5] moose,[15] wallaby,[57] and blue-white dolphin.[101]

In 1976, the low-activity, cytoplasmic isozyme, CA III, was discovered in red skeletal muscle of sheep, cat,[52] and rabbit,[64] where it comprises 1–2% of the soluble protein. It has subsequently been purified, among mammals, from skeletal muscle of the human,[20,48] gorilla,[49] pig,[91] horse,[85] ox,[20,26,113] and cat[98] and found in large amounts in male rat liver.[19] Its presence has also been detected, albeit at low levels, in human erythrocytes.[43]

3.1.1. Purification of CA I and CA II from Human Erythrocytes

Erythrocytes are removed from whole blood by centrifugation at 2,500 × *g* for 20 min, washing three times with 0.9% NaCl and recovering the cells each time by centrifugation as before.

To 250 cm³ of washed, packed cells, an equal volume of distilled water is added. The cells are shaken for 5 min and allowed to stand for 30 min in the cold. The pH is adjusted to 6.0 with dilute sulfuric acid to ensure precipitation of cell stroma, which is removed by centrifugation at 15,000 × *g* for 15 min. The lysate is then extensively dialyzed against water at pH 9.0 (adjusted with dilute NaOH) and finally against 0.1 M Tris SO_4–0.2 M Na_2SO_4 buffer, pH 9.0, over the course of a day. Any precipitate is removed by centrifugation at 15,000 × *g* for 15 min, and the supernatant is stirred gently for 3 hr (overhead paddle stirrer) with 150–200 cm³ of Prontosil–CM Sephadex affinity resin[87] equilibrated in the same buffer. Non-specifically bound protein is eluted by extensive (several liters) washing of the resin with the same buffer in a large sintered Büchner funnel until A_{280} of the eluent is close to zero. The gel is then transferred to a column (2.5 × 25 cm), where washing is completed.

CA I, the lower-activity isozyme, is then eluted by pumping 0.75 M Tris SO_4–0.2 M KI buffer, pH 7.5, through the column at a flow rate of about 40 cm³/hr. CA II remains bound under these conditions and is subsequently eluted with 0.75 M Tris SO_4–0.4 M NaN_3, pH 7.0.

Active fractions of each enzyme are pooled, promptly dialyzed extensively against 5 mM Tris SO_4 buffer, pH 8.7, and vacuum dialyzed down to a concentration of 10 mg/cm³ for storage. (N.B.: Sodium azide is extremely toxic and in acid solution will give off toxic fumes. Extreme caution must be exercised in both its use and disposal.)

The resin is regenerated for future use by washing with 500 cm³ of 1.0 M Na_2CO_3 followed by 2 liters of distilled water on a Büchner funnel and reequilibration with buffer.

This purification method is essentially the same as that described elsewhere,[87] with elution conditions similar to those of Khalifah *et al.*,[62] and should prove broadly applicable to the purification of CA isozymes from other mammalian species. It produces essentially homogeneous isozyme samples, but should other laboratories find this procedure less effective or wish to be more fastidious, a further step such as ion exchange chromatography may be used.[115] Bergenhem *et al.*,[6] who encountered problems of hemoglobin contamination when using the method of Khalifah *et al.*, innovatively developed a technique of preparative affinity electrophoresis. They removed the hemoglobin by column electrophoresis while the CA was bound to the affinity gel.

3.1.2. Purification of CA I and CA II from Other Mammals

In general, the method described above may be widely used for the purification of CA I and CA II from other mammals, although slight adjustment of elution conditions may be required for different species. For alternative methods for individual species, the reader is directed to the references given in Section 3.1.

3.1.3. Purification of CA III from Human Skeletal Muscle

Purification is performed with only minor modification to the method of Hewett-Emmett *et al.*,[50] who used autopsied psoas major muscle. However, although CA III is present only in red fibers, no human muscle comprises totally white fibers, so choice of muscle may depend on availability.

CA III is purified by using the same affinity resin as for CA I and CA II, but a lower ionic strength is essential for binding, since CA III is much less sensitive to sulfonamide inhibition. Mercaptoethanol (1 mM) is added to all buffers to minimize any possibility of dimerization.[91]

About 200 g of muscle is cut into bite-size pieces, minced twice, and homogenized in a Waring Blender with 400 cm^3 of 0.01 M Tris SO_4, pH 8.7. After being stirred overnight, the preparation is centrifuged at 10,000 rpm for 30 min. The supernatant is filtered through glass wool to remove any fat and dialyzed extensively against 0.005 M Tris SO_4, pH 8.7 (several buffer changes totaling 50 liters). Any further precipitation is removed by centrifugation as before, and 200–250 cm^3 of Prontosil–CM Sephadex affinity resin,[87] equilibrated in the same buffer, is stirred gently with the supernatant for 3 hr. Nonspecific protein is removed on a Büchner funnel, and the resin is transferred to a column (2.5 × 30 cm) as described above for CA I and CA II.

CA III is eluted with a linear gradient of 0 → 0.2 M KI in 0.005 M Tris SO_4, pH 8.7 (using 150 cm^3 of buffer without KI and 150 cm^3 of buffer containing 0.2 M KI), at a flow rate of about 40 cm^3/hr. Active fractions are pooled, dialyzed promptly against the same buffer to remove KI, and vacuum dialyzed down to about 3-cm^3 volume.

The sample is then applied to a long Sephadex G-75 column (3 × 100 cm) equilibrated with 0.005 M Tris SO_4, pH 8.7, and eluted at a rate of 40 cm^3/hr with the same buffer. Active fractions are pooled and vacuum dialyzed down to a concentration of about 10 mg/cm^3 for storage. The yield is up to 60 mg of enzyme.

3.1.4. Purification of CA III from Other Mammals

CA III from some species may bind the affinity resin less tightly than does the human isozyme. Bovine CA III provides one example of this. After slightly less exhaustive washing of the resin, this isozyme is eluted with a linear gradient of 0→ 0.2 M Na_2SO_4 in otherwise identical buffer conditions. Purification can be improved by passing it down a DEAE Trisacryl M anion exchange column (1 × 30 cm) in 0.005 M Tris Cl, pH 8.7, before the gel permeation step described for the human isozyme.[18]

Two excellent alternative methods of purification of bovine CA III, one employing cation exchange and the other employing affinity chromatography, have been developed.[26] In common with the method for purification of cat CA III[98] PAMBS–CM Bio-Gel A resin[62] is used to specifically remove any contaminating erythrocyte CAs, since this particular resin does not bind CA III from either of these two species.

Alternative methods not employing affinity chromatography include those for rabbit, pig,[91,93] and horse.[85] In our laboratory, we have purified horse CA III to homogeneity by the method described above for the human isozyme.

3.2. Membrane-Bound Carbonic Anhydrase—CA IV

It has been known for a number of years that some CA activity is membrane associated,[59,75] and it is perhaps axiomatic that this activity may well be of particular physiological importance in a number of respects. Nonetheless, the study of membrane-associated CA has, until recently, received comparatively little attention, probably because of the difficulty of purifying this form of the enzyme, which appears to be an intrinsic membrane protein, in a stable, active state and adequate yield.

Membrane-associated CA activity has been detected in a wide range of tissues.[121] Although it has been established that at least one form of CA, immunologically distinct from the cytosolic forms, exists in association with membranes,[29] and this has been tentatively designated CA IV,[120] it is as yet unclear whether the same form of the enzyme exists in membranes of different tissues.

An intriguing recent development is the discovery of CA activity associated with the sarcoplasmic reticulum.[12] Although the isozyme type has yet to be established, preliminary inhibition data do not preclude CA IV. CA activity has also been detected in association with the sarcolemma.[36] This isozyme, however, appears to be immunologically related to CA I.[81]

Purification of the membrane-bound form obviously requires special attention to isolation of the membrane fraction, solubilization of the enzyme, usually with use of a suitable detergent, and maintenance of activity thereafter. Care must be taken to avoid contamination with cytosolic isozymes.

The first purification of what appeared to be a membrane-bound CA was reported by Sapirstein and Lees,[99] who purified the enzyme from rat brain myelin by Triton extraction followed by affinity chromatography and found it to have an M_r of 30,000.[100] This isozyme, however, may well not have been CA IV but rather a membrane-associated CA II.

In contrast, both Triton X-100 and deoxycholate were apparently ineffective in solubilizing CA from human kidney.[73] This was achieved by using 3–5% SDS, in which the enzyme was surprisingly stable. Gel filtration indicated an M_r of less than 66,000. The presence of the membrane-bound enzyme in rat, rabbit, human, and bovine lung was demonstrated by Whitney and Briggle,[120] who purified the bovine enzyme to about 80% purity by affinity chromatography and gel filtration. The partially purified enzyme was remarkably stable in SDS, unlike the soluble and erythrocyte enzymes, a feature that facilitated its accurate assay. Addition of DTT, however, caused rapid inactivation, suggesting that disulfide bridges may be an important feature of its structural integrity. The data on this enzyme indicated a glycoprotein of M_r not exceeding 52,000. In contrast, the addition of 0.1% DTT actually facilitated solubilization of the rat brain myelin enzyme.[99]

Human CA IV was first purified to homogeneity by Wistrand and Knuuttila.[122] By a most elaborate method they produced a low yield of renal membrane CA of apparent molecular mass in the region of 35 kDa.

Recently Zhu and Sly[126] have purified CA IV to homogeneity from both human lung and kidney by a much simpler method which is described below. The enzymes from both tissues have an apparent molecular mass of 35 kDa and seem to be anchored to the membrane through a phosphatidylinositol–glycan linkage. They appear to differ only very slightly, perhaps as a result of posttranslational modification.

Purification of CA IV from Human Lung or Kidney[126]

The membrane fraction from 200 g human lung or kidney is prepared by ultracentrifugation from Waring blender homogenates in 0.025 M triethanolamine sulfate, pH 8.1, 0.059 M sodium sulfate, and 0.001 M benzamidine (hereafter referred to as "buffer"). It is then washed twice and resedimented in buffer to remove hemoglobin and soluble CAs.

After resuspending the membranes in buffer (10 cm^3 per g membranes), the enzyme is solubilized by the addition of solid SDS to 5% final concentration and homogenization using a Teflon glass homogenizer. Insoluble material is then removed by centrifugation (15,000 r.p.m. for 30 min). 5 cm^3 bed volume of affinity gel [PAMBS coupled to Aff-Gel-10 (Bio-Rad Laboratories) according to the manufacturer's instructions] is added to the extract which is then rocked at room temperature for 1 hr. The gel is collected on a scintered glass filter at 4°C and washed with 50 cm^3 buffer containing 1% SDS followed by 50 cm^3 buffer containing 0.1% Brij-35 (or 0.1% Triton X100). The enzyme is eluted from the gel with a mixture of 0.5 M sodium perchlorate and 0.1 M sodium acetate, pH 5.0, containing 0.1% Brij-35 and dialyzed against excess 0.05 M ammonium bicarbonate, pH 7.8.

This method produces 0.8–1.0 mg enzyme from 200 g lung and 0.4–0.5 mg from 200 g kidney.

3.3. Mitochondrial Carbonic Anhydrase—CA V

What appears to be a specific, soluble mitochondrial isozyme of CA[47] was isolated in Dodgson's laboratory from the mitochondria of guinea pig hepatocytes, where it is present in the matrix.[22,103]

CA activity has been demonstrated in kidney cortex mitochondria of the rat[23] and skeletal muscle mitochondria of the guinea pig.[104]

For purification of CA V from guinea pig liver,[22] guinea pig liver mitochondria are isolated in 225 mM mannitol–75 mM sucrose–1.0 mM ethylene glycol-bis(β-aminoethyl ether)-*N,N,N′,N′*-tetraacetate (EGTA) at 250 mosM, pH 7.0, as previously described.[78] They are then carefully washed, first in 225 mM mannitol–75 mM sucrose to remove all EGTA and then in 225 mM mannitol–75 mM sucrose–25 mM $NaHCO_3$, pH 7.4, and resuspended in the latter buffer to a final concentration of 80–120 mg of protein per cm^3. After freezing overnight and thawing to break the mitochondria, the soluble protein is separated by centrifugation at 100,000 × *g* for 1 hr.

The supernatant is loaded onto a PAMBS–CM Bio-Gel A column (100-cm^3 approximate volume) equilibrated with 0.15 M Tris SO_4, pH 9.0. The column is first washed with 250 cm^3 of equilibration buffer to remove nonspecifically bound protein, followed by 0.4 M NaN_3–0.25 M Na_2SO_4–0.1 M Bis Tris, pH 7.0, to elute the CA. Fractions containing CA activity are pooled and concentrated 20-fold by Millipore ultrafiltration.

Finally, the sample is applied to a Sephadex G-75 column equilibrated with 0.05 M Tris citrate, pH 7.2, and eluted with the same buffer.

3.4. Salivary Carbonic Anhydrase—CA VI

Since the presence of this high-molecular-mass glycoprotein was demonstrated in sheep saliva and parotid glands,[31] the rat[28], human,[83] and sheep[30] enzyme has been purified and characterized and tentatively designated CA VI.

3.4.1. Purification of CA VI from Human Saliva[83]

The flow of saliva in volunteering expectorators is stimulated by the application of 2% citric acid with gauze to the tongue. About 40 cm^3 from one individual is collected in an ice-cold bottle containing 2 cm^3 of 0.2 M benzamidine in 0.1 M Tris SO_4–0.2 M Na_2SO_4, pH 8.7, and stored at −20°C. (Benzamidine is a reversible inhibitor of serine proteases, which are known to be present in saliva.[39])

In a typical purification, 250 cm^3 of saliva is centrifuged (16,000 × *g* for 15 min), and the supernatant is diluted to 1 liter with 0.1 M Tris SO_4–0.2 M Na_2SO_4, pH 8.7 (final concentrations).

A 10-cm^3 volume of PAMBS–CM Bio-Gel A affinity gel[62] (prepared as described in Section 1.2) is added to the supernatant, and the solution is mixed gently for 24 hr. The gel is then washed in a Büchner sintered funnel with 500 cm^3 of 0.1 M Tris SO_4–0.2 M Na_2SO_4 containing 20% (vol/vol) glycerol, pH 8.7, followed by 500 cm^3 of the same buffer at pH 7.0, and transferred to a column (1.8 × 40 cm), where it is washed with a further 150 cm^3 of the latter buffer. (The addition of 20% glycerol to the buffer has been shown to very much improve the removal of nonspecifically bound protein.) The enzyme is specifically eluted with 0.1 M Tris SO_4–0.4 M NaN_3, pH 7.0 (care!), containing 1 mM benzamidine and 20% (vol/vol) glycerol.

SDS-PAGE shows that fractions of relatively pure CA VI (M_r = 42,000) are produced with less than 5% contamination of a larger protein (M_r = 60,000), which can be removed by anion exchange as follows.

Fractions of eluate containing CA activity are dialyzed twice against 2 liters 0.01 M Tris SO_4, pH 7.5, containing 1 mM benzamidine for at least 3 hr each time and applied onto a DEAE–Sephacel column (1.2 × 2.5 cm) equilibrated with the same buffer. The high-M_r contaminant does not bind to this column. After a washing with 30 cm^3 of buffer, the CA is eluted with a linear gradient of 0 → 0.1 M Na_2SO_4 in 40 cm^3 of the same buffer.

This method typically produces a yield of about 2.5 mg with a 180-fold purification.

3.4.2. Purification of CA VI from Other Mammals

Feldstein and Silverman[28] purified salivary CA from rat to homogeneity, as determined by polyacrylamide gel and SDS-PAGE, in a single-step process of affinity chromatography on PAMBS–CM Bio-Gel A.[62] This method is essentially the same as that described for the human enzyme.

In the case of the rat, saliva collection presents more of a problem and is achieved by the method of Menaker *et al.*[80] Rats are anesthetized by intraperitoneal injection of Nembutal (16.5 mg/kg), and the flow of saliva is stimulated by pilocarpine (1.5 mg/kg), also injected intraperitoneally. A tracheostomy is performed to prevent suffocation, and the saliva is collected with a Pasteur pipette and immediately transferred to vials on ice.

The saliva, which is of the order of 10 cm^3 or so from six rats, is initially diluted with 15 volumes of buffer. A 5-cm^3 volume of affinity gel is used, and the enzyme is eluted with 50 cm^3 of 0.1 M Tris SO_4–0.4 M NaN_3, pH 7.0. This eluent is then concentrated down to 1–2 cm^3 and dialyzed against five changes of 1 liter of distilled water. This produces a yield of about 0.3 mg, i.e., about three times that of the human preparation.

Fernley *et al.*[30] also purified CA VI to homogeneity from sheep parotid gland by a four-step process employing ammonium sulfate precipitation, followed by three successive chromatography steps involving a PAMBS–CH Sepharose 4B affinity resin, wheat germ lectin Sepharose, and Sepharose 6-B.

4. Avian Carbonic Anhydrases

The cytoplasmic isozymes, CA I, II, and III, have all been isolated from birds.[112] Although a partial purification of CA from duck erythrocytes was reported as early as 1962,[68] it was not until 10 years later than Bernstein and Schraer first purified an avian CA to homogeneity.[10] By a procedure involving ethanol–chloroform extraction, gel permeation on Sephadex G-75, and ion exchange chromatography on CM–cellulose, they purified a high-activity isozyme, apparently CA II, from the erythrocytes of single-comb, white Leghorn hens. Subsequently, Lemke and Graf[69] used a similar method to purify the high-activity isozyme from erythrocytes of the domestic turkey. In both cases, the presence of a reducing agent such as mercaptoethanol or DTT was found to be necessary to maintain activity.

It would appear, then, that the erythrocytes of birds, in common with that of certain mammals, lacks CA I. Absence of CA I from the red cell does not, however, preclude its expression from other tissues. Holmes[53] purified CA I, II, and III from chicken intestine, red cells, and leg muscle, respectively, by a modification of the affinity chromatography method of Osborne and Tashian[87] in the purification of the human isozymes (see Section 3.1) with different elution conditions for chicken CA III. In our laboratory, chicken CA III is purified by the method described for bovine CA III in Section 3.1.

5. Reptilian Carbonic Anhydrases

Hall and Schraer[41] purified both a low- and a high-activity CA isozyme from the erythrocyte of the diamondback terrapin; these isozymes would appear to be related to mammalian CA I and CA II.[112] The purification procedure involved separation from hemoglobin by ethanol–chloroform extraction and gel permeation on Sephadex G-75, followed by separation of the isozymes by ion exchange chromatography on a DEAE–Sephadex A-50 column. The high-activity isozyme, in common with several CA isozymes from mammalian vertebrates, required the presence of a thiol agent, such as mercaptoethanol, for activity.

6. Amphibian Carbonic Anhydrases

A high-activity CA isozyme, which appears to be related to mammalian CA II, has been purified from bullfrog erythrocytes by a procedure involving ethanol–chloroform extraction, ion exchange chromatography on DEAE–Bio-Gel A, and preparative disc gel electrophoresis.[13] This isozyme, which was the only form of CA found in this tissue, required the presence of a thiol reducing agent to maintain full activity.

7. Fish Carbonic Anhydrases

The first purification to homogeneity of a nonmammalian animal CA was achieved by Maynard and Coleman,[77] who isolated a single high-activity isozyme of CA from the erythrocyte of two species of elasmobranch, the bull shark and the tiger shark. Subsequently, Bergenham *et al.*[9] refined the purification method for the tiger shark enzyme, using ethanol–chloroform extraction followed by affinity chromatography in a modification of the method of Khalifah *et al.*,[62] and re-chromatography on DEAE–cellulose. Limited sequencing data[7] suggested that this isozyme is most closely related to CA II of amniotes.

Carlsson *et al.*[16] used ethanol–chloroform extraction followed by DEAE–cellulose chromatography to isolate from hagfish erythrocytes a single CA isozyme with activity resembling CA I but inhibition characteristics resembling CA II.

A single CA isozyme bearing considerable similarity to mammalian CA II was also isolated from erythrocytes of the teleost sheepshead by Sanyal *et al.*,[97] employing affinity chromatography.[62]

In contrast to this pattern, the isolation of both a high- and a low-activity isozyme from eel erythrocytes was reported[57] in a study which perhaps requires fuller investigation.

8. Invertebrate Carbonic Anhydrases

Despite the widespread occurrence of CA among invertebrates (reviewed in references 45, 74, and 90), little attention has been given to its purification from these animals. The first purification and partial characterization of an invertebrate CA was achieved by Nielsen and Frieden,[84] who used DEAE–Sephadex and agarose chromatography to obtain a partially purified aggregate of the oyster enzyme, which was reduced to a smaller, homogeneous preparation (M_r = 500,000) by PAGE. In common with several other CAs, the enzyme required the presence of a thiol reducing agent for full activity.

In a more recent study, Henry[46] showed the presence of CA in certain invertebrate blood cells, but these enzymes await purification.

The presence of CA has been demonstrated in various tissues of a number of insects, including the American cockroach,[3] house cricket,[25] tobacco hornworm,[56] silkmoth,[55] and locust.[102] However, to date the insect enzyme has been purified only from post-feeding larvae of the face fly,[21] which would suggest that its expression may be limited to a brief period in insect development. Using affinity chromatography, chromatofocusing, and gel filtration on Sephadex G-75-40, Darlington *et al.*[21] partially purified a CA of M_r 31,000, which preliminary data indicate may exist as more than one isozyme.

Stratakis and Linzen[105] achieved 31-fold purification of CA from hemolymph of the tarantula by anion exchange chromatography, gel permeation, and PAGE. This enzyme was inhibited by mercaptoethanol.

9. Plant and Eukaryotic Algal Carbonic Anhydrases

CAs are widely distributed throughout the plant kingdom, where they appear to exist in a range of oligomeric forms, dependent on the particular species. In general, they are strikingly different from the animal isozymes in their inhibition characteristics. Most notable, perhaps, is their resistance to sulfonamide inhibition. In this respect they are similar to mammalian CA III,[76] a point worth noting if one is contemplating purification by affinity chromatography.

In plants, the enzyme is most abundant in green leaves, especially in the chloroplasts, and appears to be virtually absent from root tissue.

The breakdown of the cell wall in plants clearly presents more of a problem than the cell membrane in animal tissues and is probably best achieved by grinding the tissue in buffer with sand. When methods that involve blending procedures are used, great care must be taken to avoid heating, since prolonged blending is generally necessary.

Detailed methods of purification to homogeneity have been described for the enzyme from parsley,[116] pea,[63] and spinach.[58,89] These variously involve ammonium sulfate extraction, ion exchange chromatography, and gel permeation. Requirement for a thiol reducing agent was not consistent. Whereas parsley CA required such an agent for full activity, spinach CA appeared to be unstable in the presence of mercaptoethanol.

For fuller details, the reader is directed to the publications cited above and to more specialist reviews.[38,66,92]

Among eukaryotic algae, the purification of CA has also been described from the green alga *Chlamydomonas reinhardtii*[14,124] and the red alga *Porphyridium cruentum*.[123] In the former, the CA appeared to be located in the periplasmic space, whereas in the latter it was inside the cell. Recently, however, an additional, distinct intracellular CA has been identified in *C. reinhardtii*.[54]

The activity of CA in both eukaryotic and prokaryotic algae (cyanobacteria) has been comprehensively reviewed.[2]

10. Prokaryotic Carbonic Anhydrases

CA activity has been found in a wide variety of prokaryotic organisms.[2,60] Adler *et al.*[1] purified CA to homogeneity from *Neisseria sicca* by three consecutive ion exchange steps. The enzyme was released into solution by simply shaking the cells with buffer, suggesting that it was located in the periplasmic space. A yield of 15–25 mg of enzyme was produced from a bacterial culture of 20 liters under optimized growth conditions.

11. Carbonic Anhydrase Production from cDNA

Forsman *et al.*[34] produced active human CA II in *Escherichia coli* from the cDNA clone, and this system has been used to express several mutants of the enzyme produced by site-directed mutagenesis.[33] Details of the enzyme purification from the bacterial culture have yet to be published.

A cDNA clone for human CA II has also been inserted into the mammalian Cos-7 cell line and expressed.[82] The use of mammalian cell culture in this way is of particular importance, since it would be likely to accomplish any posttranslational modifications that a mammalian protein may normally undergo during synthesis *in vivo*.

Undoubtedly, systems such as these will be increasingly used in the production of CA isozymes.

ACKNOWLEDGMENTS. I am particularly indebted to Jenny Shelton for her skilled practical assistance and much valuable discussion and to Dr. Richard Tashian for his sage counsel. I am also most grateful to Dr. David Hewitt-Emmett and Dr. Barry Davis for their advice and to Marguerite Lyons for her patient assistance in preparing the manuscript.

References

1. Adler, L., Brundell, J., Falkbring, S. O. and Nymen, P. O., 1972, *Biochim. Biophys. Acta* **284**:298–310.
2. Aizawa, K., and Miyachi, S., 1986, *FEMS Microbiol. Rev.* **39**:215–233.
3. Anderson, A. D., and March, R. B., 1956, *Can. J. Zool.* **34**:68–74.
4. Armstrong, J. McD., Myers, D. V., Verpoorte, J. A., and Edsall, J. T., 1966, *J. Biol. Chem.* **241**(21):5137–5149.
5. Ashworth, R. B., Brewer, J. M., and Stanford, R. L., Jr., 1971, *Biochem. Biophys. Res. Commun.* **44**:667–674.
6. Bergenhem, N., Carlsson, U., and Hansson, C., 1983, *Anal. Biochem.* **134**:259–263.
7. Bergenhem, N., 1989, Ph.D. thesis, University of Umeå, Umeå, Sweden.
8. Bergenhem, N., Carlsson, U., and Klasson, K., 1985, *J. Chromatogr.* **319**:59–65.
9. Bergenhem, N., Carlsson, U., and Strid, L., 1986, *Biochim. Biophys. Acta* **871**:55–60.
10. Bernstein, R. S., and Schraer, R., 1972, *J. Biol Chem.* **247**:1306–1322.
11. Brundell, J., Falkbring, S. O., and Nymann, P. O., 1972, *Biochim. Biophys. Acta* **284**:311–323.
12. Bruns, W., Dermietzel, R., and Gros, G., 1986, *J. Physiol.* **371**:351–364.

13. Bundy, H. F., and Cheng, B., 1976, *Comp. Biochem. Physiol.* **55B**:265–271.
14. Bundy, H. F., and Coté, S., 1980, *Phytochemistry* **19**:2531–2534.
15. Carlsson, V., Hannestad, V., and Lindskog, S., 1973, *Biochim. Biophys. Acta* **327**:515–527.
16. Carlsson, U., Kjellstrom, B., and Antonsson, B., 1980, *Biochim. Biophys. Acta* **612**:160–170.
17. Carter, M. J., and Parsons, D. S., 1970, *Biochem. J.* **120**:797–808.
18. Carter, N. D., Chegwidden, W. R., Hewett-Emmett, D., Jeffery, S., Shiels, A., and Tashian, R. E., 1984, *FEBS Lett.* **165**(2):197–200.
19. Carter, N. D., Hewett-Emmett, D., Jeffery, S., and Tashian, R. E., 1981, *FEBS Lett.* **128**: 114–118.
20. Carter, N. D., Shiels, A., and Tashian, R. E., 1978, *Biochem. Soc. Trans.* **6**:552–553.
21. Darlington, M. V., Meyer, H. J., and Graf, G., 1984, *Ann. N.Y. Acad. Sci.* **429**:219–221.
22. Dodgson, S. J., 1987, *J. Appl. Physiol.* **63**:2134–2141.
23. Dodgson, S. J., and Contino, L. C., 1988, *Arch. Biochem. Biophys.* **260**(1):334–341.
24. Duff, T. E., and Coleman, J. E., 1966, *Biochemistry* **5**:2009–2019.
25. Edwards, L. J., and Patton, R. L., 1969, *Insect Physiol.* **13**:1333–1341.
26. Engberg, P., Millqvist, E., Pohl, G., and Lindskog, S., 1985, *Arch. Biochem. Biophys.* **241**(2):628–638.
27. Falkbring, S. O., Göthe, P. O., Nyman, P. O., Sundberg, L., and Porath, J., 1972, *FEBS Lett.* **24**(2):229–235.
28. Feldstein, J. B., and Silverman, D. N., 1984, *J. Biol. Chem.* **259**(9):5447–5453.
29. Fernley, R. T., 1988, *Trends Biochem. Sci.* **13**:356–359.
30. Fernley, R. T., Coghlan, J. P., and Wright, R. D., 1988, *Biochem. J.* **249**:201–207.
31. Fernley, R. T., Wright, R. D., and Coghlan, J. P., 1979, *FEBS Lett.* **105**(2):299–302.
32. Ferrell, R. E., Stroup, S. K., Tanis, R. J., and Tashian, R. E., 1978, *Biochim. Biophys. Acta* **533**:1–11.
33. Forsman, C., Behravan, G., Jonsson, B.-H., Liang, Z. Lindskog, S., Ren, X., Sandström, J., and Wallgren, K., 1988, *FEBS Lett.* **229**(2):360–362.
34. Forsman, C., Behravan, G., Osterman, A., and Jonsson, B.-H., 1988, *Acta Chem. Scand.* **B42**(5):314–318.
35. Furth, A. J., 1968, *J. Biol. Chem.* **243**:4832–4841.
36. Geers, C., Gros, G., and Gärtner, A., 1985, *J. Appl. Physiol.* **59**(2): 548–558.
37. Girard, J. P., and Istin, M., 1975, *Biochim. Biophys. Acta* **381**:221–232.
38. Graham, D., Reed, M. L., Patterson, B. D., and Hockley, D. G., 1984, *Ann. N.Y. Acad. Sci.* **429**:222–237.
39. Gresik, E. W., 1980, *J. Histochem. Cytochem.* **28**:860–870.
40. Gros, G., and Dodgson, S. J., 1988, *Annu. Rev. Physiol.* **50**:669–694.
41. Hall, G. E., and Schraer, R., 1978, *Comp. Biochem. Physiol.* **63B**:561–567.
42. Hames, B. D., and Rickwood, D. (eds.), 1981, *Gel Electrophoresis of Proteins—a Practical Approach*, IRL Press, Oxford.
43. Heath, R., Carter, N. D., Hewett-Emmett, D., Fincanci, E., Jeffery, S., Shiels, A., and Tashian, R. E., 1983, *Fed. Proc.* **42**:2180.
44. Henriksson, D., Tanis, R. J., and Tashian, R. E., 1980, *Biochem. Biophys. Res. Commun.* **96**(1):135–142.
45. Henry, R. P., 1984, *Ann. N.Y. Acad. Sci.* **429**:544–546.
46. Henry, R. P., 1987, *J. Exp. Zool.* **242**:113–116.
47. Hewett-Emmett, D., Cook, R. G., and Dodgson, S. J., 1986, *Isozyme Bull.* **19**:13.
48. Hewett-Emmett, D., and Tashian, R. E., 1979, *Am. J. Hum. Genet.* **31**:50A.
49. Hewett-Emmett, D., and Tashian, R. E., 1981, *Am. J. Phys. Anthropol.* **54**:232–233.
50. Hewett-Emmett, D., Welty, R. J., and Tashian, R. E., 1983, *Genetics* **105**:409–420.
51. Hoare, D. G., and Koshland, D. E., 1967, *J. Biol. Chem.* **242**:2447–2453.
52. Holmes, R. S., 1976, *J. Exp. Zool.* **197**:289–295.

53. Holmes, R. S., 1977, *Eur. J. Biochem.* **78**:511–520.
54. Husic, H. D., Kitayama, M., Togasaki, R. K., Moroney, J. V., Morris, K. L., and Tolbert, N. E., 1989, *Plant Physiol.* **89**:904–909.
55. Johnston, J. W., and Jungreis, A. M., 1979, *Comp. Biochem. Physiol.* **62B**:465–469.
56. Johnston, J. W., and Jungries, A. M., 1981, *J. Exp. Biol.* **91**:255–269.
57. Jones, G. L., and Shaw, D. C., 1982, *Biochim. Biophys. Acta* **709**:284–303.
58. Kandel, M., Gornall, A. G., Cybulsky, D. L., and Kandel, S. I., 1978, *J. Biol. Chem.* **253**(3):679–685.
59. Karler, R., and Woodbury, D. M., 1960, *Biochem. J.* **75**:538–543.
60. Karrasch, M., Bott, M., and Thauer, R. K., 1989, *Arch. Microbiol.* **151**:137–142.
61. Keilin, D., and Mann, T., 1940, *Biochem. J.* **34**:1163–1176.
62. Khalifah, R. G., Strader, D. J., Bryant, S. H., and Gibson, S. M., 1977, *Biochemistry* **16**:2241–2247.
63. Kiesel, W., and Graf, G., 1972, *Phytochemistry* **11**:113–117.
64. Koester, M. K., Register, A. M., and Noltmann, E. A., 1977, *Biochem. Biophys. Res. Commun.* **76**:196–204.
65. Laemmli, U. K., and Farre, M., 1973, *J. Mol. Biol.* **80**:575–599.
66. Lamb, J. E., 1977, *Life Sci.* **20**:393–406.
67. Laurent, G., Charrel, M., Castay, E., Nahon, C., Marriq, C., and Darrien, Y., 1962, *C. R. Soc. Biol.* **154**:1461–1464.
68. Leiner, M., Beck, H., and Eckert, H., 1962, *Physiol. Chem.* **327**:144–165.
69. Lemke, P. R., and Graf, G., 1974, *Mol. Cell. Biochem.* **4**(2):141–147.
70. Lindskog, S., 1960, *Biochim Biophys. Acta* **39**:218–226.
71. McIntosh, J. E. A., 1969, *Biochem. J.* **114**:463–476.
72. McIntosh, J. E. A., 1970, *Biochem. J.* **120**:299–310.
73. McKinley, D. N., and Whitney, P. L., 1976, *Biochim. Biophys. Acta* **445**:780–790.
74. Maren, T. H., 1967, *Physiol. Rev.* **47**:595–665.
75. Maren, T. H., 1980, *Ann. N.Y. Acad. Sci.* **341**:246–258.
76. Maren, T. H., and Sanyal, G., 1983, *Annu. Rev. Pharmacol. Toxicol.* **23**:439–459.
77. Maynard, J. R., and Coleman, J. E., 1971, *J. Biol. Chem.* **246**(14):4455–4464.
78. Mela, L., and Seitz, S., 1979, *Methods Enzymol.* **55**:39–46.
79. Meldrum, N. U., and Roughton, F. J. W., 1933, *J. Physiol.* **80**:113–142.
80. Menaker, L., Sheetz, J. H., Cobb, C. M., and Naria, M. J., 1974, *Lab. Invest.* **30**:341–349.
81. Moyle, S., Jeffery, S., and Carter, N. D., 1984, *J. Histochem. Cytochem.* **32**:1262–1264.
82. Murakami, H., Marelich, G. P., Grubb, J. H., Kyle, J. W., and Sly, W. S., 1987, *Genomics* **1**: 159–166.
83. Murakami, H., and Sly, W. S., 1987, *J. Biol. Chem.* **262**(3):1382–1388.
84. Nielsen, S. A., and Frieden, E., 1972, *Comp. Biochem. Physiol.* **41B**:875–889.
85. Nishita, T., and Deutsch, H. F., 1981, *Biochem. Biophys. Res. Commun.* **103**:573–580.
86. Nyman, P. O., 1961, *Biochim. Biophys. Acta* **52**:1–12.
87. Osborne, W. R. A., and Tashian, R. E., 1975, *Anal. Biochem.* **64**:297–303.
88. Pihar, O., 1965, *Collect. Czech. Chem. Commun.* **30**:3220–3223.
89. Pocker, Y., and Ng, J. S. Y., 1973, *Biochemistry* **12**:5127–5134.
90. Polya, J. B., and Wirtz, A. J., 1965, *Enzymologia* **28**:355–366.
91. Pullan, L. M., and Noltmann, E. A., 1984, *Biochem. Pharmacol.* **33**:2641–2645.
92. Reed, M. L., and Graham, D., 1981, *Prog. Phytochem.* **7**:47–94.
93. Register, A. M., Koester, M. K., and Noltmann, E. A., 1978, *J. Biol. Chem.* **253**:4143–4152.
94. Reid, M. S., and Bieleski, R. L., 1968, *Anal. Biochem.* **22**:374–381.
95. Rickli, E. E., and Edsall, J. T., 1962, *J. Biol. Chem.* **237**:PC258–260.
96. Rickli, E. E., Ghazanfar, S. A. S., Gibbons, B. H., and Edsall, J. T., 1964, *J. Biol. Chem.* **239**(4):1065–1078.

97. Sanyal, G., Pessah, N. J., Swenson, E. R., and Maren, T. H., 1982, *Comp. Biochem. Physiol.* **73B**:937–944.
98. Sanyal, G., Swenson, E. R., Pessah, N. J., and Maren, T. H., 1982, *Mol. Pharmacol* **22**:211–220.
99. Sapirstein, V. S., and Lees, M. B., 1978, *J. Neurochem.* **31**:505–511.
100. Sapirstein, V. S., Strocchi, P., Wesolowski, M., and Gilbert, J. M., 1983, *J. Neurochem.* **40**:1251–1261.
101. Shimizu, C., and Matsuura, F., 1962, *Bull. Jpn. Soc. Sci. Fish.* **28**:924–929.
102. Spencer, I. M., Hargreaves, I., and Chegwidden, W. R., 1988, *Biochem. Soc. Trans.* **16**: 973–974.
103. Storey, B. T., Dodgson, S. J., and Forster, R. E., II, 1984, *Ann. N.Y. Acad. Sci.* **429**:210–211.
104. Storey, B. T., Lin, L. C., Tompkins, B., and Forster, R. E., II, 1989, *Arch. Biochem. Biophys.* **270**(1):144–152.
105. Stratakis, E., and Linzen, B., 1984, *Hoppe-Seyler's Z. Physiol. Chem.* **365**:1187–1198.
106. Tanis, R. J., and Tashian, R. E., 1971, *Biochemistry* **10**:4852–4858.
107. Tanis, R. J., Tashian, R. E., and Yu, Y.-S. L., 1970, *J. Biol. Chem.* **245**:6003–6009.
108. Tashian, R. E., 1969, in: *Biochemical Methods in Red Cell Genetics* (J. J. Yunis, ed.), Academic Press, New York, pp. 307–336.
109. Tashian, R. E., 1989, *BioEssays* **10**(6):186–192.
110. Tashian, R. E., and Carter, N. D., 1976, *Adv. Hum. Genet.* **7**:1–56.
111. Tashian, R. E., Hewett-Emmett, D., Dodgson, S. J., Forster, R. E., and Sly, W., 1984, *Ann. N.Y. Acad. Sci.* **429**:262–275.
112. Tashian, R. E., Hewett-Emmett, D., and Goodman, M., 1983, in: *Isozymes: Current Topics in Biological and Medical Research*, Vol. 7 (M. C. Rattazzi, J. G. Scandalios, and G. S. Whitt, eds.), Alan R. Liss, New York, pp. 79–100.
113. Tashian, R. E., Hewett-Emmett, D., Stroup, S. K., Goodman, M., and Yu, Y.-S. L., 1980, in: *Biophysics and Physiology of* CO_2 (C. Bauer, G. Gros, and H. Bartels, eds.), Springer-Verlag, Berlin, pp. 165–176.
114. Tashian, R. E., Shreffler, D. C., and Shows, J. B., 1968, *Ann. N.Y. Acad. Sci.* **151**:64–77.
115. Tibell, L., Forsman, C., Simonsson, I., and Lindskog, S., 1984, *Biochim. Biophys. Acta* **789**:302–310.
116. Tobin, A. J., 1970, *J. Biol. Chem.* **245**:2656–2666.
117. Venta, P. J., Montgomery, J. C., and Tashian, R. E., 1987, in: *Isozymes: Current Topics in Biological and Medical Research*, Vol. 14 (M. C. Rattazzi, J. G. Scandalios, and G. S. Whitt, eds.), Alan R. Liss, New York, pp. 59–72.
118. Watts, D. C., and Moreland, B., 1970, *Exp. Physiol. Biochem.* **3**:1–30.
119. Whitney, P. L., 1974, *Anal. Biochem.* **57**:467–476.
120. Whitney, P. L., and Briggle, T. V., 1982, *J. Biol. Chem.* **257**(20):12056–12059.
121. Wistrand, P. J., 1984, *Ann. N.Y. Acad. Sci.* **429**:195–206.
122. Wistrand, P. J., and Knuuttila, K.-G., 1989, *Kidney Int.* **35**:851–859.
123. Yagawa, Y., Muto., S., and Miyachi, S., 1987, *Plant Cell. Physiol.* **28**(7):1253–1262.
124. Yang, S.-Y., Tsuzuki, M., and Miyachi, S., 1985, *Plant Cell. Physiol.* **26**(1):25–34.
125. Johansen, J. T., 1976, *Carlsberg Res. Commun.* **41**(2):73–80.
126. Zhu, X. L., and Sly, W. S., 1990, *J. Biol. Chem.* **265**:8795–8801.

CHAPTER 8

Techniques for Measuring Carbonic Anhydrase Activity in Vitro

The Electrometric Delta pH and pH Stat Methods

RAYMOND P. HENRY

1. Introduction

The isozymes of carbonic anhydrase (CA) (EC 4.2.1.1.) catalyze the reversible hydration–dehydration of carbon dioxide and water, with H^+ ions being transferred between the active site of the enzyme and a surrounding buffer.[3,16,17] This results in a change in pH as the reaction proceeds toward equilibrium. With the advent of rapid-responding pH and reference electrodes coupled to sensitive pH meters, these changes in H^+ ion concentration could be accurately measured. That measurement, which can be performed either directly or indirectly, forms the underlying principle upon which the electrometric methods for the assay of CA activity are built.

The electrometric assay methods measure either a change in pH (the delta pH method) or the addition of an acid or base while pH is kept constant (the pH state method). These two assays share a number of common advantages: they are easy to use, they employ relatively inexpensive equipment, and the results are accurate and quantitative.[4]

RAYMOND P. HENRY • Department of Zoology and Wildlife Science, Auburn University, Auburn, Alabama 36849-5414.

2. The Delta pH Assay

2.1. Description

In its original design, this method measured the time needed for a color change to occur in a pH-sensitive dye, with or without CA present.[1,14,15] That approach suffered numerous drawbacks, including pH-dependent changes in the reaction rate and direct inhibition by the dye.[2,4,11,18] The modern delta pH assay obviates those potential artifacts by using a sensitive recording pH meter to directly measure the rate of change in pH in a buffer containing CA upon the addition of substrate (e.g., CO_2 for hydration or HCO_3^- for dehydration).

The hydration of carbon dioxide, for which the delta pH assay is most commonly used, it quite rapid[3,6] and therefore hard to measure. Because the rate constants for the uncatalyzed and catalyzed reactions are very much slower at low temperatures, the delta pH assay is usually carried out at between 0 and 5°C to slow the reaction rates. This is not an absolute requirement, however, as Maren and Couto[12] have successfully used this method at 24°C. The reaction vessel is thermostatted to the desired temperature by a surrounding water jacket connected to a constant-temperature circulating water bath (Fig. 1).

The pH electrodes must have a rapid response time. Response times of 100 msec or less ensure that the speed of measurement does not lag behind the actual speed of the reaction[4] (e.g., the MEPH-1 pH and MERE reference electrodes made by World Precision Instruments, New Haven, Conn., work well). Electrometric measurements are vulnerable to outside electrical interference; thus, all electrical equipment needs proper grounding or shielding.

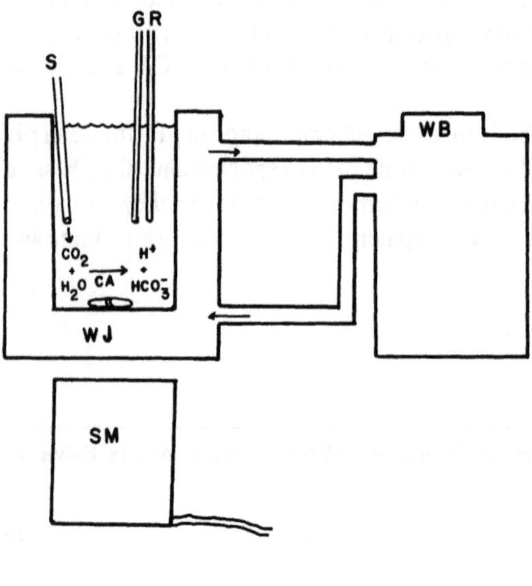

Figure 1. Schematic representation of the reaction vessel used to measure CO_2 hydration via the delta pH method. G and R represent the pH and reference electrodes, respectively. S represents the delivery system for CO_2 (substrate); this can either be a small air stone through which CO_2 gas is bubbled or a syringe that delivers a specific volume of CO_2-saturated water. WJ and WB are the water jacket around the vessel and a refrigerated circulating water bath, respectively, for controlling the temperature of the reaction mixture. SM is the magnetic stir motor, which drives a small magnetic stir bar in the vessel for continuous mixing of the vessel's contents.

Buffer concentration in the reaction mixture is also important. Too low a buffer concentration (<10 mM) decreases the specific activity of CA,[17] probably as a result of decreased proton transfer from the active site during the hydration reaction.[16] High buffer concentration dampens any pH change and thus also reduces measured CA activity. Phosphate, Tris–phosphate, and barbital buffers at concentrations of between 10 and 25 mM are routinely used and appear to avoid either problem,[5,8,12] although there are reports that high concentrations of phosphate can directly inhibit the CA II isozyme. Continuous stirring is essential to prevent CO_2 or H^+ gradient formation at the pH electrode.

2.2. Experimental Protocol and Apparatus

A wide variety of procedures has been reported in the literature for the delta pH assay since it was described in its modern form by Wilbur and Anderson.[19] The method of choice should give an accurate and quantitative measure of CA activity, and it should be suitable for kinetic studies. Two such methods are described in this section.

The first is taken from the work of Maren and Couto.[12] Their instrumentation consisted of a reaction vessel, rapid-reacting pH electrode (Radiometer GK 2322), and pH meter (Fisher Accumet 420). For measuring hydration, CO_2 gas is bubbled through 5 ml of solution in either the presence or absence of CA. The reaction is initiated by the addition of 2 ml of barbital buffer (25 mM, pH 7.8), and the time needed for the pH of the reaction mixture to drop 0.6 units was measured. From that value, plus the value for the amount of buffer titrated during that time interval, the rate constant for the reaction (in units of $seconds^{-1}$) can be calculated (see reference 12 for details). Reaction velocity can also be reported as enzyme units (with 1 enzyme unit being the amount of CA needed to double the uncatalyzed rate or, in this case, to halve to time of reaction), or, if the buffer capacity of the reaction mixture is determined, as standard international units (micromoles of CO_2 catalyzed per unit time). These values can then be standardized according to tissue weight or protein concentration. This procedure gave reliable results at 24°C for CO_2 concentrations of between 2.8 and 17.5 mM[12]; it is also one of the few delta pH methods to be used for the dehydration reaction as well.

A second approach involves the addition of a chart recorder to measure the rate of change in pH.[5,8] An isotonic Tris–phosphate buffer (10 mM, pH 7.40, 6-ml volume) is held at 4°C in the reaction vessel (Fig. 1). The reaction is initiated by the addition of a volume (e.g., 100 μl) of CO_2-saturated water via a gas-tight syringe. The drop in pH is measured continuously by a null-point pH meter (100 mV per pH unit output; Biological Instrument Group, University of Pennsylvania) and recorded on a chart recorder (linear, 100-mV full scale). The initial slope (chart divisions per unit time) is equivalent to the initial rate of hydration. The system is calibrated via the addition of aliquots of a known concentration of strong acid (e.g., 10 μl of 0.1 N HCl). Using the slope, buffer capacity, and CO_2 concentration, results can be reported as a rate constant (calculations from reference 12), as enzyme units, or as standard international units. To obtain quantitative results, calibration of the buffer and subtraction of the uncatalyzed rate are necessary.

The null-point feature on the pH meter gives zero suppression of the background electrical output of the meter at the initial pH. The subsequent change in pH is fed through a DC amplifier, allowing very small changes in pH (0.05–0.15 units) to be accurately measured. Any pH meter can be converted to a null-point meter by the addition of an external bucking circuit between the meter and recorder (for details, see Fig. 2; diagram and functional information provided by Dr. J. N. Cameron, University of Texas Marine Science Institute, Port Aransas).

2.3. Troubleshooting

Although this technique is simple and reliable, some problems crop up from time to time. Drift in baseline pH values over time is usually the fault of a clogged reference junction, and it can be eliminated by either cleaning or replacing the reference electrode. Noisy baseline, reduced sensitivity, and nonlinear slopes can also be caused by dirty electrodes (e.g., protein adsorption); it is advisable to routinely clean electrodes with an enzymatic solution. Large and sudden fluctuations in pH are caused by electrical noise; sources include the magnetic stir motor, water bath condenser/heater, and static electricity of the operator's clothes. These spikes can be avoided by grounding all components and by not allowing the stir motor to contact the reaction vessel. The inability to calibrate the system usually signifies a failed pH electrode which needs replacing.

3. The pH Stat Assay

The major advantage of the pH stat assay is that both the hydration and dehydration reactions can be measured at constant pH. Dehydration is measured by the rate of addition of an acid necessary to maintain constant pH as H^+ is consumed by CA; conversely, hydration is determined by the rate of base addition in the face of H^+ production. Acid or base additions are performed via an autotitrator system (Fig. 3).[7,9,10,13]

Both measurements use a water-jacketed vessel whose bottom is a sintered glass disk through which gas can be bubbled into the reaction medium (Fig. 4).[7,9] For dehydration, nitrogen is used as a carrier gas to remove CO_2 from solution as it is produced. The vessel initially contains buffer plus enzyme, and the reaction is started by the addition of $NaHCO_3$. The pH of the reaction is fed to a control module, which maintains a constant value by adding acid to the reaction medium. The volume of acid added per unit time is recorded graphically on a chart recorder. Since the stoichiometry of H^+ consumption to CO_2 production is 1:1, the rate of acid addition (micromoles of H^+ per minute) is a measure of the rate of CO_2 production.

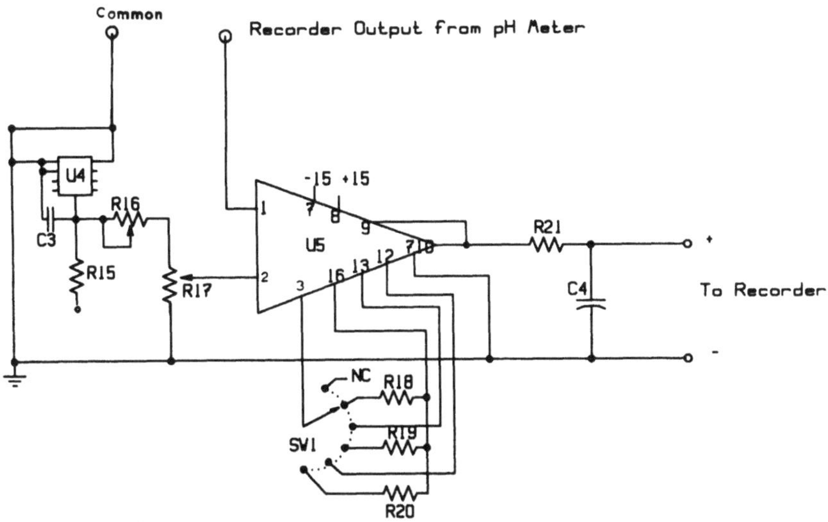

Figure 2. Circuit diagram for conversion of a research-grade pH meter to a null-point meter. This circuit provides zero suppression (null-point conversion), voltage gain (amplification), and filtering. Scaling of zero suppression, which depends on the output of the specific meter (e.g., Radiometer pH meters have an output of −500 mV per pH unit), is controlled by selection of R16 and R17. Amplification is controlled by R17, gain by SWL, and filtering by selection of the values for R21 and C4. The circuit requires a regulated ±15 VDC power supply (not shown). The components needed are as follows:

1. C3 0.11 μF, 100V polystyrene or Mylar film
2. C4 1 μF, 25V electrolytic (varies with filtering desired)
3. R15 2.4K, 5% 1/4W carbon resistor
4. R16 15 turn Cermet trimmer (optional)
5. R17 10 turn pot; Clarostat 73JA (value depends on pH voltage)
6. R18 20.0K, 1%, 1/4W metal film resistor
7. R19 2.105K 1%, 1/4W metal film resistor
8. R20 201 ohm, 1%, 1/4W metal film resistor
9. R21 100K, 1% (provides time constant of approximately 0.1 sec)
10. U4 AD584JH, Analog Devices
11. U5 AD524KD, Analog Devices
12. SW1 1P5T wafer switch (e.g., TRW T205)

Diffusion of CO_2 out of solution can become the rate-limiting factor unless one uses a rate of nitrogen flow through the vessel that is fast enough to strip CO_2 out of solution as it is formed. This rate is determined by increasing the gas flow until the reaction rate becomes independent of the gas flow rate; rates of between 400 and 900 ml/min have been used.[7,9] Because high rates of gas flow can result in excessive foaming, the assay is sometimes run at low temperatures and reduced flow rates.

The pH stat procedure can be modified to measure the rate of CO_2 hydration as well. Carbon dioxide gas (e.g., 5% in air) is bubbled through the reaction vessel to

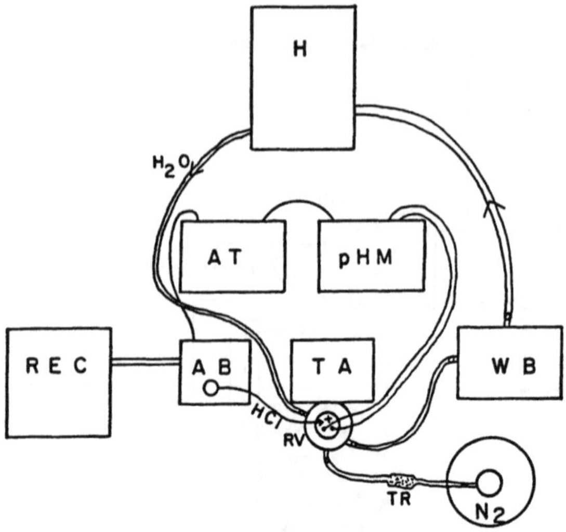

Figure 3. Representative diagram of the components of a typical autotitration system used in the pH stat assay for the dehydration reaction. RV is the reaction vessel, N_2 is a nitrogen source, TR is a CO_2 trap (e.g., Ascarite). TA represents the titrator assembly, which holds the vessel in place and houses the pH and reference electrodes, the stir mechanism, and the delivery system for the addition of HCl. The pHM is a research-grade pH meter connected to an autotitragraph module (AT), which monitors and maintains a constant pH through the addition of acid from the autoburette (AB). A trace of the amount of acid added per unit time is obtained by the recorder (REC). H represents a circulating refrigerated–heated water bath for temperature control, and WB is a water bath for maintaining reagents at the desired temperature. For the hydration reaction, N_2 is replaced by a CO_2 source and the trap is eliminated. HCl is replaced by NaOH.

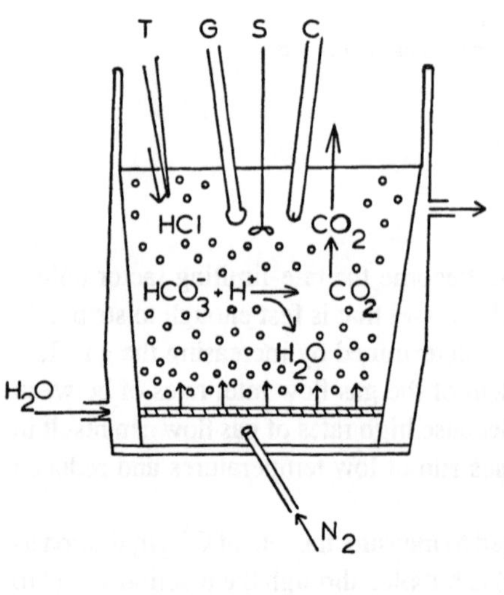

Figure 4. Schematic representation of the reaction vessel for measuring the dehydration in the pH stat assay. N_2 is a stream of nitrogen gas that strips CO_2 out of solution as it is formed. G and R are the pH and reference electrodes, respectively. T represents the delivery system for the addition of titrant (e.g., HCl) to hold a constant pH.

provide substrate for the enzyme.[7,10] The reaction is initiated by the addition of buffer plus enzyme; pH is held constant by the automated addition of base (e.g., NaOH), and substrate concentration is held constant by the continuous flow of CO_2. Substrate can also be provided in the form of CO_2-saturated water.[13] For kinetic studies and for determining rate constants, the molar concentration of CO_2 can be calculated from partial pressure and solubility values for the experimental conditions used.

The autotitration system are available commercially as a package (e.g., Radiometer and Brinkmann). The reaction vessel must be custom made. This assay has been used successfully for the quantitative determination of CA activity in tissue homogenates and for kinetic studies on crude and purified CA. Results appear to be valid over a wide range of temperature, pH, and substrate concentrations.

References

1. Brinkman, R., 1933, *J. Physiol.* **80**:171–173.
2. Carter, M. J., 1972, *Biol. Rev.* **47**:465–513.
3. Coleman, J. E., 1980, in: *Biophysics and Physiology of Carbon Dioxide* (C. Bauer, G. Gros, and H. Bartels, eds.), Springer-Verlag, New York, pp. 133–150.
4. Davis, R. P., 1963, *Methods Biochem. Anal.* **11**:307–327.
5. Dodgson, S. J., Forster, R. E., Schwed, D. A., and Storey, B. T., 1983, *J. Biol. Chem.* **258**:7696–7701.
6. Edsall, J. T., 1968, in: *CO_2: Chemical, Biochemical, and Physiological Aspects* (R. E. Forster, J. T. Edsall, A. B. Otis, and F. J. W. Roughton, eds.), NASA SP #188, Washington, D.C., pp. 15–28.
7. Hansen, P., and Magid, E., 1966, *Scand. J. Clin. Lab. Invest.* **18**:21–32.
8. Henry, R. P., and Kormanik, G. A., 1985, *J. Crust. Biol.* **5**:234–241.
9. Henry, R. P., and Camerion, J. N., 1982, *J. Exp. Zool.* **221**:309–321.
10. Liebman, K. C., Alford, D., and Boudet, R. A., 1961, *J. Pharmacol. Exp. Ther.* **131**:271–274.
11. Maren, T. H., 1967, *Physiol. Rev.* **47**:595–781.
12. Maren, T. H., and Couto, E. O., 1979, *Arch. Biochem. Biophys.* **196**:501–510.
13. McIntosh, J. E., 1968, *Biochem. J.* **109**:203–207.
14. Philpot, F. J., and Philpot, J. St. L., 1936, *Biochem. J.* **30**:2191–2194.
15. Roughton, F. J. W., and Booth, V. H., 1946, *Biochem. J.* **40**:319–330.
16. Silverman, D. N., Tu, C. K., and Wynnns, G. C., 1980, in: *Biophysics and Physiology Carbon Dioxide* (C. Bauer, G. Gros, and H. Bartels, eds.), Springer-Verlag, New York, pp. 254–261.
17. Tu, C. K., and Silverman, D. N., 1975, *J. Am. Chem. Soc.* **97**:5935–5936.
18. Van Goor, H., 1948, *Enzymologica* **13**:73–164.
19. Wilbur, K. M., and Anderson, N. C., 1948, *J. Biol. Chem.* **176**:147–154.

CHAPTER 9

Modified Micromethod for Assay of Carbonic Anhydrase Activity

WOLFGANG BRUNS and GEROLF GROS

The introduction of an assay for microliter samples of carbonic anhydrase (CA) (EC 4.2.1.1.) by Dr. T. H. Maren provided a simple, rapid method by which CA activity could be detected in 0.05 μl of dog blood.[3] Quantitation results from the change of pH caused by the hydration of CO_2, continuously supplied as gas, following the addition of alkaline buffer; the time needed for the indicator phenol red to change color from red to yellow depends on CA activity. We have modified the assay first by blowing CO_2 on the surface of the enzyme solution (instead of through it; this allows us to use samples that tend to foam), second by adding a magnetic stirrer, third by reducing the reaction volume 50%, thus increasing the sensitivity of the assay. Another modification is the use of barbital buffer—as an alternative to the original CO_3^{2-}/HCO_3^- buffer—which further increases the assay's sensitivity, especially in the case of anion-sensitive CA isozymes such as CA I or CA III, which are inhibited by HCO_3^- but not by barbital.[5] The technique cannot be used to determine Michaelis--Menten constants since the initial velocities of substrate turnover cannot be obtained.

1. Equipment

The glass reaction vessel (Fig. 1) has one short sidearm (connected to a CO_2 supply), one long sidearm (for cleaning with vacuum and drying with air), and a reaction chamber (total volume of 15 ml). The entire vessel is mounted in a glass beaker filled with ice water, with the ice occupying the upper half of the beaker's

WOLFGANG BRUNS and GEROLF GROS • Zentrum Physiologie, Medizinische Hochschule Hannover, 3000 Hannover 61, Federal Republic of Germany.

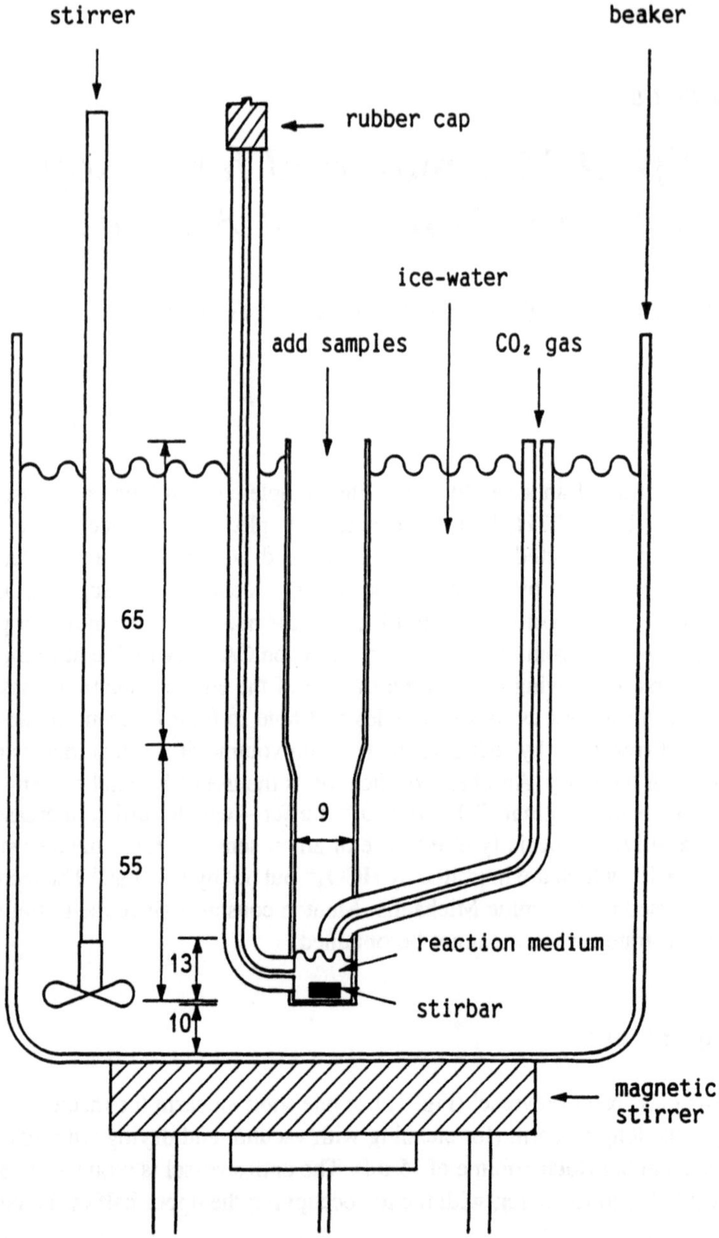

Figure 1. Schematic drawing of a longitudinal section through the setup for CA measurement. The numbers indicate dimensions in millimeters. The internal diameter of the sidearm tubings is 1 mm. For further details, see text.

height. A magnetic stir bar (7 mm × 3 mm) spins on the flat bottom of the reaction vessel, and convection in the ice water is achieved by using a motor-driven stirrer.

2. Solutions

For the CO_3^{2-}/HCO_3^- system: Indicator solution (12.5 mg/liter phenol red, 2.6 mM $NaHCO_3$, in distilled H_2O) and buffer solution (300 mM Na_2CO_3, 206 mM $NaHCO_3$). For the barbital system: Indicator solution (12.5 mg/liter phenol red in 35 μM NaOH) and buffer solution (64 mM Na-barbital titrated to pH 9.0 at room temperature with 1 M H_2SO_4). The solutions are stored on ice until use.

3. Assay

Before the addition of solutions, CO_2 is blown through the reaction chamber at approximately 500 ml/min; this rate must be constant throughout the experiment. A flowmeter is helpful but not essential. The large opening of the reaction chamber is stoppered for a few seconds to ensure filling of the long sidearm with CO_2. Immediately after the stopper has been removed, the external opening of the long sidearm is covered with a soft rubber cap. Solutions are then added with micropipettes (Eppendorf): first, 200 μl of indicator solution, then the sample (with or without CA), and then H_2O to bring the fluid volume to 350 μl. The solutions then equilibrate, with the magnetic stirrer set low. After 2 min, the magnetic stirrer is set to maximum speed; then 50 μl buffer is added, and simultaneously the stopwatch is started. Immediately thereafter, the rubber cap covering the long sidearm is gently squeezed to push the solution that has entered it during equilibration back into the reaction chamber, thus allowing the buffer just added to mix with the entire volume of the other solutions. When the indicator is yellow, the watch is stopped and the time is noted.

4. Results

In this laboratory, employing the CO_3^{2-}/HCO_3^- buffer system, the uncatalyzed reaction time, t_0, is 59.3 ± 0.9 sec ($n = 22$; $\bar{x} \pm$ SD). With CA present, the reaction time, t, is shortened. The reaction is accelerated by a factor t_0/t. CA activity is defined as:

$$A = t_0/t - 1 \tag{1}$$

Thus, $A = 1$ if the reaction time is half the uncatalyzed time.[4] In order to compare the values obtained by other assay techniques, A must be related to enzyme concentration. CA activity of 1 unit is defined as that concentration of CA in the reaction chamber required to halve t_0. Division by [protein] in the reaction chamber gives the specific activity in units per milliliter per milligram of protein.[1]

Figure 2A is a plot of values of t versus ratios of packed erythrocyte volume to total assay volume, volume fractions (VF), for diluted, hemolyzed rabbit erythrocytes (CO_3^{2-}/HCO_3^- buffer system). The uncatalyzed time, 58.5 sec, is halved by VF $= 5.65 \cdot 10^{-5}$, equivalent to adding 0.0225 μl of packed erythrocytes. CA activity is thus calculated as $A = (5.65 \cdot 10^{-5})^{-1} = 18{,}000$. Figure 2B shows the same data, with the straight line representing the theoretical equation derived from Eq. 1:

$$1/t = (1/t_0)A - 1/t_0 \tag{2}$$

where $t_0 = 58.5$ s, and $A = \mathrm{VF}/(5.65 \cdot 10^{-5})$, from Fig. 2A. It is apparent that there is excellent agreement between data points and the prediction of Eq. 2 over the entire range of data. From linear regression analysis of the experimental data the y-intercept, equivalent to $1/t_0$, is 0.0168 $\sec^{-1}$ with $r = 0.999$. This gives $t_0 = 59.5$ sec.

In the barbital system, t_0 is 38.0 ± 0.3 sec ($n = 11, \bar{x} \pm$ SD). A plot similar to Fig. 2B obtained with different amounts of rabbit erythrocytes gives a linear regression line with a y-intercept equivalent to 40.5 sec ($r = 0.995$). The CA activity of packed red cells is $A = 43{,}500$. This increase in activity upon changing the buffer system is in line with the observation of Maren *et al.*[4] In both systems measurements are accurate over a > five-fold increase in CA concentration.

CA inhibitor (I) constants may be determined by this technique. To constant

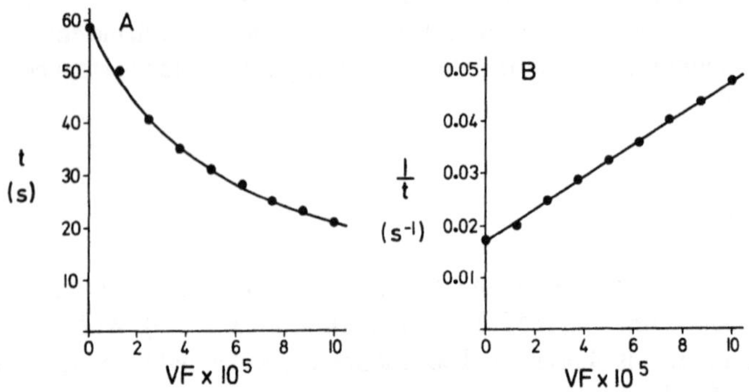

Figure 2. Calibration curves obtained from the measurement of reaction times upon addition of different amounts of rabbit red blood cells lysed and diluted 1:2000 in distilled water. The points represent means of duplicate determinations. (A) Reaction times (t) plotted against the volume fractions (VF) of packed erythrocytes in the final assay volume. (B) Reaction times (t) plotted reciprocally against the theoretical relationship expressed in Eq. 2, if $t_0 = 58.5$ sec and if a volume fraction of $5.65 \cdot 10^{-5}$ yields an activity of $A = 1$, as determined from Fig. 2A.

[CA] in the reaction chamber, with $A = 2$,[3] increasing [I] is added. From [I] and the fractional inhibition of CA, the inhibition constant K_i can be calculated.[2,4] The K_i of acetazolamide of rabbit hemolysate was thus determined to be 1.7×10^{-8} M,[1] in agreement with another report in which a 7-ml chamber was used[7] and one in which a pH stop-flow technique was used at 25°C.[6]

When samples are dissolved in buffers and sucrose solutions, t_0 can be prolonged (often to a greater extent than t), resulting in an overestimation of CA activity (Eq. 1). The problem can be overcome by constructing calibration curves with different [CA] in the presence of the same amount of interfering solute.

References

1. Bruns, W., Dermietzel, R., and Gros, G., 1986, *J. Physiol.* **371**:351–364.
2. Easson, L. H., and Stedman, E., 1936–1937, *Proc. Roy. Soc.* **121**:142–164.
3. Maren, T. H., 1960, *J. Pharmacol. Exp. Ther.* **130**:26–29.
4. Maren, T. H., Parcell, A. L., and Malik, M. N., 1960, *J. Pharmacol. Exp. Ther.* **130**:389–400.
5. Maren, T. H., Rayburn, C. S., and Liddell, N. F., 1976, *Science* **191**:469–472.
6. Siffert, W., and Gros, G., 1982, *Biochem. J.* **205**:559–566.
7. Skipski, I. A., and Scott, W. N., 1980, *Comp. Biochem. Physiol.* **65B**:583–593.

CHAPTER 10

Localization of Carbonic Anhydrase by Chemical Reactions

YVONNE RIDDERSTRÅLE

The first useful procedure to visualize carbonic anhydrase (CA) in tissues was devised by Häusler[5] after an early attempt by Kurata.[7] Further improvements were later made by Hansson[3] and Ridderstråle.[18] Criticisms of this cobalt–phosphate precipitation technique[1,16,17] have been convincingly refuted,[21] especially by Lönnerholm,[9,10] and the technique seems to be generally accepted after the validation by Maren.[15]

This technique gives rise to a precipitate resulting from the activity of all isozymes present in the tissue section. Two variants are described below, one using freeze sectioning[3] the other using resin embedding[18,19] prior to incubation.

1. Processing of the Tissue

1.1. Fixation

Fixatives inactivate the enzyme to different degrees,[3,5,22] and the use of unfixed tissue may appear advantageous. This can, however, lead to a loss of the soluble isozymes to the incubation medium, and unfixed tissue is thus rarely used. The staining intensity is somewhat lower after fixation with glutaraldehyde than with formaldehyde. However, the improved morphology with glutaraldehyde is a great advantage, and fixatives including glutaraldehyde are therefore usually preferred.

A fixative containing 2.5% glutaraldehyde in 0.1 M phosphate buffer or 0.1 M cacodylate buffer gives good results, as does Karnovsky's fixative.[6] In most cases,

YVONNE RIDDERSTRÅLE • Department of Animal Physiology, Swedish University of Agricultural Sciences, S-750 07 Uppsala, Sweden.

4–6 hr of fixation at 4°C is sufficient. There appears to be no advantage of a very prolonged or shortened fixation time.

After fixation, the tissue is thoroughly rinsed in buffer solution (three to four changes and 10 min in each). It can be stored in the buffer in the refrigerator for some days.

1.2. The Cobalt Precipitation Technique[3]

Fixed (or unfixed) tissue is frozen, and sections 10–20 μm thick are cut. Thinner sections disintegrate during the process and must be carried on a Millipore filter during the procedure. (This has been called the Hansson Method.)

1.2.1. Incubation Reagents

Incubation Medium. Mix 17 ml of solution I with 40 ml of freshly prepared solution II (enough for one 9.5-cm petri dish).

Solution I: 1 ml 0.1 M $CoSO_4$, 6 ml 0.5 M H_2SO_4, 1–10 ml 1⁄15 M KH_2PO_4, and distilled water to a final volume of 17 ml.

Solution II: Dissolve 0.75 g $NaHCO_3$ in 40 ml distilled water.

Blackening Solution. $(NH_4)_2S$ 0.5% in distilled water is filtered through a filter paper on which some CoS has been deposited. Some grains of CoS are added to the filtrate to ensure saturation.

Rinsing Solution. Prepare 100 ml of physiological saline buffered with 1 ml 1⁄15 M phosphate buffer, pH 5.9.

1.2.2. Incubation Procedure

Put the sections on the surface of the incubation medium immediately after it has been mixed and incubate them for 1–10 min at room temperature. With the higher phosphate concentration in the medium, shorter times are needed. Make sure that the sections do not dip under the surface of the medium.

Move the sections onto the rinsing solution for 1–3 min.

Transfer the sections onto the blackening solution for 3 min.

Rinse by floating the sections for 1 min on each of the three successive physiological saline dishes.

Dry sections onto slides, counterstain if desired, and mount with Canada balsam or similar agent.

1.2.3. Modifications

Two similar variants of the cobalt precipitation technique have been developed.[8,20] These authors fasten freeze sections on glass slides; the first method incubates the slides while gently agitating a reaction vessel, where the medium covers the sections with a height of no more than 1 mm. The second keeps the slides vertical and dips them in the incubation medium every 30 sec. In this way, the fragile freeze sections do not disintegrate as easily as when floating on the surface of the medium, and the technique can also be useful when serial sections are needed. According to the authors, the results appear comparable to those yielded by the original cobalt precipitation technique.

1.3. Ridderstråle's Resin Method[18,19]

Different types of water-soluble glycol methacrylates polymerizable at room temperature have been found suitable for this purpose, e.g., JB-4 (Sorvall) and Historesin (LKB), as well as the water-soluble Lowicryl K4M. These resins are commercially available as embedding kits.

1.3.1. Dehydration and Infiltration

The fixed and rinsed tissue is dehydrated in ethanol, acetone, or the infiltration solution of the resin used. The more lipid soluble acetone is preferred when the tissue is rich in lipids, e.g., brain and active mammary gland. The following time schedule it suitable for tissue pieces not thicker than 2 mm. Gentle shaking during the steps including resin is advantageous. With Lowicryl K4M, the infiltration and embedding takes place at 4°C. For 50, 70, 90, and 100% dehydrating solution, 30 min; for 1 part resin mixture + 1 part dehydrating solution, 1 hr; for 2 parts resin mixture + 1 part dehydrating solution, 1 hr; for pure resin mixture with one change, 2–3 hr (overnight for Lowicryl K4M).

1.3.2. Embedding

The tissue is placed with fresh resin in embedding molds such as Beem capsules or polyethylene embedding molds fitting the microtome to be used.

Skin contact should be avoided with all kinds of resin solutions. Sometimes small amounts of unpolymerized resin may be found on the surface of the resin blocks, which should be wiped off carefully.

JB-4 and Historesin. After addition of the accelerator, the resin starts to polymerize within 30 min at room temperature, and the polymerization is completed after a couple of hours. When making blocks larger than 3 cm^3, it is advisable to facilitate heat dissipation by placing the mold in a water bath of about 15°C or to use the special Teflon molds available for this purpose. Heating is no problem with smaller blocks.

Lowicryl K4M. The resin is polymerized for 24 hrs at 4°C by indirect UV irradiation (360 mm, 2 × 15 W; Philips TLAD 15W/05 or equivalent) at a distance of 30–40 cm. At this temperature (4°C or warmer), benzoin ethyl ether is recommended as accelerator. Lowicryl K4M infiltration and polymerization can also take place at temperatures down to −35°C, with probably better survival of enzyme activity.

1.3.3. Storage

The embedded tissue may be stored at −20°C or lower with undiminished stainability for at least 1 year and often considerably longer. In the refrigerator, the lifetime is substantially shorter.

1.3.4. Sectioning

Sectioning is performed on a microtome suitable for sectioning with glass knives. Large sections with a thickness of 0.5–2 μm can easily be cut.

The sections are immediately transferred from the glass knife with tweezers and put onto a water surface in a petri dish to stretch. They are transferred on to the different solutions by lifting them from below with a metal loop or a glass rod.

1.3.5. Incubation Reagents

Incubation Medium. Mix 8.5 ml of solution I with 20 ml of solution II and pour it into a Petri dish.

Solution I: Prepare a stock solution by mixing 10 ml 0.2 M $CoSO_4$ + 60 ml 0.5 M H_2SO_4 + 100 ml $\frac{1}{15}$ M KH_2PO_4 (higher cobalt concentration than in the cobalt precipitation technique). This solution can be stored in the refrigerator a couple of weeks, but give it time to reach room temperature before use.

Solution II: Dissolve 3.75 g $NaHCO_3$ in 200 ml distilled water. This solution should be freshly prepared.

Rinsing Solution. A stock solution of $\frac{1}{15}$ M phosphate buffer, pH 5.9, is diluted 100× with distilled water. The rinsing solution should be exchanged for fresh solution after five or six incubation series.

Blackening Solution. $(NH_4)_2S$ (0.5%) in distilled water. This is filtered through a filter paper on which CoS has been deposited. [The CoS is made by mixing small amounts of $CoSO_4$ and $(NH_4)_2S$ in water and pour it onto the filter paper to be used.]

1.3.6. Incubation Procedure

Incubate the sections for 1–8 min on the surface of the incubation medium immediately after mixing of the reagents.

Float the sections for 1 min on the rinsing solution.

Transfer the sections to the blackening solution for 3 min.

Transfer the sections to two successive baths with distilled water, 1 min in each.

Bring the sections onto a glass slide either with the metal loop or by pushing the slide obliquely into the water and move the section onto the glass with a fine hair brush.

1.3.7. Counterstaining and Mounting

If desired, the sections can be counterstained after drying. Many conventional stains adapted for this type of resin, such as hematoxylin-eosin and others, can be used. Azure blue is convenient for general staining.

Azure Blue Solutions. Stock solution I: 1% azure I in 1% sodium borate solution.

Stock solution II: 1% azure II in distilled water.

Staining solution: Mix equal parts of I and II with two parts of distilled water. If this gives too strong staining, the solution can be diluted with distilled water.

Staining Procedure. Warm the slide (about 60°C), cover the section with a drop of the staining solution, rinse the solution off with water after 1 or 2 sec and let the slide dry.

Permanent mounting of the sections can be performed with Eukitt, Canada balsam, or any similar mounting medium. Epon, the same mixture as is used for embedding in electron microscopy, is an excellent mounting medium for histological sections. After polymerization for 1 hr at 90°C, the resin is hard and the sections are ready to be viewed in the microscope. Unpolymerized epon for this purpose can be stored in syringes in the freezer over long periods.

With all of the above-mentioned mounting media, some of the preparations may fade within a year, whereas others remain unaffected for several years.

1.4. Electron Microscopy

There are two different approaches to electron microscopy. The first one is to embed the stained section made according to Hansson in epon and then make ultrathin sections.[12] Sugai and Ito have improved this technique considerably.[22] The other is to incubate thin sections made with Ridderstråle's resin method and study them in the electron microscope without further treatment. With this method, the thinnest sections are achieved with the Lowicryl resin, but sections of about 0.1–0.2 μm can be cut with JB-4 or Historesin. During sectioning of these resins, it is necessary to use a relatively high cutting speed and the water level in the trough should be as low as possible. The rinsing following the incubation of such thin sections should be kept short, about 15–30 sec, to avoid possible loss of the initial precipitate during washing. Despite this precaution thin sections sometimes are unstained, though light microscopy shows CA to be present. In this case, it is thus advisable not to use too thin sections for electron microscopy.

The sections are collected on uncoated grids, since the Formvar film prevents the staining. The grids are transferred onto the different incubation baths with a metal loop. Avoid copper grids, since the copper reacts with the ammonium sulfide.

Another, often more convenient way is to incubate sections with an area larger than a grid. While the sections are still floating on the rinsing water, one or more grids provided with a supporting film are placed on top of the sections with the film facing downward. The grids are picked up from the water surface with a piece of Parafilm.

The sections made in either of these ways are not as stable as Epon sections under the electron beam but can be stabilized with carbon. Since they are thicker than traditional ultrathin sections, it is an advantage when studying them to use 100 kV or still higher accelerating voltage.

1.5. Controls

1.5.1. Specificity

Acetazolamide is a well-known specific inhibitor of CA[14] that abolishes the staining due to CA activity. A stock solution of 1.14×10^{-3} M acetazolamide is prepared and can be stored for several months in the refrigerator. It is advisable to preincubate the sections for 10 min on a 10^{-5} M acetazolamide solution, at least with high concentrations of CA in the sections. Before incubation, the inhibitor is included in solution II to give a final concentration of the inhibitor in the medium of 10^{-5} M. The sections are incubated with corresponding times as the uninhibited sections and are processed in an identical manner. Higher concentrations of inhibitor has to be used when the more insensitive CA III is present.[10]

Heating the sections before incubation[3] is an alternative control method that can be used when acetazolamide-insensitive isoenzymes are encountered.

1.5.2. Test for Preformed Crystallization Nuclei

The sections are preincubated for 30 min on a solution containing 8×10^{-5} M acetazolamide. This results in total inhibition of CA, except possibly CA III. The sections are transferred to the incubation medium containing the same concentration of inhibitor and incubated for 10–20 min. The sections are then processed as usual.

2. Interpretation

2.1. Physicochemical Basis

This discussion follows the terminology and principles used by van Duijn.[2]

After mixing, the incubation solution will have a pH slightly below 6, and the carbon dioxide tension is such that the pH will not change appreciably unless CO_2 disappears. When this occurs, H_2CO_3 will decompose spontaneously into CO_2 and water. The pH will thus increase continuously, most rapidly at the surface. Carbonate ions and phosphate ions will increase in concentration as a result of the pH increase. At a certain pH, the solution will become saturated with regard to the reaction product consisting of cobalt ions, carbonate ions, and phosphate ions.[3] At this time (between 8 and 10 min in the incubation solution used in the resin method), precipitate may start to form provided crystallization nuclei exist in the solution. Without such nuclei, no precipitation occurs until at a later time the pH has risen to a level where the concentration of the reaction product has increased to that necessary for rapid nucleation.[2] At this time, precipitate forms rapidly at the surface of the solution and increases in amount with time.

When a tissue section is floating on the surface of the incubation solution, similar processes occur in the section. At locations containing CA activity, carbonic acid decomposition occurs more rapidly and the pH rises more quickly than in other parts of the tissue. Supersaturation, and later the concentration of rapid nucleation and ensuing precipitate formation, is reached earlier at these locations than in other parts of the section. If the section contains structures that can act as crystallization nuclei (the so-called preformed crystallization nuclei) for the reaction product, precipitation may occur at these sites as soon as the saturation pH has been reached.

2.2. Sources of Error

With the cobalt precipitation methods, in most cases some cell nuclei are more strongly stained than the cytoplasm, whereas others are unstained. In general, there is a close correlation between cytoplasmic staining and strong nuclear stain. An example of this is shown in Fig. 1, where the periportal hepatic cells have unstained cytoplasm and nuclei. The hepatic cells closer to the central vein show cytoplasmic stain and marked nuclear staining. This marked difference in staining must reflect a corresponding difference in CA localization. This is shown by the fact that they are unstained in the acetazolamide control (Fig. 2).

The high intensity of staining of some cell nuclei apparently is related to the fact that cell nuclei in general act as preformed crystallization sites. This is shown by the consistent and fairly strong staining of all cell nuclei in the test for preformed crystallization nuclei (Fig. 3). In this picture, several other cell components are stained (though much more weakly than the cell nuclei), indicating that they also contain preformed crystallization nuclei.

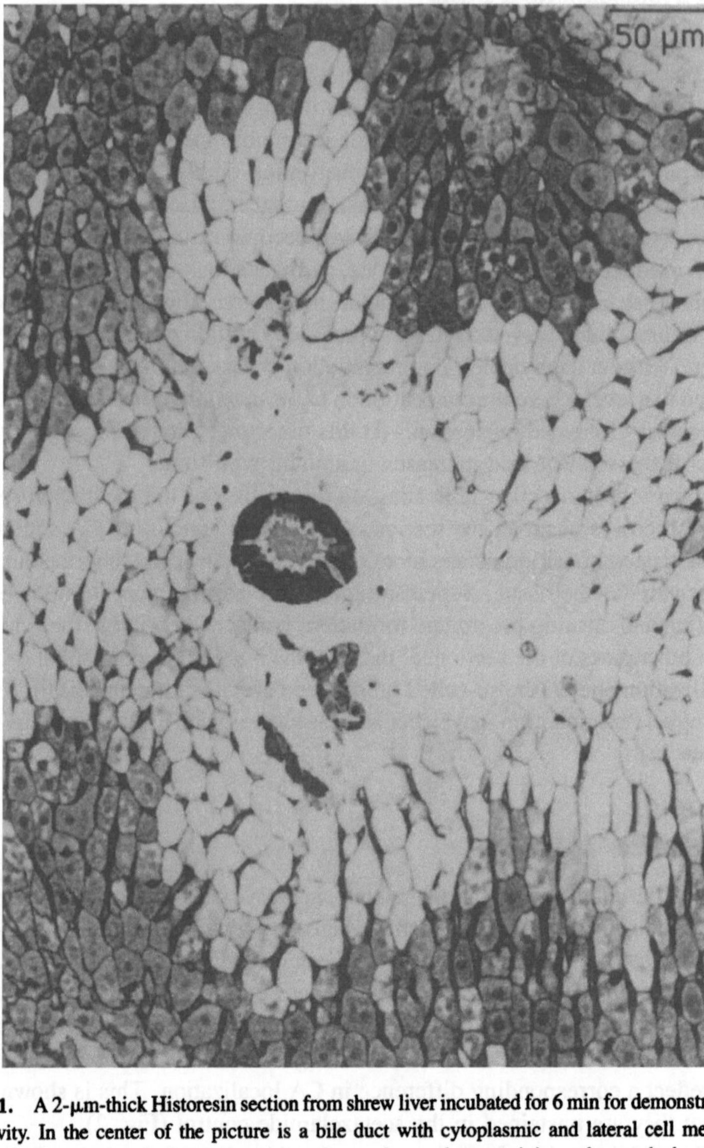

Figure 1. A 2-µm-thick Historesin section from shrew liver incubated for 6 min for demonstration of CA activity. In the center of the picture is a bile duct with cytoplasmic and lateral cell membrane staining. The hepatocytes in the periportal area show only membrane staining, whereas the hepatocytes in the vicinity of the central vein have strong cytoplasmic and membrane stain. A large part of the sinusoids is also stained.

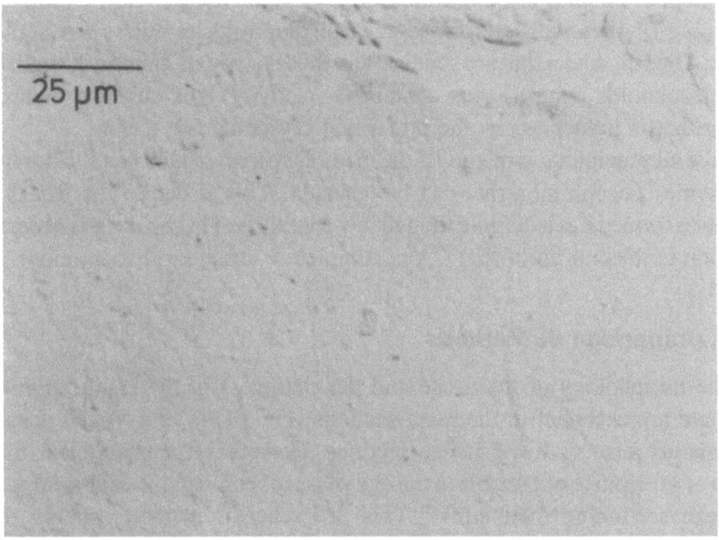

Figure 2. Acetazolamide-inhibited 2-μm-thick Historesin section from shrew liver incubated for 6 min, showing no precipitate. The picture is taken with slight interference contrast.

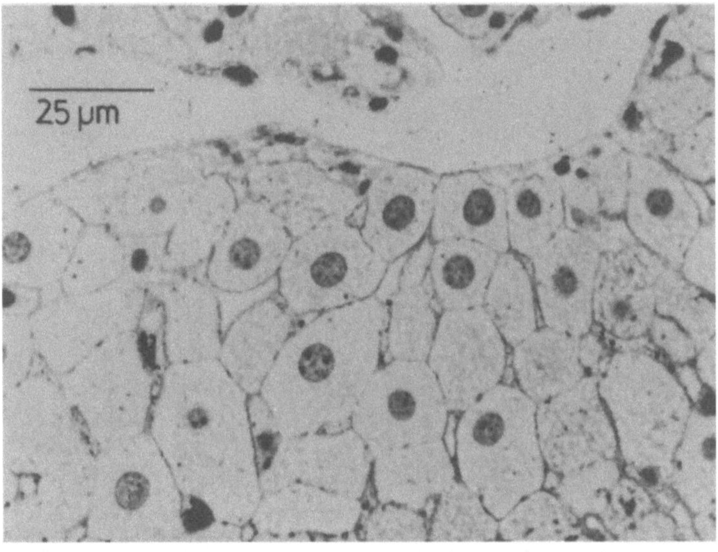

Figure 3. Test for preformed crystallization nuclei performed on a 2-μm-thick Historesin shrew liver section incubated for 10 min. See Section 3.2.

It must be emphasized that with short incubation times (below 8 min) there is no staining of sites with preformed crystallization nuclei unless CA activity is present. The cell nuclei that are stained (and those in which staining is abolished in the acetazolamide control) thus contain CA activity. It is only the intensity of staining that is influenced by the preformed crystallization nuclei.

Like all precipitation methods, the method described here has a lower limit to the enzyme concentration that can be detected. This is due to the fact that the precipitate formed has a definite though low solubility. That no stain is observed in a location is thus no proof that CA is completely absent at that location.

2.3. Comparison of Methods

The morphology of the tissue and the sharpness of the localization of the precipitate are excellent in the resin sections (Figs. 1, 4, and 5) and usually of better quality than with the freeze sections. However, it appears clear that the amount of precipitate obtainable in the cytoplasm of comparable cells is larger with the freeze sectioning method.[13,19] This difference is almost certainly due to reduction of cytoplasmic CA activity during the embedding procedure.

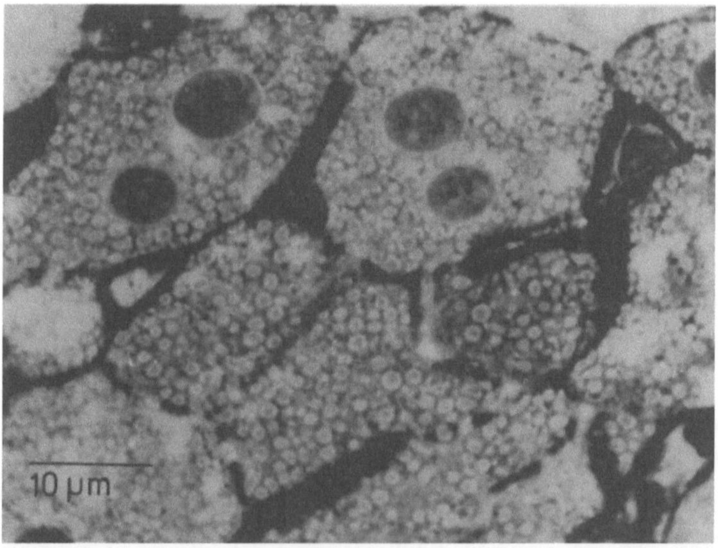

Figure 4. A 0.5-μm-thick Historesin section incubated for 6 min, showing the intensely stained hepatocytes close to the central vein of the shrew liver. The nuclear membrane appears to be unstained. In the stained cytoplasm, there are many mitochondria with a stained central core. At the upper right is a stained sinusoid with a stained erythrocyte.

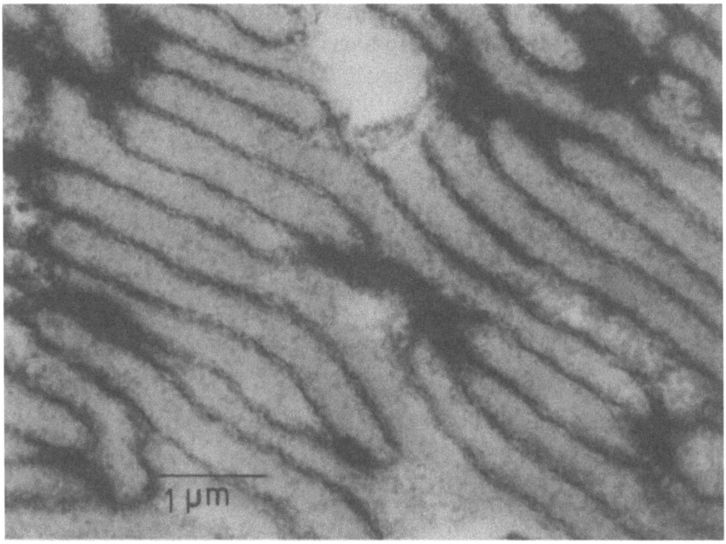

Figure 5. Electron micrograph from a thin Lowicryl section incubated for 6 min, showing distinct membrane staining in the shrew kidney thick tubule.

Methods using freeze sections do not in general show staining of cell membranes or capillary endothelium. Strong but selective staining at these locations, on the other hand, is characteristic of the resin method (Figs. 1, 4, and 5).

This difference may be explained as follows. In the resin section, the cell components are embedded in a three-dimensional network of polymerized resin that restricts the diffusion some 15–30 times.[19] If we consider a small amount of CA activity at a point which is topographically isolated from other CA activity, this will in the resin give rise to a sharply localized pH rise and a correspondingly small area of precipitate.

In a similar situation in a freeze section, the ions produced by the same amount of CA activity will diffuse out into a much larger volume. The resulting pH increase is consequently much smaller and will result in a more widespread and much fainter stain, or often in no precipitate at all. When the CA is more homogeneously distributed in larger areas, as is most often the case with the cytoplasmic CA, the effect of the differences in diffusion rate will of course be much less important and is usually overshadowed by differences in residual activity in the two methods.

The low diffusion rate in the resin is also undoubtedly the reason for the high resolution in the localization of the precipitate characteristic of the resin method, as the sharpness of localization is much improved by a reduction of the diffusion rate.[4]

Acknowledgments. Invaluable technical assistance was given by Mrs. Ingrid Wennerberg. For constructive discussions, Prof. Ivar Sperber and Prof. P. G. Knutsson are gratefully acknowledged.

REFERENCES

1. Anderson, R. E., Gay, C. V., and Schraer, H., 1982, *J. Histochem. Cytochem.* **30**:1135–1145.
2. Duijn, P. van, 1974, in: *Electron Microscopy and Cytochemistry* (E. Wisse, W. T. Daems, I. Molenaar, and P. van Duijn, eds.), North-Holland Publishing Co., Amsterdam, pp. 3–23.
3. Hansson, H. P. J., 1967, *Histochemie* **11**:112–128.
4. Holt, S. J., and O'Sullivan, D. G., 1958, *R. Soc. Proc. London Ser. B.* **148**:465–480.
5. Häusler, G., 1958, *Histochemie* **1**:29–47.
6. Karnovsky, M. J., 1965, *J. Cell Biol.* **27**:137A.
7. Kurata, Y., 1953, *Stain Technol.* **28**:231–233.
8. Loveridge, N., 1978, *Histochem. J.* **10**:361–372.
9. Lönnerholm, G., 1974, *Acta Physiol, Scand. Suppl.* **418**:1–43.
10. Lönnerholm, G., 1980, *J. Histochem. Cytochem.* **28**:427–433.
11. Lönnerholm, G., 1984, *Ann. N.Y. Acad. Sci.* **429**:369–381.
12. Lönnerholm, G., and Ridderståle, Y., 1974, *Acta Physiol. Scand.* **90**:764–778.
13. Lönnerholm, G., and Ridderstråle, Y., 1980, *Kidney Int.* **17**:162–174.
14. Maren, T. H., 1967, *Physiol. Rev.* **47**:595–781.
15. Maren, T. H., 1980, *Histochem. J.* **12**:183–190.
16. Muther, T. F., 1972, *J. Histochem. Cytochem.* **20**:319–330.
17. Muther, T. F., 1977, *J. Histochem. Cytochem.* **25**:1043–1050.
18. Ridderstråle, Y., 1976, *Acta Physiol. Scand.* **98**:465–469.
19. Ridderstråle, Y., 1980, *Acta Physiol. Scand. Suppl.* **488**:1–23.
20. Riley, D. A., Ellis, S., and Bain, J., 1982, *J. Histochem. Cytochem.* **30**:1275–1288.
21. Rosen, S., and Musser, G. L., 1972, *J. Histochem. Cytochem.* **20**:951–954.
22. Sugai, N., and Ito, S., 1980, *J. Histochem. Cytochem.* **28**:511–525.

CHAPTER 11

Localization of the Carbonic Anhydrases by Immunocytochemistry

CAROL V. GAY

The purpose of this review is to discuss new technical developments that have enhanced immunocytochemical localization of the isozymes of carbonic anhydrase (CA) (EC 4.2.1.1.).

The advantages of immunocytochemistry, high sensitivity and high specificity, are ensured by selection of preparative methods to retain antigenicity and by demonstrating antibody specificity for the various CA isozymes. Immunodiffusion and dot blot methods are useful for establishing specificity. General strategies described by Sternberger *et al.*[63] serve well for production and application of antibodies to CA; the CA enzymes have sufficient molecular mass (approximately 30,000 daltons) for antigenicity.

Fixation is critical for immunocytochemistry. Soluble forms of antigen must be anchored in their correct microanatomical positions without masking or destroying antigenic sites. The possibility that fixation may alter the antigen so that it will cross-react with antibodies to other CA isozymes is another concern.[16,73] Frémont and colleagues[16] have noted that after denaturation with sodium dodecyl sulfate or mercaptoethanol, antibodies to CA III will cross-react with CA I and CA II.

Penetration of the large antibody molecules (usually an IgG; molecular weight 146,000) has to be considered. The use of frozen sections or water-permeable embedding media seems to provide adequate access of antibodies to section surfaces so that antigenic sites close to a surface will be registered.

CAROL V. GAY • Departments of Molecular and Cell Biology and Poultry Science, The Pennsylvania State University, University Park, Pennsylvania 16802.

1. Light Microscopic Studies

The approximately 60 CA immunolocalization studies of the past 5 years reveal that the CA isozymes as antigens are robust proteins. There are many tissue preparative methods that provide definitive localizations; no single method stands out as a clear best choice. Numerous studies use strong fixatives such as Carnoy's, a mixture of ethanol, chloroform, and glacial acetic acid in the ratio of 6:3:1 (see, for example, references 26, 31, 39, 46, 54, 55, and 71). Other fixatives used include Bouin's fixative, which is picric acid, formalin, and glacial acetic acid in the ratio of 15:5:1,[54,55] 4–20% buffered formalin,[10,11,23,26,33,39,46,71] or formalin–ethanol–acetic acid–water in the ratio of 4:69:5:22, which has been particularly useful for neural studies.[8,9,11,15,42] Another fixative that retains antigenicity is 6% $HgCl_2$–1% sodium acetate–0.1% glutaraldehyde.[25,26]

Fixation is often followed by dehydration in ethanol, clearing in xylene, and embedment in paraffin. Adequate results have been achieved with these standard methods, possibly because of maintaining the tissues at a cool temperature throughout processing except for brief heating at modest temperatures (52–58°C) for paraffin embedding.

Perfusion fixation is useful, particularly with delicate tissues such as brain and spinal cord.[10,11,64] After brief fixation, tissues may be cut into 20–50-μm slices for application of antibodies before embedding and thin sectioning.[12,29,34] Instruments available for this include the Vibratome (Technical Products International, St. Louis, Mo.), Tissue Chopper (Sorvall, Norwalk, Conn.), and Bioslicer (Dosaka EM Co. Ltd., Kyoto, Japan). Another approach is to dehydrate, clear, mount, and view the 20–50-μm sections with Normarski differential interference optics[64] without further sectioning.

All fixation, dehydration, and embedding solutions may be avoided by making cryostat sections. There is the possibility that small, unbound molecules such as the CAs (molecular weight approximately 30,000) may diffuse. However, in a study from Carter's group,[30] rapid refreezing of thaw-mounted cryostat sections resulted in retention of CA II and CA III exclusively in type I fibers.

For investigators interested in examining skeletal tissues, it is useful to note that CA antigenicity also survives decalcification with EDTA.[14,39]

A useful gauge for monitoring CA retention during processing for immunocytochemistry is demonstration of the presence of soluble CA in a glandular lumen, as, for example, in salivary glands.[29]

Within the past 5 years, immunohistochemical studies have been performed on amphibian epidermis[75]; brain tissue and cells and peripheral nerves[4,6,8–12,15,23,34,36,42–44,64,66]; gastrointestinal tract[48]; gill[60]; invertebrate spicule-forming structures[35]; kidney[25,26,47]; macrophages[65]; muscle[28,30,49,52,61,71]; myoepithelial cells[68]; reproductive tissues[22,24]; retinal tissue[45,46,58,74]; salivary glands[29,32,53–55]; skeletal tissues such as bone, cartilage, and giant-cell tumors of bone and tendon[13,21,31,33,37,39–41,64,67]; and a variety of other tumors.[50,51,56,57,59]

2. Ultrastructural Studies

Immunocytochemistry for CA isozymes at the fine-structural level has been applied to a number of tissues. Light microscope studies have provided evidence for the relative ease in preserving the antigenicity of the CA isozymes. Thus, it comes as no surprise that both fine-structural preservation and antigenicity can be accomplished simultaneously. Ultrastructural studies include investigation of muscle,[16,70] bone,[2,13,39,41] fish gill,[60] salivary gland,[29] kidney,[7] and cultured retinal cells.[74]

The remainder of the discussion focuses on several protocols that have produced excellent ultrastructural results. Frémont *et al.*[16] employed perfusion fixation and used protein A–gold as a probe. The hindlimb of adult rats was perfused with buffer consisting of 10 mM 3-(*N*-morpholino)propanesulfonic acid (MOPS) at pH 7.4, 2 mM $MgCl_2$, 0.15 M NaCl, 3.0 mM $NaHC0_3$, 2 mM EGTA, 5 mM KCI, 0.4 mM KH_2PO_4, 0.6 mM Na_2HPO_4, and 10 U of heparin at a flow rate of 4 ml/min until the effluent was clear. Fixative (3% glutaraldehyde, 2 mM $MgCl_2$, 2 mM EGTA in 100 mM sodium cacodylate buffer, pH 7.2) was then introduced at the same flow rate for 30 min. Small pieces (1 × 3 mm) of muscle were excised for further fixation in the same fixative, washed in cacodylate buffer (15 min × 2), and then treated with 0.5 mg/ml sodium borohydride in cacodylate buffer for 30 min to block free aldehyde groups. A final buffer wash preceded embedding in 20% (wt/vol) polyvinyl alcohol (Vinol 203; M_r 11,000–31,000; Air Products and Chemicals, Allentown, Pa.). Thin sections (gold interference color) were cut onto glycerine and floated on grids on 20 mM phosphate buffer, pH 7.4, and 150 mM NaCl (PBS) overnight to extract embedding medium and allow antibody penetration. The grids were then rinsed with distilled water, treated with 5% dry milk solids in PBS to minimize nonspecific staining for 15 min, and reacted with CA III antiserum (in 50 mM PBS) for 1 hr at room temperature. This was followed by rinsing in PBS (15 min × 2) and a 30-min incubation with a protein A–gold suspension. Two more PBS rinses, followed by distilled water and contrast staining with 2% aqueous uranyl acetate and lead citrate, completed the process.

The procedure of Frémont *et al.*[16] is attractive because quantification of CA III cytoplasmic extracts on nitrocellulose with ^{125}I-labeled antibody revealed similar proportions of CA III in three types of muscles as found by counting gold particles on tissue sections and by measuring CA activity. The finding of CA III solely in type I fibers and not in adjacent structures and the finding of no staining in CA I- and CA II-rich red blood cells adds further confidence.

A similar distribution of CA III was also found in muscle by Väänänen's group,[72] who used a less complicated procedure. Small pieces of tissue were block fixed in 2% paraformaldehyde–0.1% glutaraldehyde in 10 mM phosphate buffer, pH 7.4, for 4 hr. After an overnight phosphate buffer wash, dehydration at 4°C in a graded alcohol series was followed by Lowicryl K4M embeddment. Thin sections were treated with 1% normal goat serum for 1 hr and then reacted with antibodies, using the immunogold–IgG conjugate method. A similar approach was used for fish gill.[60] Although the use of water-soluble embedding media, such as polyvinyl alcohol and Lowicryl, is necessary to provide assess of antibodies to the antigens, it should be noted that these materials are not easy to section.

Application of CA antibodies to ultrathin cryosections is an alternative approach to use of fixatives and embedding media. Good structural detail and retention of antigenicity have been reported.[7] Nuclei were sometimes stained, however, and some unwanted diffusion may have occurred.

Another alternative for ultrastructural immunocytochemistry is use of pre-embedding staining on whole cells or semithin sections of tissue. Detergents, such as 0.3% saponin, are used to promote antibody penetration into lightly fixed cells.[2,13,29] After the final step of the immunocytochemical reactions, the tissue is postfixed and embedded in one of the standard epoxies.

3. Immunolocalization of CA in Skeletal Tissues

Immunolocalization of CA in skeletal tissues is an area where particularly significant progress has been made.

3.1. Osteoclasts

The progressive development of immunocytochemical techniques has allowed inferences on CA function in osteoclasts to be made. Localization of CA with indirect immunofluorescence[19,33,67,69] corroborated the identification of CA in cytosol of osteoclasts shown previously by autoradiographic detection of the specific inhibitor, ^{3}H-acetazolamide.[18] The pursuit of CA localization in osteoclasts at the ultrastructural level has revealed both cytoplasmic CA and the presence of the enzyme in association with the ruffled border.[2,13,33] This membrane association is reversed by treatment with calcitonin, a calciostatic hormone that inhibits osteoclast activity. The high-activity CA II exists in osteoclasts; in both avian[19] and mammalian[67] cells, antibodies to CA II are reactive.

The discovery of CA along the cytoplasmic surface of the ruffled border of osteoclasts has lead to the postulate that CA provides H^+ to a proton pumping ATPase[17] that is situated in the ruffled border membrane.[1,5] CA is necessary for osteoclast acidification, as shown in cultured osteoclasts that have been treated with 10^{-7} M acetazolamide, osteoclast acidification, as detected by acridine orange fluorescence, is blocked.[3,27] Additional discussion of the functioning of CA in osteoclasts is reviewed elsewhere.[62,70]

3.2. Giant-Cell Tumors of Bone and Tendon

An ultrastructural study of giant-cell tumor of human bone[67] indicates that these multinucleate cells, which are similar in morphology to osteoclasts, contain

CA in the cytoplasm, membranes of some vesicles, and microvilli-like projections of the plasma membrane. This localization of CA further implicates the osteoclast-like nature of the multinucleate giant cell found in giant-cell tumors. Macrophage-like stromal cells also contain CA.[67] A difference between giant cells and osteoclasts is the presence of both CA I and CA II isozymes in giant cells,[39,67] whereas osteoclasts contain only CA II.

Kuwahara[39] purified CA from crude extracts of giant-cell tumor of bone and found two 30,000-dalton forms, each cross-reacting with anti-CA I and anti-CA II, respectively. In addition, a 56,000-dalton membrane-bound form is present.

The finding of membrane-associated and vesicle-associated CA in osteoclast-like giant cells from human bone is noteworthy; similar localizations have been reported for avian osteoclasts.[2]

The distribution of CA in multinucleate cells of a benign giant-cell tumor of bone appears to be similar to osteoclasts and to stromal cells of the tumor.[41]

In a giant-cell tumor of tendon,[67] the osteoclast-like cells react to CA II antibodies but not to CA I antibodies.

3.3. Cartilage

A study of human fetal growth plate[67] confirms the distribution found in growth plate of rats[38] and chicks.[20,21] Late proliferative and early hypertrophic chrondrocytes and hypertrophic extracellular matrix have been consistently found reactive to antibodies against CA II but not CA I. No ultrastructural studies of CA in cartilage have been reported.

ACKNOWLEDGMENTS. Editorial assistance with preparing the manuscript by Nancy L. Kief and Virginia R. Gilman is gratefully acknowledged.

References

1. Akisaka, T., and Gay, C. V., 1986, *Cell Tissue Res.* **245**:507.
2. Anderson, R. E., Schraer, H., and Gay, C. V., 1982, *Anat. Rec.* **204**:9–20.
3. Anderson, R. E., Jee, W., and Woodbury, D. M., 1985, *Calcif. Tissue Res.* **37**:646–650.
4. Banks, D. A., Anderson, R. E., and Woodbury, D. M., 1986, *Epilepsia* **27**:510–515.
5. Baron, R., Neff, L., Louvard, D., and Courtoy, P. J., 1985, *J. Cell Biol.* **101**:2210–2222.
6. Benjelloun, S., Delaunoy, J. P., Gomes, D., DeVitry, F., Langui, D., and Dupouey, P., 1986, *Dev. Neurosci.* **8**:150–159.
7. Brown, D., and Kumpulainen, T., 1985, *Histochemistry* **83**:153–158.
8. Cammer, W., and Tansey, F. A., 1986, *J. Neurol. Sci.* **73**:299–310.
9. Cammer, W., and Tansey, 1987, *J. Histochem. Cytochem.* **35**:865–870.
10. Cammer, W., and Tansey, F. A., 1988, *J. Neurosci. Res.* **20**:23–31.
11. Cammer, W., Sacchi, R., and Sapirstein, V., 1985, *J. Histochem. Cytochem.* **33**:45–54.
12. Cammer, W., Sacchi, R., Kahn, S., and Sapirstein, V., 1985, *J. Neurosci. Res.* **14**:303–316.
13. Cao, H., and Gay, C. V., 1985, *Experientia* **41**:1472–1473.
14. Carey, J. T., and Tonna, E. A., 1988, *J. Dent. Res.* **67**:401.

15. Delaunoy, J. P., Langui, D., Ghandour, S., Labourdette, G., and Sensenbrenner, M., 1988, *Int. J. Dev. Neurosci.* **6**:129–136.
16. Frémont, P., Charest, P. M., Côté, C., and Rogers, P. A., 1988, *J. Histochem. Cytochem.* **36**: 775–782.
17. Gay, C. V., 1988, *Rev. Poult. Biol.* **1**:197–210.
18. Gay, C. V., and Mueller, W. J., 1973, *Science* **183**:432–434.
19. Gay, C. V., Faleski, E. J., Schraer, H., and Schraer, R., 1974, *J. Histochem. Cytochem.* **22**: 819–825.
20. Gay, C. V., Anderson, R. E., Schraer, H., and Howell, D. S., 1982, *J. Histochem. Cytochem.* **30**:391–394.
21. Gay, C. V., Andersen, R. E., and Leach, R. M., 1985, *Avian Dis.* **29**:812–821.
22. Ge, Z. H., and Spicer, S. S., 1988, *Biol. Reprod.* **38**:439–452.
23. Griot, C., and Vandevelde, M., 1988, *J. Neuroimmunol.* **18**:333–340.
24. Härkönen, P. L., and Väänänen, H. K., 1988, *Biol. Reprod.* **38**:377–384.
25. Holthöfer, H., 1987, *Pediatr. Res.* **22**:504–508.
26. Holthöfer, H., Schulte, B. A., Pasternack, G., Siegel, G. J., and Spicer, S. S., 1987, *Lab. Invest.* **57**:150–156.
27. Hunter, S. J., Schraer, H., and Gay, C. V., 1988, *J. Bone Miner. Res.* **3**:297–303.
28. Ibi, T., Haimoto, H., Nagura, H., Sahashi, K., Kato, K., Mokuno, K., Sugimura, K., and Matsuoko, Y., 1985, *Acta Neuropathol.* **68**:74–76.
29. Ikejima, T., and Ito, S., 1984, *J. Histochem. Cytochem.* **32**:625–635.
30. Jeffery, S., Carter, N. D., and Smith, A., 1986, *J. Histochem. Cytochem.* **34**:513–516.
31. Jilka, R. L., Rogers, J. I., Khalifah, R. G., and Väänänen, H. K., 1985, *Bone* **6**:445–449.
32. Kadoya, Y., Kuwahara, H., Shimazaki, M., Ogawa, Y., and Yagi, T., 1987, *Osaka City Med. J.* **33**:99–109.
33. Kadoya, Y., Nagahama, S., Kuwahara, H., Shimazaki, M., Shimazu, A., Ogawa, Y., and Yagi, T., 1987, *Osaka City Med. J.* **33**:111–119.
34. Kahn, S., Tansey, F. A., and Cammer, W., 1986, *J. Neurochem.* **47**:1061–1065.
35. Kingsley, R. J., and Watabe, N., 1987, *J. Exp. Zool.* **241**:171–180.
36. Komoly, S., Jeyasingham, M. D., Pratt, O. E., and Lantos, P. L., 1987, *J. Neurol. Sci.* **79**:141–148.
37. Kumasa, S., Mori, H., Tsujimura, T., and Mori, M., 1987, *J. Cutan. Pathol.* **14**:181–187.
38. Kumpulainen, T., and Väänänen, H. K., 1982, *Calcif. Tissue Res.* **34**:428–430.
39. Kuwahara, H., 1986, *J. Osaka City Med. Cent.* **35**:619–636.
40. Kuwahara, H., Chanoki, Y., Masuda, H., Shimazaki, M., Shimazu, A., Ogawa, Y., and Yagi, T., 1986, *Acta Histochem. Cytochem.* **19**:392.
41. Kuwahara, H., Shimazaki, M., Kadoya, Y., Chanoki, Y., Mitsuhashi, T., Ogawa, Y., Yagi, T., Ishida, T., and Shimagu, A., 1986, *Osaka City Med. J.* **32**:89–97.
42. Langui, D., Delaunoy, J. P., Ghandour, M. S., and Sensenbrenner, M., 1985, *Neurosci. Lett.* **60**:151–156.
43. Levine, S. M., and Macklin, W. B., 1988, *Brain Res.* **444**:199–203.
44. Linser, P. J., 1985, *J. Neurosci.* **5**:2388–2396.
45. Linser, P. J., Sorrentino, M., and Moscona, A. A., 1984, *Dev. Brain Res.* **13**:65–71.
46. Linser, P. J., Smith, K., and Angelides, K., 1985, *J. Comp. Neurol.* **237**:264–272.
47. Lönnerholm, G., and Wistrand, P. J., 1984, *Kidney Int.* **25**:886–898.
48. Lönnerholm, G., Selking, O., and Wistrand, P. J., 1985, *Gastroenterology* **88**:1151–1161.
49. Moyle, S., Jeffery, S., and Carter, N. D., 1984, *J. Histochem. Cytochem.* **32**:1262–1264.
50. Nakagawa, Y., Perentes, E., and Rubinstein, L. J., 1986, *Acta Neuropathol.* **72**:15–22.
51. Nakagawa, Y., Perentes, E., and Rubinstein, L. J., 1987, *J. Neuropathol. Exp. Neurol.* **46**: 451–460.
52. Nishita, T., Matsushita, H., and Kai, M., 1987, *Equine Vet. J.* **19**:509–513.
53. Noda, Y., Sumitomo, S., Hikosaka, N., and Mori, M., 1986, *J. Oral Pathol.* **15**:187–190.

54. Noda, Y., Sumitomo, S., Orito, T., and Mori, M., 1986, *Arch. Oral Biol.* **31**:795–800.
55. Noda, Y., Takai, Y., Hikosaka, N., Meenaghan, M. A., and Mori, M., 1986, *Arch Oral Biol.* **31**:441–447.
56. Noda, Y., Takai, Y., Iwai, Y., Meeaghan, M. A., and Mori, M., 1986, *Virchows Arch. A* **408**: 449–460.
57. Noda, Y., Oosumi, H., Morishima, T., Tsujimura, T., and Mori, M., 1987, *J. Cutan. Pathol.* **24**:285–290.
58. Nork, T. M., Ghobrial, M. W., Peyman, G. A., and Tso, M. O. M., 1986, *Arch. Ophthalmol.* **104**:1383–1389.
59. Okada, Y., Mochizuki, K., Sugimura, M., Noda, Y., and Mori, M., 1987, *Pathol. Res. Pract.* **182**:647–657.
60. Rahim, S. M., Delaunoy, J.-P., and Laurent, P., 1988, *Histochemistry* **89**:451–459.
61. Shima, K., 1984, *Hokkaido J. Med. Sci.* **59**:98–116.
62. Silverton, S. F., this volume, Chapter 33.
63. Sternberger, L. A., 1986, *Immunocytochemistry,* 3rd ed., John Wiley & Sons, New York.
64. Sternberger, N. H., Del Cerro, C., Kies, M. W., and Herndon, R. M., 1985, *J. Neuroimmunol.* **7**:355–363.
65. Sundquist, K. T., Leppilamp, M., Jarvelin, K., Kumpulainen, T., and Väänänen, H. K., 1987, *Bone* **8**:33–38.
66. Tansey, F. A., Thampy, K. G., and Cammer, W., 1988, *Dev. Brain, Res.* **43**:131–138.
67. Väänänen, H. K., 1984, *Histochemistry* **81**:485–487.
68. Väänänen, H. K., and Autio-Harmainen, H., 1987, *J. Histochem. Cytochem.* **35**:683–686.
69. Väänänen, H. K., and Parvinen, E.-K., 1983, *Histochemistry* **78**:481–485.
70. Väänänen, H. K., and Parvinen, E.-K., this volume, Chapter 32.
71. Väänänen, H. K., Paloniemi, M., and Vuori, J., 1985, *Histochemistry* **83**:231–235.
72. Väänänen, H. K., Takala, T., and Morris, D. C., 1986, *Histochemistry* **86**:175–179.
73. Van Der Sluis, P. J., and Boer, G. J., 1986, *Cell Biochem. Funct.* **4**:1–17.
74. Wakakura, M., and Foulds, W. S., 1988, *Invest. Ophthalmol. Visual Sci.* **29**:892–900.
75. Zaccone, G., Fasulo, S., Lo Cascio, P., and Licata, A., 1986, *Histochemistry* **84**:5–9.

CHAPTER 12

Immunological Methods for Detection and Quantification of Carbonic Anhydrase Isozymes

STEPHEN JEFFERY

1. Raising Antisera

The quality of the antibody is the most important factor in any immunological detection system, qualitative or quantitative. The ideal antibody has high affinity for its antigen and monospecificity. The response to a given protein varies greatly between animals, and proteins differ in immunogenicity. In our laboratory, we have found it easy to raise antisera against rat carbonic anhydrase (CA II) but much more difficult to produce high-affinity antibodies to rat CA III.

An initial injection of 100 μg of protein should be sufficient if rabbit is to be used to raise the antibodies. A higher concentration can result in immunotolerance or an antiserum more likely to cross-react with trace contaminating protein. The protein is dissolved in 500 μl of distilled water (d.w.), an equal volume of Freund complete adjuvant is added, and the mixture is homogenized to produce a viscous solution that will release antigen slowly into the animal's bloodstream.

The Freund mixture is injected subcutaneously in three or four sites on the back of the rabbit, using a wide-bore needle. After 5 or 6 weeks, a further 50 μg of antigen is injected in incomplete Freund adjuvant. After another week, serum can be taken for testing, and further subcutaneous injections of 50–100 μg of antigen can be given at intervals of 2–3 weeks until the desired titer is obtained.

To test the specificity of the antibody, Western blotting is the method of choice (see Section 2.2).

STEPHEN JEFFERY • **Department of Child Health, St. George's Hospital Medical School, University of London, London SW17 ORE, United Kingdom.**

2. Immunological Detection Methods

2.1. Ouchterlony Plates

The Ouchterlony plate method is the simplest to test for the presence of a CA isozyme in a complex mixture, but it is purely qualitative. A 30-ml amount of 1% agar (hand hot) is poured into a 9-cm petri dish. A central well for the antibody (2–4 mm in diameter) is cut in the plate, surrounded by other wells of similar diameter. The distance between the central well and those surrounding it should be about 6 mm. To test whether a sample contains the CA isozyme against which the antibody was raised, the antibody is placed in the central well, the sample is placed in one of the surrounding wells, and a pure sample of the CA isozyme is placed in an adjacent well. The plate can be incubated for a few hours at 37°C or overnight at 4°C. Precipitation lines form in the agar between the antibody and sample wells if there is CA in the sample that the antibody recognizes. If the two lines formed by the sample and pure isozyme meet and are continuous, the isozyme in question may be present in the sample. If the lines meet but spur, the sample may contain a related antigen. Figure 1 shows the situation for rat CA III and bovine CA III on an Ouchterlony plate in which anti-bovine CA III is in the center well.

2.2. Western Blotting

Western Blotting is a sensitive, semiquantitative method for detecting the presence of a given CA isozyme in a complex mixture. There are numerous protocols for this technique. The one described below is that used routinely in our laboratory for the soluble rat isozymes, CA II and CA III.

Twelve percent acrylamide gels are used to separate the proteins, with a 6.8% stacking gel. A discontinuous buffer system is used.[4] Samples are boiled for 2 min in buffer containing SDS and mercaptoethanol and then loaded onto the gel, with bromophenol blue added as a tracking dye. When the dye has reached the end of the gel, CA isozymes are approximately at the middle. The gel is then removed and soaked in Tris–glycine buffer (20 mM Tris–250 mM glycine–20% methanol, pH 8.3) for 20 min. Either nitrocellulose or nylon membranes can be used for protein transfer.

For blotting, a Scotch pad is soaked in the Tris–glycine–methanol buffer and placed in a shallow tray. A piece of 1-mm filter paper is put on the pad, followed by the presoaked SDS–gel. A nylon or nitrocellulose membrane, cut to the size of the gel, is placed on the gel, and air bubbles are excluded. A piece of Whatman MM paper soaked in buffer is placed on the membrane. Another Scotch pad is placed on the filter paper, and the entire sandwich of pads, gel, and membrane is held together in a perspex clamp with holes to allow passage of electric current. The clamp and contents are put in a transfer tank filled with buffer, with the membrane toward the anode and the gel facing the cathode. The transfer is completed by using overnight

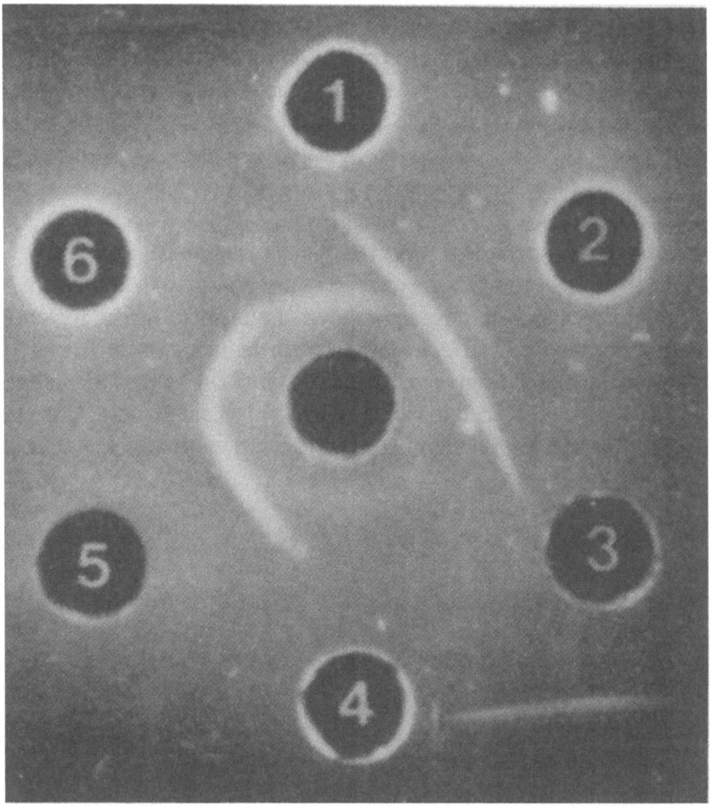

Figure 1. Ouchterlony plate with bovine anti-CA III in the center well. Outer wells: (1) human muscle CA III, (2) bovine muscle CA III, (3) rat red cell CA II, (4) rat red cell CA I, (5) rat liver CA III, (6) rat muscle CA III. (Taken from reference 2.)

electrophoresis at 30 V and 100 mA without cooling. The membrane is washed in Tris-buffered saline (TBS) for 10 min and then transferred to blocking solution, which varies depending on the type of membrane used. After blocking, the membrane is placed directly into primary antibody diluted in TBS containing 1% gelatin. The primary antibody concentration we use for our antibodies against rat CA II and CA III for Western blotting is 1/200. After 1–2 hr of incubation at room temperature in primary antibody, the membrane should be washed twice in TBS for 10 min each wash. If background is a problem, 0.05% Tween can be added to the washes.

After washing, the membrane is incubated for 1 hr in peroxidase- or alkaline phosphatase-conjugated goat anti-rabbit IgG second antibody (1:1000) in TBS plus 1% gelatin. If greater sensitivity is required, a biotin–avidin amplification step can be introduced at this point. Two 10-min washes as before are carried out before the membrane is put into substrate for the enzyme.[3,5] Colored bands are produced wherever the antibody has bound to protein on the membrane. Figure 2 shows a Western blot of CA III for rat liver homogenate transferred to nitrocellulose, together with a duplicate stain of the SDS–gel from which the blot was prepared.

2.3. Laurell Rocket Electrophoresis

Rocket electrophoresis is a quantitative technique for determining concentrations of an antigen in a complex mixture. It is less sensitive than Western blotting and requires more antibody. Figure 3 shows the result of using rocket electrophoresis to determine CA III concentrations in human muscle from fetuses of different ages. The method can detect in the region of 1–10 μg/ml in our hands. For human CA III, there is no need to carbamylate the protein to increase the negative charge on the molecule. For human erythrocytes, with which rocket electro-

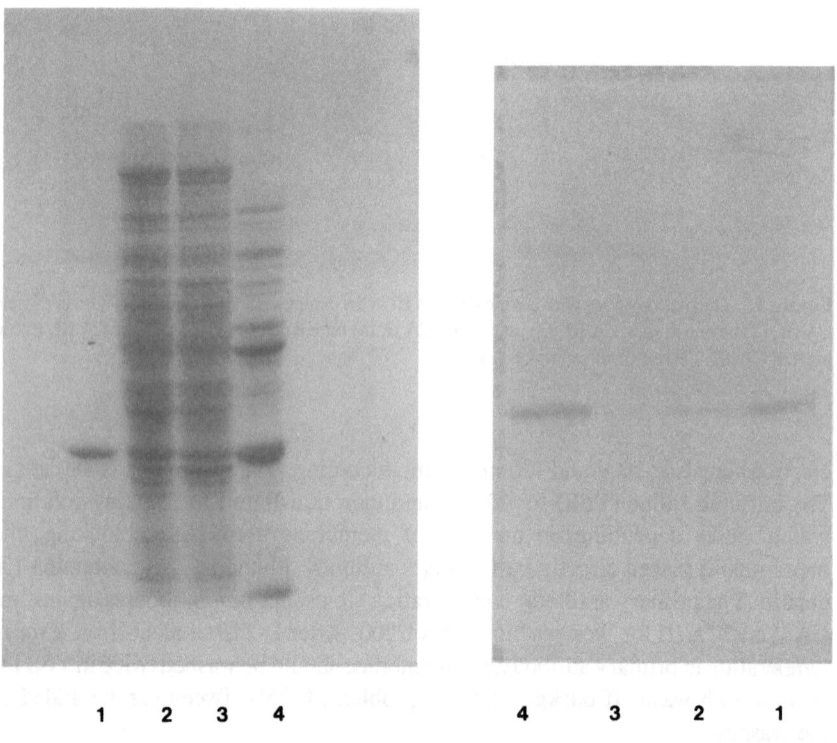

Figure 2. (Left) SDS–gel of (1) pure rat CA III, (2) male rat liver homogenate supernatant, (3) female rat liver supernatant, and (4) soleus muscle homogenate supernatant. (Right) Western blot of the same samples (numbers are of equivalent lanes), using specific anti-rat CA III antiserum at 1:200 dilution.

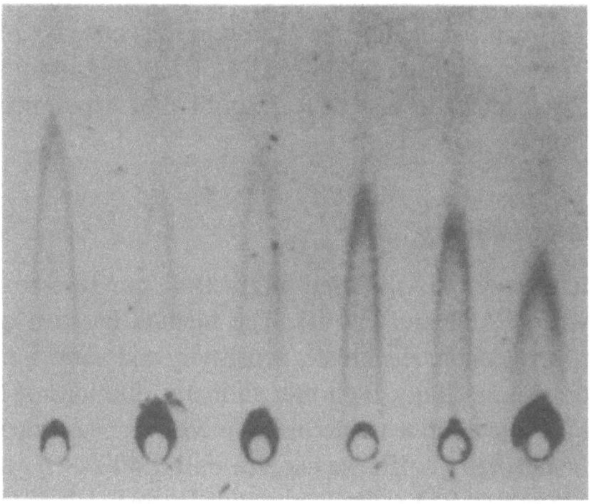

Figure 3. Laurell rocket electrophoresis of fetal muscle homogenates with anti-human CA III in the gel. Samples are 1:2 (w:v) skeletal muscle homogenate supernatants from fetuses of different gestational ages: (a) 25 weeks, (b) 23 weeks, (c) 20.5 weeks, (d) 18.5 weeks, (e) 15.5 weeks, (f) 12 weeks (1:1 homogenate).

phoresis has also been used,[6] it is not possible to detect CA II without carbamylation.

One percent agarose in 20 mM sodium barbitone–7 mM diethylbarbituric acid, containing 120 μl of anti-human CA III, is poured onto glass plates (0.27 × 7 × 10 cm). When the agarose is set, 5-μl wells are cut at about 1.5 cm from one end of the plate. Electrophoresis can be carried out overnight at 5–6 V/cm or for 3 hrs at 12–15 V/cm, at 10°C. Gels are washed twice in 0.1 M NaCl and then in distilled water, blotted dry by using 1-mM filter paper, and finally air dried. Staining is by 0.5% Coomassie blue in glacial acetic acid–ethanol–water (20:9:9), with the same solution minus dye as a destain.

2.4. Enzyme-Linked Immunosorbent Assays

Enzyme-linked immunosorbent assays (ELISAs) have not been used very often to measure CA concentrations; see reference 7 for ELISA estimation of human CA I concentrations in hemolysates. In these studies, antisera against human CA I were raised in rabbit and goat. Rabbit anti-human CA I IgG was used to coat the wells in a microtiter plate. The wells were washed, and a test sample or standard CA I was then applied to the plate. After incubation, the plates were washed and the goat anti-human CA I was applied and incubated. Unbound IgG was washed off, and then alkaline phosphatase-labeled anti-goat IgG was added.

Excess labeled IgG was washed off, and substrate was added. Color reaction was stopped after 20 min by addition of 3 M NaOH (50 μl). Sensitivity for human CA I in hemolysates was found to be about 5 ng/ml. For experimental details, see reference 7.

2.5. Radioimmunoassay

Radioimmunoassay (RIA) is used routinely in our laboratory, primarily to detect rat isozymes CA II and CA III. The method has two advantages over previously described techniques: greater sensitivity and a much reduced demand for antibody. For a high-affinity, high-titer antibody, dilutions of 1:60,000 can be used in RIA. This is the dilution that achieves a good standard curve for our CA II antibody. For the anti-rat CA III, we can use only 1:4000.

The principle behind RIA is the competition between labeled protein (usually tagged with ^{125}I) and unlabeled protein for a limiting quantity of antibody. The unlabeled protein is either a known amount of standard or an unknown amount in an experimental sample. The amount of labeled protein (tracer) is constant in each assay tube, and the more unlabeled protein there is in the sample, the less tracer is bound to the antibody at equilibrium. Antibody–antigen complexes are removed for counting by using a second antibody against the primary antibody.

Labeling of CA with ^{125}I can be achieved with a variety of methods. Iodogen,[8] lactoperoxidase,[9] and Bolton–Hunter[1] have all been used in our laboratory, but we routinely use a modification of the chloramine-T method, utilizing L-cysteine as a reductant instead of sodium metabisulfite. This procedure gives good incorporation and yields a stable tracer. The methods we use in our assays are outlined below.

To an Eppendorf tube at room temperature, add 10 μl of 0.25 M sodium phosphate, pH 7.5, 5 μl of 1 mg/ml CA in 50 mM sodium phosphate, pH 7.5, and 0.5 mCi of Na^{125}I. Add 16 μg of chloramine-T in 10 μl of d.w., let stand for 10 sec, and then add 56 μg of L-cysteine in 60 μl of d.w., 100 μg of potassium iodide in 10 μl of d.w., and 250 μl of 50 mM sodium phosphate, pH 7.5, containing 1 g/liter bovine serum albumin (BSA).

Once the reaction has been stopped, the mix is put through a G-50 column preequilibrated with column buffer (50 mM sodium phosphate, pH 7.5, with 1 g/liter BSA and 0.02% with respect to sodium azide). Fractions of 1 ml are collected at a flow rate of 15 ml/hr. The iodinated protein peak is first to elute, followed by free Na^{125}I. Fractions are stored at 4°C after separation.

Ideally, excess antibody should bind all of the counts in the tube; in practice, the percentage bound depends on the isozyme. For rat CA II, over 90% of the counts are bound by excess antibody; for CA III, the binding is around 30%. When the maximum binding has been determined, the antibody dilution to be used in the assay must be calculated from an antibody dilution curve. Serial dilutions of antibody are incubated with 20,000 cpm/100 μl of tracer, with 100 μl of buffer A (50 mM Tris, pH 7.5–0.25% BSA) replacing the unlabeled CA present in the actual assay. Many references quote antibody concentrations that achieve 50% binding (i.e., 50% of counts bound with excess antibody) as the amount of antibody to use in the assay. In fact, 1500 cpm above nonspecific binding (NSB) can be used; the procedure both achieves greater sensitivity and conserves antibody. Second antibody plus normal rabbit serum is added in buffer B (50 mM Tris–10 mM EDTA–0.1% BSA) and incubated at 4°C overnight before centrifugation.

When the desired antibody dilution has been determined, a standard curve can be established. Known standards are diluted in buffer A, and 100 μl of each is used in the assay tubes. Standards (and samples) are used in triplicate. Results are plotted as $B/B_0 \times 100$ against log of concentration. B_0 is the counts bound by antibody in the absence of competing unlabeled antigen (with 100 μl of buffer A in the tube to make up the volume), and B is the counts bound for each standard. Before calculation of B or B_0, the nonspecific binding (NSB) has to be subtracted. NSB is given by incubating second antibody with labeled CA, with 200 μl of buffer A in the tube to make up the volume. A typical standard curve is sigmoid, and dilutions of samples should be such that B/B_0 lies in the linear part of the curve. A 90% value for B/B_0 is considered the limit of detection for the assay.

To validate an assay, all substances liable to cross-react (e.g., other CA isozymes) must be tested, quality controls must be put in every assay, and samples must dilute in parallel with the standard curve. Once these criteria are satisfied, the assay can be used experimentally.

References

1. Bolton, A. E., and Hunter, W. M., 1973, *J. Endocrinol.* **55**:xxx–xxxi.
2. Carter, N. D., Hewett-Emmett, D., Jeffery, S., and Tashian, R. E., 1981, *FEBS Lett.* **128**:114–118.
3. Graham, R. C., and Karnovsky, M. J., 1966, *J. Histochem. Cytochem.* **14**:291–295.
4. Laemli, Y. N., 1979, *Nature* (London) **227**:680–685.
5. Leary, J. J., Brigati, D. J., and Ward, D. C., 1983, *Proc. Natl. Acad. Sci. USA* **80**:4045–4049.
6. Norgaard-Pedersen, B., 1973, *Scand J. Immunol.* **2**(Suppl. 1):125–131.
7. Shepherd, J. N., Spencer, N., Bidwell, D. E., and Voller, A., 1982, *Clin. Biochem.* **15**:248–253.
8. Sraker, S. J., and Speck, J. C., 1978, *Biochem. Biophys. Res. Commun.* **80**:849–854.
9. Thorell, J. J., and Johansson, B. G., 1971, *Biochim. Biophys. Acta* **251**:363–369.

PART III

Carbonic Anhydrases in Clinical Medicine

CHAPTER 13

Carbonic Anhydrase Inhibitors in Ophthalmology

MARIANNE E. FEITL and THEODORE KRUPIN

1. The Anterior Segment of the Eye

1.1. Carbonic Anhydrase and Aqueous Humor Production

Aqueous humor is secreted by the ciliary processes into the posterior chamber (Fig. 1). Intraocular pressure (IOP), normally less than 20 mm Hg, results from the balance between the production (approximately 2 μl/min) and the drainage (approximately 0.20 μl/min per mm Hg) of aqueous humor. Elevated IOP due to reduced drainage of aqueous humor, open-angle glaucoma, causes optic nerve damage; therefore, medical therapy is directed to pharmacologically increasing the drainage or decreasing the production of aqueous humor.[6,32,44]

Formation of aqueous humor involves the active transport of solutes across the ciliary processes (approximately 70) which project into the posterior chamber (Fig. 2).[5] Aqueous humor production results from osmotic pressure differences across epithelial layers which are developed by active electrolyte transport. Inhibition of cell-mediated processes lowers IOP in laboratory animals and humans; ouabain reduces sodium influx,[8] and carbonic anhydrase (CA) inhibitors reduce bicarbonate and sodium movement,[61,62] with a reduction of aqueous humor production.

CA is present in the anterior uvea,[55] including the ciliary epithelium,[28,57] within the membranes of the nonpigmented epithelium. In primate eyes, CA is also present in the capillary endothelium adjacent to the ciliary epithelium.[33] CA II is

MARIANNE E. FEITL • Department of Ophthalmology, Geisinger Medical Center, Danville, Pennsylvania 17822. ***THEODORE KRUPIN*** • Department of Ophthalmology, Northwestern University Medical School, Chicago, Illinois 60611.

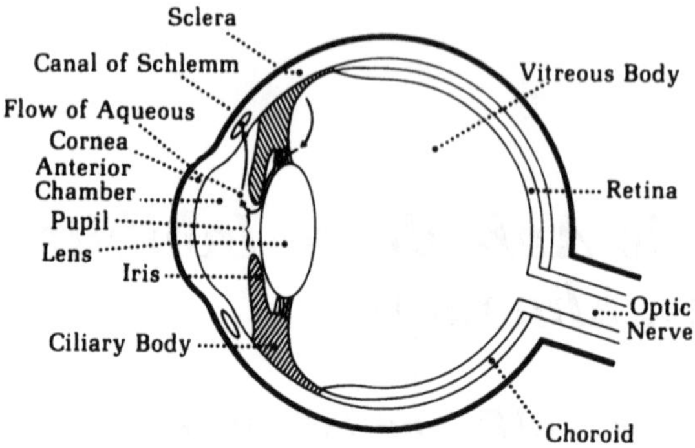

Figure 1. Schematic drawing of the globe.

the only isozyme found in the ciliary epithelium,[28,57] at a concentration of approximately 0.3 μM.

1.2. CA Inhibitors and IOP

Preliminary reports in 1952 on the sulfonamide acetazolamide as a new diuretic[34] and the discovery of CA in the anterior uvea[55] led these and other researchers to study the effect of this drug on IOP.[2,3,15] Acetazolamide was found to lower eye pressure in patients with glaucoma. Clinical studies demonstrated that orally administered acetazolamide did not alter the outflow of aqueous humor from the eye; the reduced IOP resulted from reduced aqueous humor inflow. Noninvasive fluorophotometric studies indicate that acetazolamide administration reduces aqueous humor formation by approximately 40–50%.[11]

CA inhibitors produce systemic acidosis, which by itself lowers IOP by reducing the rate of aqueous humor formation.[25,17,58]

Intravenous administration of a low dose of acetazolamide (125 mg) to acidotic patients lacking erythrocyte CA II[49] did not lower IOP, whereas it did reduce IOP of control subjects without causing systemic acidosis.[27] This finding was consistent with the hypothesis that CA II is present in the ciliary epithelium.

The effects of CA inhibitors have been determined in various species by measuring the rate of radioactive isotope movement from the plasma to the posterior chamber.[37,39] In dogs and monkeys, bicarbonate concentration in the posterior chamber aqueous humor normally is three-fold greater than in the plasma. After administration of CA inhibitors, sodium accession into the posterior chamber aqueous humor decreases from 2.4 to 1.3 mmol/min, bicarbonate accession decreases from 1.1 to 0.4 mmol/min, and chloride movement is only slightly changed (from 1.6 to 1.2 mmol/min).[61,62] The rate of water entry is also reduced.

The ciliary body has been mounted in Ussing–Zerahan-type chambers so that the ciliary body (blood) side faces one chamber half and the ciliary process

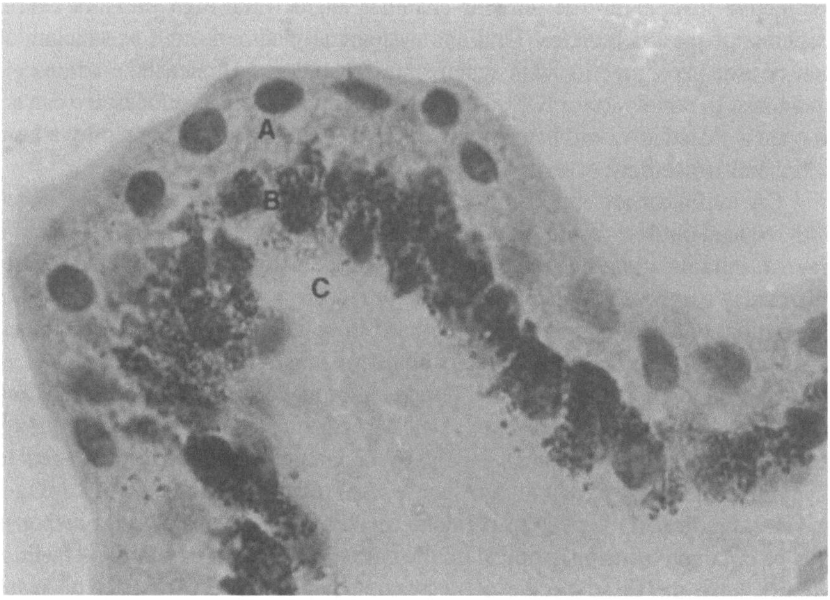

Figure 2. Histology of the ciliary process: inner nonpigmented ciliary epithelium (A); outer pigmented ciliary epithelium (B); vascular core (C).

(aqueous) side faces the other chamber half.[22,23,26] In this preparation with identical bathing solutions in both chambers, the ciliary process side is consistently negative with respect to the ciliary body side. Short-circuit current is dependent on Na^+, K^+, and HCO_3^- in the bathing solution. Deletion of HCO_3^- results in a reversal of the electrical parameters; the ciliary process side becomes positive in relation to the blood side. The addition of acetazolamide to the bathing solution results in a reduction of the transmembrane electrical potential and the short-circuit current.[26]

It is concluded that CA inhibitors directly affect transport mechanisms of the ciliary epithelium.[38] Several hypotheses could explain the decrease of active transport in this tissue: (1) a change in intracellular pH, which may inhibit Na^+–K^+ ATPase and sodium movement, (2) a decreased production of H^+ with reduced H^+–K^+ exchange, and (3) a reduced Cl^-–HCO_3^- exchange.

1.3. Glaucoma Therapy

CA inhibitors lower IOP in both acute and chronic situations. Because of their systemic side effects,[56] CA inhibitors are not the first drug of choice for long-term therapy.

In the treatment of acute closed-angle glaucoma, topical miotic agents are ineffective in opening the anterior chamber angle when high pressure causes ischemia of the iris muscles. Oral and systemically administered acetazolamide may be used in conjunction with hyperosmotic agents and a topical beta-adrenergic antagonist to rapidly lower IOP so that a topical miotic such as pilocarpine can act to open a closed anterior chamber angle. Further attacks are prevented by a laser peripheral iridectomy.

CA inhibitors are useful in the therapy of secondary glaucomas associated with reduced outflow facility. These conditions include elevated IOP due to ocular trauma, inflammation, and the early postoperative period after laser treatment or intraocular surgery. Treatment may be required for a relatively short interval in these situations. CA inhibitors may be used in combination with a topical beta-antagonist that also decrease aqueous humor production.

Chronic open-angle glaucoma requires prolonged therapy to lower IOP and control progression of the disease. Unless medically contraindicated (e.g., by respiratory or cardiac problems), a topical beta-antagonist is often the first agent in therapy. These drops have a high response rate, require only twice-daily dosage, and are usually well tolerated by patients. An additive pressure-lowering response may be obtained by using a topical miotic agent, which increases outflow facility. Miotics have the disadvantage of decreasing pupil size, which may decrease vision. Epinephrine compounds, which primarily act by increasing outflow facility, may also be used in the medical management of open-angle glaucoma. If topical therapy does not lower IOP sufficiently (or is poorly tolerated), a CA inhibitor may be added.

The usual dose of acetazolamide tablets is 250 mg every 6 hr. The sustained release capsule (500 mg) peaks at 8 hr and lasts for 12–24 hr.[48] A more uniform plasma concentration, with a lower plasma peak and fewer side effects, is obtained with the capsules. The dose for methazolamide tablets is 25–50 mg every 12 hr. Some patients who cannot tolerate acetazolamide may be able to take methazolamide successfully.[30] Many patients respond adequately to much lower doses of either methazolamide or acetazolamide, so therapy should be titrated to the lowest effective dose to minimize side effects and increase tolerance.[4,50]

Acetazolamide may be given intravenously in acute or preoperative situations. This treatment is particularly important when the patient may have nausea or vomiting associated with elevated IOP. An almost immediate pressure-lowering effect begins after intravenous administration. The intravenous dose is 250 mg every 6 hr.

1.4. Systemic Side Effects

Caution is warranted in the administration of CA inhibitors in patients with a history of sensitivity to other sulfonamides (e.g., sulfa antibiotics), since CA inhibitors may make them prone to dermatitis or anaphylaxis. Parasthesias of the fingers, toes, and the area around the mouth occur transiently in most patients.

Abdominal discomfort, nausea, and decreased appetite are common. These symptoms may be reduced by taking the medication with meals. Carbonated beverages may take on a metallic taste, probably because of CA inhibition in the saliva[13] or in the taste buds. Some of these side effects decrease with time, but they often necessitate discontinuation of the medication. Transient myopia has also been reported.[1,18] Since acetazolamide penetrates the lens capsule poorly, it is unlikely that the myopia is related to inhibition of CA in the lens.[16]

The more serious side effects include blood dyscrasias, kidney stones, and mental changes, including excitation, confusion, and depression.[24,30] CA inhibitors increase kidney stone formation by depressing renal citrate and magnesium excretion and increasing calcium availability for insoluble salts.[9,46] If a patient develops a stone while on CA therapy, the recurrence rate when the drug is continued is much higher than in patients with calculi not associated with CA therapy. The incidence of kidney stones is higher with acetazolamide than with methazolamide.[20]

The diuretic effect of acetazolamide is due to its action in the kidney. The resulting renal loss of HCO_3^- carries out Na^+, water, and K^+. The resulting hypokalemia is usually mild and transient. However, caution is mandated with concurrent use of chlorothiazide diuretics, digitalis, and systemic corticosteroids and in patients with adrenal insufficiency or hepatic cirrhosis. Plasma potassium should be monitored in these patients, and supplemental therapy should be given for significant hypokalemia.[4]

Methazolamide is metabolized mainly in the liver, and only 20–25% is excreted unchanged.[35] In the presence of hepatic disease, methazolamide is contraindicated. Metabolic acidosis occurs with higher doses of all of the agents as a result of bicarbonate depletion. Methazolamide has a weaker renal inhibitory effect than does acetazolamide, which results in less induced acidosis.[36] Patients with underlying medical conditions that prevent compensation for alterations in acid–base balance should not receive CA inhibitors. These medical states include renal failure, severe pulmonary obstructive disease, hepatic insufficiency, hyperchloremic acidosis, and adrenocortical insufficiency. Even the normal degree of induced metabolic acidosis may trigger the symptom complex of malaise, fatigue, weight loss, anorexia, depression, and decreased libido.[51] Systemic sodium bicarbonate treatment may reduce the induced acidosis and the symptom complex.[12]

Blood dyscrasias are a very rare but potentially fatal adverse effect.[54] Aplastic anemia has been reported as late as 5 or 6 years after initiation of therapy; however, most cases occurred within the first 3 months of therapy.[14,43] Although the mechanism appears to be idiosyncratic, blood dyscrasias may be reversible upon discontinuation of the drug when symptoms occur. A complete blood count should perhaps be done before the initiation of therapy and at periodic intervals during treatment[14]; this procedure is, however, controversial.[29,43]

CA inhibitors should be avoided during pregnancy and breast feeding of infants, since teratological effects have been demonstrated.[19,35,47,60]

2. Posterior Segment of the Eye

2.1. Carbonic Anhydrase and Retinal Pigment Epithelium

CA has been identified in human, mammalian, and vertebrate retina,[33,45,59] in the apical and basal membranes of the retinal pigment epithelium (Fig. 3),[7] in the capillary endothelial cells, and in Müller's cells.[31]

2.2. CA Inhibitors and the Retina

Inorganic anions, including the dye fluorescein, are actively removed from the vitreous cavity. Acetazolamide increases the rate of fluorescein disappearance.[52] Acetazolamide also increases the rate of subretinal fluid reabsorption in experimental retinal detachment[41] and increases the adhesion between the retina and pigment epithelium.[40] In isolated pigment epithelial preparations, acetazolamide reduces the transepithelial potential difference.[21,42]

2.3. CA Inhibitors and Cystoid Macular Edema

Cystoid macular edema (CME), or accumulation of extracellular fluid in the center of the macula, occurs as a secondary event after intraocular surgery and with

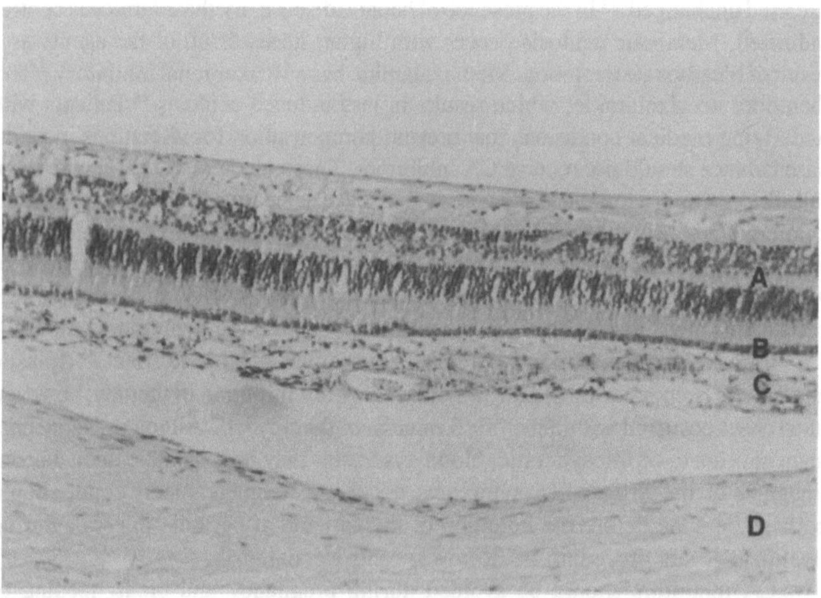

Figure 3. Histology of retina (A) and retinal pigment epithelium (B) with underlying vascular charotid (C) and sclera (D). The space between the choroid and sclera is a fixation artifact.

vascular, inflammatory, hereditary, and other diseases. CME, which can be transient or chronic, reduces visual acuity. Histopathologically, the cystoid spaces contain eosinophilic exudate and are found in the outer plexiform and inner nuclear layers of the retina.[53] The largest cysts develop around the fovea.

A recent prospective and unmasked study suggests that acetazolamide can modify or cause resolution of chronic macular edema in patients with inherited outer retinal disease or intraocular inflammation. In patients with primary vascular disorders, acetazolamide was of no benefit. The maintenance dose required to sustain the therapeutic response with improved vision was smaller than doses used to reduce IOP.[10]

References

1. Beasley, F. J., 1962, *Arch Ophthalmol.* **68**:490–491.
2. Becker, B., 1954, *Am. J. Ophthalmol.* **37**:13–15.
3. Becker, B., 1955, *Am. J. Ophthalmol.* **39**(Pt. 2):177–183.
4. Berson, F. G., and Epstein, D. L., 1980, *Perspect. Ophthalmol.* **4**:91–95.
5. Bill, A., 1973, *Exp. Eye Res.* **16**:287–296.
6. Caprioli, J., 1987, in: *Adler's Physiology of the Eye: Clinical Application*, 8th ed. (R. A. Moses and W. M. Hart, Jr., eds.), The C. V. Mosby Co., St. Louis, pp. 204–222.
7. Cohen, A. I., 1987, in: *Adler's Physiology of the Eye: Clinical Application*, 8th ed. (R. A. Moses and W. M. Hart, Jr., eds.), The C. V. Mosby Co., St. Louis, pp. 458–490.
8. Cole, D. F., 1960, *Br. J. Ophthalmol.* **44**:739–750.
9. Constant, M. A., and Becker, B., 1960, *Am. J. Ophthalmol.* **49**:929–934.
10. Cox, S. N., Hay, E., and Bird, A. C., 1988, *Arch. Ophthalmol.* **106**:1190–1195.
11. Dailey, R. A., Brubaker, R. F., and Bourne, W. M., 1982, *Am. J. Ophthalmol.* **93**:232–237.
12. Epstein, D. L., and Grant, W. M., 1977, *Arch. Ophthalmol.* **95**:1378–1382.
13. Fernley, R. T., this volume, Chapter 34.
14. Fraunfelder, F. T., Meyer, S. M., Bagby, G. C., Jr., and Dreis, M. W., 1985, *Am. J. Ophthalmol.* **100**:79–81.
15. Friedenwald, J. S., 1955, *Am. J. Ophthalmol.* **39**:59–64.
16. Friedland, B. R., and Maren, T. H., 1981, *Exp. Eye Res.* **33**:545–561.
17. Friedman, Z., Krupin, T., and Becker, B., 1982, *Invest. Ophthalmol. Visual Sci.* **23**:209–213.
18. Galin, M. A., Baras, I., and Zweifach, P., 1962, *Am. J. Ophthalmol.* **54**:237–240.
19. Hallsey, D. W., and Layton, W. M., 1969, *Proc. Soc. Exp. Biol.* **126**:6–12.
20. Kass, M. A., Kolker, A. E., Gordon, M., Goldberg, I., Gieser, D. K., Krupin, T. D., and Becker, B., 1981, Acetazolamide and urolithiasis, *Ophthalmology* **88**:261–265.
21. Kawasaki, K., Mukoh, S., Yonemura, D., Fujii, S., and Segawa, Y., 1986, *Doc. Ophthalmol.* **63**:375–381.
22. Kishida, K., Miwa, Y., and Wata, C., 1986, *Exp. Eye Res.* **43**:981–995.
23. Kishida, K., Sasabe, T. K., Iizuka, S., Manabe, R., and Otori, T., 1982, *Curr. Eye Res.* **2**: 149–157.
24. Kristinsson, A., 1967, *Br. J. Ophthalmol.* **51**:348–349.
25. Krupin, T., Oestrich, C. J., Bass, J., Podos, S. M., and Becker, B., 1977, *Invest. Ophthalmol. Visual Sci.* **16**:997–1001.
26. Krupin, T., Reinach, P. S., Candia, O. A., and Podos, S. M., 1984, *Exp. Eye Res.* **38**:115–123.
27. Krupin, T. D., Sly, W. S., Whyte, M. P., and Dodgson, S. J., 1985, *Am. J. Ophthalmol.* **99**: 396–399.
28. Kumpulainen, T., 1983, *Histochemistry* **77**:281–284.

29. Lichter, P. R., 1988, *Ophthalmology* **95**:711–712.
30. Lichter, P. R., Newman, L. P., Wheeler, N. C., and Beall, O. P., 1978, *Am. J. Ophthalmol.* **85**: 495–502.
31. Linser, P. J., and Cohen, J. L., this volume, Chapter 26.
32. Lütjen-Drecoll, E., 1984, in: *Glaucoma: Applied Pharmacology in Medical Treatment* (S. M. Drance and A. H. Neufeld, eds.), Grune & Stratton, Inc., Orlando, Fla., pp. 23–33.
33. Lütjen-Drecoll, E., Lonnerholm, G., and Eichhorn, M., 1983, *Graefes Arch. Clin. Exp. Ophthalmol.* **220**:285–291.
34. Maren, T. H., 1952, *Trans. N.Y. Acad. Sci.* **15**(Ser. II):53–55.
35. Maren, T. H., 1967, *Physiol. Rev.* **47**:595–781.
36. Maren, T. H., 1977, *Exp. Eye Res.* **25**(Suppl.):245–247.
37. Maren, T. H., 1984, *Ann. N.Y. Acad. Sci.* **429**:10–17.
38. Maren, T. H., 1987, *Dev. Drug. Res.* **10**:255–276.
39. Maren, T. H., Maywood, J. R., Chapman, S. K., and Zimmerman, T. J., 1977, *Invest. Ophthalmol. Visual Sci.* **16**:730–742.
40. Marmor, M. F., and Maack, T., 1986, *Invest. Ophthalmol. Visual Sci.* **23**:121–124.
41. Marmor, M. F., and Negi, A., 1986, *Arch. Ophthalmol.* **104**:1674–1677.
42. Miller, S. S., and Steinberg, R. H., 1977, *Exp. Eye Res.* **25**:235–248.
43. Mogk, L. G., and Cyrlin, M. N., 1988, *Ophthalmology* **95**:768–771.
44. Moses, R. A., 1987, in: *Adler's Physiology of the Eye: Clinical Application*, 8th ed. (R. A. Moses and W. M. Hart, Jr., eds.), The C. V. Mosby Co., St. Louis, pp. 223–245.
45. Musser, G. I., and Rosen, S., 1973, *Exp. Eye Res.* **15**:105–119.
46. Parfitt, A. M., 1969, *Arch. Int. Med.* **124**:736–740.
47. Samples, J. R., and Meyer, S. M., 1988, *Am. J. Ophthalmol.* **106**:616–622.
48. Shields, M. B. (ed.), 1982, *A Study Guide for Glaucoma*, Williams & Wilkins, Baltimore, pp. 428–434.
49. Sly, W. S., Hewett-Emmett, D., Whyte, M. P., Yu, Y.-S. L., and Tashian, R. E., 1983, *Proc. Natl. Acad. Sci. USA* **80**:2752–2756.
50. Stone, R. A., Zimmerman, T. J., Shin, D. H., Becker, B., and Kass, M. A., 1977, *Am. J. Ophthalmol.* **83**:674–679.
51. Stoudemire, A., and Houpt, J. L., 1984, *Psych. Med.* **1**:353–355.
52. Tsuboi, S., and Pederson, J., 1985, *Arch. Ophthalmol.* **103**:1557–1558.
53. Tso, M. O. M., 1981, *Ophthalmologica* **183**:46–54.
54. Wisch, N., Fischbein, F. I., Siegel, R., Glass, J. L., and Leopold, I., 1973, *Am. J. Ophthalmol.* **75**:130–132.
55. Wistrand, P. J., 1951, *Acta Physiol. Scand.* **24**:144–148.
56. Wistrand, P. J., 1984, *Ann. N.Y. Acad. Sci.* **429**:609–619.
57. Wistrand, P. J., and Garg, L. C., 1979, *Invest. Ophthalmol. Visual Sci.* **18**:802–806.
58. Wistrand, P. J., and Maren, T. H., 1960, *Am. J. Ophthalmol.* **50**:291–297.
59. Wistrand, P. J., Schenholm, M., and Lonnerholm, G., 1986, *Invest. Ophthalmol. Visual Sci.* **27**:419–428.
60. Worsham, G. F., Beckman, E. N., and Mitchell, E. M., 1978, *J. Am. Med. Assoc.* **240**:251–253.
61. Zimmerman, T. J., Garg, L. C., Vogh, B. P., and Maren, T. H., 1976, *J. Pharmacol. Exp. Ther.* **196**:510–516.
62. Zimmerman, T. J., Garg, L. C., Vogh, B. P. and Maren, T. H., 1976, *J. Pharmacol. Exp. Ther.* **199**:510–517.

CHAPTER 14

The Eye

Topical Carbonic Anhydrase Inhibitors

ERIK A. LIPPA

Although systemically administered carbonic anhydrase (CA) inhibitors are effective in lowering intraocular pressure (IOP) in patients with glaucoma, patient acceptance and compliance are limited by side effects such as fatigue, anorexia, gastrointestinal disturbances, paresthesias, renal calculi, and loss of libido, and there are concerns regarding the rare association of CA inhibitors with occurrences of aplastic anemia. The goal of producing an efficacious topical CA inhibitor for the treatment of glaucoma without the side effects caused by systemic CA inhibition has proved elusive.

Investigations of topical administration of the marketed systemic agents (acetazolamide, methazolamide, ethoxzolamide, and dichlorphenamide) have demonstrated at most a minimal effect on IOP in animal models and in humans.[6,9,13,16,17,19,21,38,43,44,47] It is probable that this lack of IOP-lowering activity is due to insufficient ocular penetration of these compounds, since their mechanism of action is generally believed to depend on almost total inhibition of CA in the ciliary processes of the eye,[37,38,42,66] the site of aqueous humor formation.

Later, when topical CA inhibitors effective in lowering IOP in rabbits were finally developed, a new difficulty arose. Compounds that had demonstrated marked activity in rabbits showed minimal activity in humans. The degree to which this disparity is related to a slower blink rate and different tear dynamics between species or to other *in vivo* ocular penetration determinants, such as tissue-specific or formulation-dependent parameters, is unknown. Although a monkey model may be somewhat more predictive of results in humans, disparities also occur.

ERIK A. LIPPA • Clinical Research, Merck Sharp & Dohme Research Laboratories, West Point, Pennsylvania 19486, Department of Ophthalmology, Jefferson Medical College of Thomas Jefferson University, Philadelphia, Pennsylvania 19107, and Department of Ophthalmology, University of Pennsylvania, Philadelphia, Pennsylvania 19104.

For many years, the development of effective, novel topical CA inhibitors has focused on improving transcorneal penetration rates by achieving an appropriate balance of lipid–water solubility. Recent evidence[45] suggests that noncorneal penetration is also involved. Moreover, the degree of importance of properties such as formulation pH, tonicity, and viscosity, pigment binding, and CA isozyme selectivity can only now be fully explored since the advent of topically effective CA inhibitors in humans.

1. Compounds Marketed for Systemic Use

Significant IOP-lowering activity in animals has been observed only at high concentrations (5–10%) achievable only at an alkaline pH unsuitable for clinical use.[15,35,57,58] Limited ocular penetration in humans was observed when the concentration of methazolamide in aqueous humor collected 1 hr after administration of a 6 mM methazolamide solution was less than 0.6 μM, substantially below that anticipated to be necessary for IOP-lowering activity.[14]

It became increasingly apparent over the past decade that a topically effective CA inhibitor depended on the development of drugs with physicochemical properties (especially differential lipid–water solubility) different from those of the available drugs.[39,43] These agents would have to achieve greater ocular penetration as well as allow stable formulation at a sufficiently high concentration at a pH consistent with ocular tolerability. The first of these new compounds shown to decrease IOP in normotensive rabbits was trifluoromethazolamide, chosen over methazolamide for increased lipophilic properties and high water solubility while still retaining potency.[43]

2. Trifluoromethazolamide

In 1983, Maren *et al.* showed that corneal contact with a 3% trifluormethazolamide solution (pH 7.8) for 10 min in normotensive rabbits induced an IOP decrease that averaged 1.7 mm Hg over the subsequent 5 hr.[43] Topical administration of 5 consecutive drops of a 2.5% suspension (pH 6.5) reduced IOP in normotensive albino rabbits.[57] Later, a single drop of 2% suspension was shown to decrease IOP of normotensive rabbits by 3.1 mm Hg over that seen in the contralateral untreated eye at 1 hr postdose.[44] Unfortunately, trifluormethazolamide is rapidly hydrolyzed to a much weaker inhibitor.

Compounds were subsequently designed and developed expressly for topical ocular administration in an attempt to enhance ocular penetration over that of known compounds while maintaining or improving CA inhibition activity. These novel compounds, which have been investigated in animals and are related to studies in humans, can best be categorized as benzothiazole-2-sulfonamides, benzothiophene-2-sulfonamides, thiophene-2-sulfonamides, and thienothiopyran-2-sulfonamides. Their chemical structures are given in Table I. The earliest of these, benzothiazole-2-sulfonamides, are structurally related to ethoxzolamide.

3. Benzothiazole-2-Sulfonamides

3.1. L-645,151

L-645,151 is the *O*-pivaloyl ester of L-643,799. During ocular penetration, L-645,151 is hydrolyzed to the active metabolite L-643,799. The concentrations inhibiting enzymatic activity of human CA II by 50% *in vitro* (IC_{50}) are 4 and 6 nM for L-645,151 and L-643,799, respectively.[54,67] One drop of 2% L-645,151 given to normotensive albino rabbits produced a 2.2-mm-Hg decrease in IOP in comparison with the contralateral untreated eye.[3] A single drop of L-645,151 at concentrations as low as 0.25% lowered the elevated IOP of alpha-chymotrypsinized rabbits, and after administration of a 2% suspension of L-645,151, a decrease of greater than 9 mm Hg at 4 hr postdose was observed.[58] However, L-645,151 was noted to be a potential contact sensitizer, as is the case with several benzothiazole-2-sulfonamides.[55]

3.2. 6-Hydroxyethoxzolamide

Slight activity was found in normotensive rabbits when 3 consecutive drops of a 1% suspension was given; this activity was potentiated somewhat when the compound was formulated in a gel.[30] The activity of the gel formulation was confirmed in laser-induced ocular hypertensive monkeys.[11]

3.3. Aminozolamide

Both a 3% suspension and a 3% gel formulation of aminozolamide were active in alpha-chymotrypsinized rabbits, with peak decreases of 16 and 16.7 mm Hg at 8 and 10 hr postdose, respectively.[50] This activity was hypothesized to be secondary to prolonged retention of an active intraocular metabolite (6-acetamido-2-benzothiazole sulfonamide) in the iris plus ciliary body. Activity was also demonstrated in laser-induced ocular hypertensive monkeys.[11]

Some degree of activity in patients with ocular hypertension was reported after administration of 3% aminozolamide gel.[31] A single dose induced a peak decrease of 6.5 mm Hg at 4 hr postdose versus a decrease of 1.1 mm Hg in the contralateral untreated eye; however, treatment with vehicle alone resulted in a decrease of 3.5 mm Hg versus a decrease of 3.4 mm Hg in the fellow untreated eye. Moreover, single and multiple doses of a 2.5% suspension of aminozolamide did not prove effective in lowering IOP in patients,[29] even though activity of the

TABLE I

Carbonic Anhydrase Inhibitors Recently Investigated for Topical Activity[a]

Compound	Chemical structure	IC_{50} (nM) to CA II	pKa	Partition coefficient (pH 7.4, octanol buffer)
L-643,799	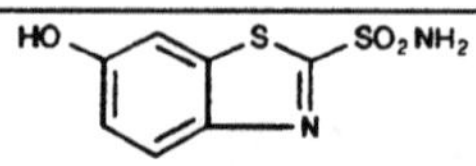	6.0	7.10, 8.54	13
L-645,151	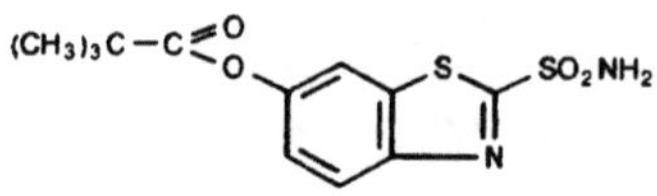	4.0	7.75	280
L-650,719	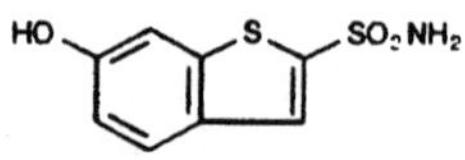	8.8	8.48, 9.90	19
L-651,465	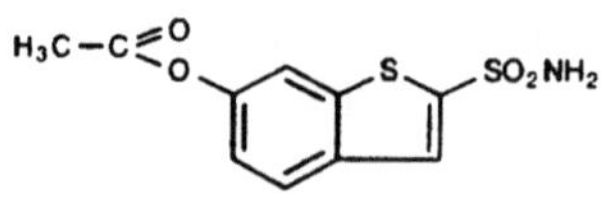	4.0	8.85	19
L-662,583	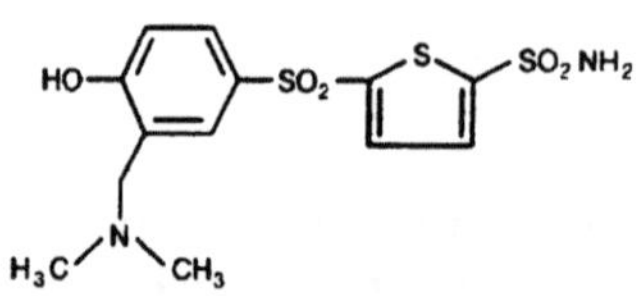	1.4	5.89, 8.94, 10.32	0.49
L-654,230	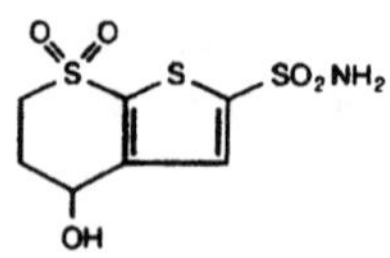	13.0	8.3	0.45
L-671,152	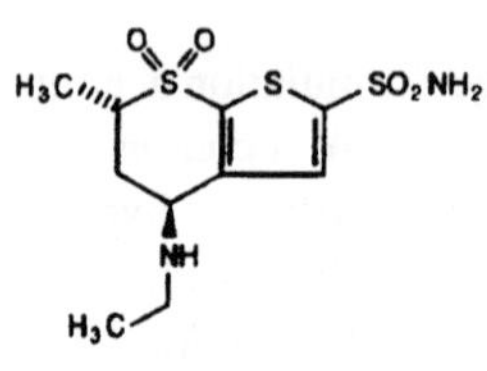	2.2	6.16, 8.60	1.5
MK-927	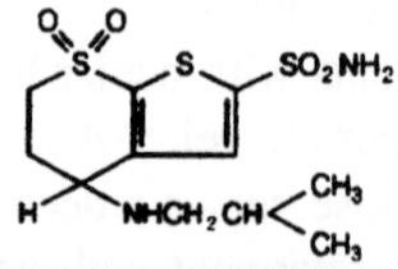	13.0	5.6, 8.6	8

TABLE I (*Continued*)

Compound	Chemical structure	IC_{50} (nM) to CA II	pKa	Partition coefficient (pH 7.4, octanol buffer)
MK-417	*S*-enantiomer of MK-927	7.7 (52 for *R*-enantiomer)	5.6, 8.6	8
Aminozolamide	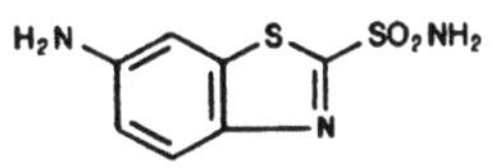	7.0	7.3	4
Acetazolamide	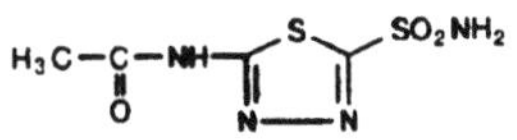	9.9	7.2, 8.8	0.14

[a]L-662,583, L-671,152, MK-927, and MK-417 were tested as the hydrochloride salt. IC_{50} values were determined by the procedure in reference 48 except that preincubation with 1 nM CA II was 1 min at 3°C. (Acetazolamide pK_a, from reference 32; other values, from reference 51.)

suspension had been demonstrated in both rabbits and monkeys. A single dose of the 3% gel caused only a small decrease in aqueous humor formation in normal volunteers, a 9% greater decrease than that of the contralateral placebo-treated eye.[26] A multiple-dose aqueous humor flow study was discontinued because of ocular side effects, including marked conjunctival injection and follicular conjunctivitis, after four doses of drug.

4. Benzothiophene-2-Sulfonamides

4.1. L-650,719

L-650,719 was chosen for further study *in vivo* among a group of novel benzothiophene-2-sulfonamides being investigated as potential topical CA inhibitors. The concentration inhibiting enzymatic activity of human CA II by 50% *in vitro* is 8.8 nM.[20,56] A single drop of suspensions in concentrations up to 8% of L-650,719 given to normotensive rabbits resulted in a peak decrease in IOP of 2.6 mm Hg, with duration of activity through 4 hr postdose. Maximal effect was achieved at the 2% concentration.[4,40] A single drop of a 0.1% L-650,719 suspension caused a peak decrease in IOP of 7 mm Hg in alpha-chymotrypsinized rabbits.[20]

Despite the substantial activity demonstrated in rabbits, multiple doses of a 2% suspension of L-650,719 did not lower IOP in normal volunteers.[65] Moreover, in a study of patients with ocular hypertension, single-dose 2% L-650,719 induced only a slight decrease in IOP at 3 hr postdose, a 2.7 mm Hg greater decrease than that observed in the contralateral untreated eye. In addition, the 4.2-mm-Hg decrease (12.2%) observed post 2% L-650,719 was less than half the decrease observed the next day after 250 mg of acetazolamide given orally (8.7 mm Hg, 26.1%).[46]

The reason for this disparity in topical activity between rabbits and humans was then pursued. Pharmacologic activity of L-650,719 in humans was demonstrated when a single oral dose of 100 or 200 mg of L-650,719 given to normal volunteers produced a decrease in IOP of 4.7 mm Hg (23.6%) at 4 hr postdose and an increase in urine pH and potassium and sodium excretion.[10] Next, ocular penetration in humans was investigated by collecting aqueous humor samples at the start of cataract surgery 20–100 min after 1–3 drops of 2% L-650,719. The results revealed a maximum aqueous concentration of 3.4 μM,[28] less than would be anticipated to be compatible with substantial IOP-lowering activity. An attempt was made to enhance ocular penetration by investigating esters of L-650,719.

4.2. L-651,465

L-651,465 is the acetate ester of L-650,719. During ocular penetration, L-651,465 is hydrolyzed to the active metabolite L-650,719 *in vitro*.[56] A 2% suspension of L-651,465 was investigated in patients with ocular hypertension.[22] Only limited activity was demonstrated. Moreover, a 2.6-mm-Hg decrease (9.3%) from baseline was observed 3 hr after treatment with a 2% L-651,465 suspension, compared with a 3.5-mm-Hg decrease (13.8%) in the same eyes 3 hr after administration of 250 mg of acetazolamide orally. Attention was then directed to developing classes of water-soluble compounds more potent than acetazolamide.

5. Thiophene-2-Sulfonamide (L-662,583)

L-662,583 at 0.7 nM inhibits enzymatic activity of human CA II by 50% *in vitro*; the corresponding concentrations were 10.8 and 21.2 nM for acetazolamide and methazolamide, respectively. One drop of a 2% solution of L-662,583 was shown to decrease the IOP of normotensive albino rabbits by up to 2.3 mm Hg.[60] In the alpha-chymotrypsinized rabbit model, 1 drop of 0.5% L-662,583 decreased IOP by up to 9.8 mm Hg; in laser-induced ocular hypertensive monkeys, 2% L-662,583 caused a peak decrease of 8.3 mm Hg at 3 hr postdose. A slightly hypotonic formulation was chosen for clinical trials with this compound on the basis of enhanced ocular penetration in rabbits.[53] Unfortunately, excellent activity in both rabbit and monkey models of ocular hypertension was not predictive of activity in humans as multiple doses of 2% L-662,583 (pH 5.2) did not lower IOP in normal volunteers.[52]

Recently, in a further attempt to develop potent CA inhibitors that are water soluble to at least 1–2% within a physiologically tolerable pH range, novel thienothiopyran-2-sulfonamides were investigated.[48]

6. Thienothiopyran-2-Sulfonamides

6.1. L-654,230

L-654,230 at 13 nM inhibits enzymatic activity of human CA II by 50% *in vitro*.[48] However, L-654,230 has limited water solubility, and neither a single dose nor three doses of a 0.8% solution of L-654,230 resulted in a significant lowering of IOP in patients with ocular hypertension.[46] Further alterations in the chemical structure (see Table I) produced enhanced water solubility so that the second thienothiopyran-2-sulfonamide administered to humans could be formulated at concentrations up to 2%.

6.2. MK-927

The racemic compound MK-927 was found both to be highly water soluble and to demonstrate excellent activity, with a concentration inhibiting the enzymatic activity of human CA II by 50% *in vitro* of 13 nM.[1,49] One drop of a 2% MK-927 solution induced an IOP decrease of 3 mm Hg in normotensive albino rabbits at 2 hr postdose.[41] Single doses of 0.01, 0.1, and 0.5% solutions of MK-927 reduced the IOP of alpha-chymotrypsinized rabbits by 2.5, 4.8, and 6.5 mm Hg, respectively, with a duration of activity for 0.5% MK-927 of approximately 6 hr.[59] Of interest is the suggestion that MK-927 gains access to the ciliary processes in part through a noncorneal route.[45] Furthermore, MK-927 binds to ocular pigment to a limited degree and demonstrates greater activity in homogenates of iris plus ciliary body in pigmented rabbits than in albino rabbits.[36]

The multiple-dose activity of 2% MK-927 was investigated in laser-induced ocular hypertensive monkeys, showing a peak decrease in IOP at 3 hr postdose of 11.7 mm Hg on treatment day 1. The effect increased after multiple doses, with a decrease from baseline IOP of 14.8 mm Hg at 3 hr postdose after 5 days of twice-daily therapy.[63]

Three consecutive drops of 2% MK-927, the maximum available concentration at pH 5.2, administered to normal volunteers resulted in a 29.6% decrease in IOP measured 4 hr after the first dose from that determined 20 hr predose.[34]

The effect of 3 drops of 2% MK-927 (1 drop every 10 min) was studied in patients with glaucoma or ocular hypertension.[33] A peak decrease in IOP of 26.7% from that immediately predose, representing a decrease of 7.7 mm Hg, occurred at 6 hr postdose. A contralateral effect was not observed in this crossover study, implying that the effect was due to direct ocular penetration of the compound. Single doses were then shown to be active in patients, with a peak reduction in IOP of 33.1% from that immediately predose occurring at 4.5 hr postdose (peak decrease of 10.5 mm Hg).[23] Activity was still present at 8 hr postdose.

A single-dose dose–response study of 2, 1, and 0.5% MK-927 in patients indicated that a clear dose–response relationship was present, with 0.5% MK-927 being a single-dose minimal-effect or no-effect dose, though the activity of even the 2% solution appeared to be waning substantially by 10 to 12 hr postdose.[24,27] A multiple-dose study of once-daily versus twice-daily therapy in patients indicated that MK-927 possesses a small degree of residual activity even at 24 hr postdose.[25]

A 2-week multiple-dose dose–response study of 2, 1, and 0.5% MK-927 given every 12 hr to patients demonstrated a significant dose–response relationship.[62] MK-927 at 0.5% was a minimal-effect dose. Both 2 and 1% MK-927 were active through 12 hr postdose, with peak decreases of 20.4 and 17.7%, respectively, observed at 2 hr postdose on treatment day 14 from time-matched IOP prestudy. A substantial increase in activity from day 1 to day 14 was not observed, and the degree of morning trough activity seen at 12 hr postdose (an 11% decrease in IOP for 2% MK-927) would be insufficient for twice-daily monotherapy treatment of patients.

To further increase the clinical activity demonstrated by MK-927, the compound was resolved into its *S*- and *R*-enantiomers.

6.3. MK-417

MK-417, the *S*-enantiomer of MK-927, has been shown to be more active *in vitro* than the *R*-enantiomer, with the concentrations inhibiting enzymatic activity of human CA II by 50% *in vitro* being 7.7 and 52 nM for the *S*- and *R*-enantiomers, respectively.[1] Moreover, MK-417 demonstrates a high degree of selectivity for CA II versus CA I.

MK-417 at 1% is more potent than 1% MK-927 in patients with glaucoma or ocular hypertension, producing a peak decrease in IOP of 23.9% at 6 hr postdose versus a decrease of 19.8% for 1% MK-927.[12] The multiple-dose efficacies of 1.8% MK-417 (the maximum available concentration at pH 5.2) and 2% MK-927 given every 12 hr were then compared in patients.[18] MK-417 at 1.8% was slightly more active than MK-927 at 2%, inducing a peak decrease of 19.9% at 2 hr postdose on treatment day 14 from the time-matched IOP prestudy; for MK-927, a decrease of 17.8% was observed.

No systemic biochemical effects were observed when normal volunteers were given 1.8% MK-417 or 2% MK-927 four times daily in each eye for 2 weeks.[8] A pilot pharmacokinetics study indicated half-lives in red blood cells of 0.57 and 300 hr for *R*-enantiomer and *S*-enantiomer (MK-417), respectively, with the extrapolated blood levels at steady state anticipated to be too low to induce the metabolic effects seen with oral CA inhibitors.[7] Further clinical studies are ongoing.

6.4. L-671,152

Recent chemical structural refinement around MK-417 led to the development of L-671,152, which is both more active and more CA II selective than MK-417[2] *in vitro* and possesses greater ocular hypotensive activity in both rabbits and monkeys than does MK-927.[61,64] Clinical studies on this promising compound have been initiated.

7. Summary

Thus investigations during the past several years have brought us ever closer to the goal of an efficacious topical CA inhibitor for the treatment of glaucoma. Several novel CA inhibitors have been synthesized and have been specifically designed and developed for ocular hypotensive activity after topical ocular administration. Pharmacologic studies have helped to elucidate many of the factors critical to enhancing both ocular penetration and IOP-lowering activity. Clinical trials have provided the vital feedback necessary for compound optimization. Future clinical research with these or other compounds will be required to establish whether the full potential of a topical CA inhibitor can be met, with maximal local activity and minimal or no systemic effect.

References

1. Baldwin, J. J., Ponticello, G. S., Sugrue, M. F., Mallorga, P. J., Randall, W. C., Schwam, H., Springer, J. P., Smith, G. M., and Murcko, M., 1988, abstract 95, Medicinal Chemistry Section, 3rd Chemical Congress of North America, Toronto, Canada, June 5, 1988.
2. Baldwin, J. J., Ponticello, G. S., Murcko, M., Springer, J. P., Randall, W. C., Schwam, H., and Sugrue, M. F., 1989, *Invest. Ophthalmol. Visual Sci.* **30**(Suppl.):374.
3. Bar-Ilan, A., Pessah, N. I., and Maren, T. H., 1986, *J. Ocular Pharmacol.* **2**:109–120.
4. Bar-Ilan, A., Pessah, N. I., and Maren, T. H., 1989, *J. Ocular Pharmacol.* **5**:99–110.
5. Becker, B., 1954, *Am. J. Ophthalmol.* **37**:13–15.
6. Becker, B., 1955, *Am. J. Ophthalmol.* **39**:177–184.
7. Biollaz, J., Lippa, E. A., Buclin, T., Winchell, G., Matuszewski, B. K., Munafo, A., Piguet, M., Brunner-Ferber, F., and Schelling, J. L., 1989, *Eur. J. Clin. Pharmacol.* **36**(Suppl.):A182.
8. Buclin, T., Lippa, E. A., Biollaz, J., Brunner-Ferber, F., Schoeneich, M., Faggionni, R., Munafo, A., and Schelling, J. L., 1989, *Eur. J. Clin. Pharmacol.* **36**(Suppl.):A188.
9. De Feo, G., Piccinelli, D., Putzolu, S., and Silvestrini, B., 1975, *Arzneim. Forsch.* (*Drug Res.*) **25**:806–809.
10. Calissendorff, B., personal communication.
11. DeSantis, L., Sallee, V., Barnes, G., Schoenwald, R., Barfknecht, C., Duffel, M., and Lewis, R., 1986, *Invest. Ophthalmol. Visual Sci.* **27**(Suppl.):179.
12. Diestelhorst, M., Béchetoille, A., Lippa, E. A., Brunner-Ferber, F., and Krieglstein, G. K., 1989, *Invest. Ophthalmol. Visual Sci.* **30**(Suppl.):23.
13. Duffel, M. W., Ing, I. S., Segarra, T. M., Dixson, J. A., Barfknecht, C. F., and Schoenwald, R. D., 1986, *J. Med. Chem.* **29**:1488–1494.
14. Edelhauser, H. F., and Maren, T. H., 1988, *Arch. Ophthalmol.* **106**:1110–1115.
15. Flach, A. J., Peterson, J. S., and Seligmann, K. A., 1984, *Am. J. Ophthalmol.* **98**:66–72.
16. Foss, R. H., 1955, *Am. J. Ophthalmol.* **39**:336–339.
17. Friedman, Z., Allen, R. C., and Ralph, S. M., 1985, *Arch. Ophthalmol.* **103**:963–966.
18. George, J.-L., Sirbat, D., Lesure, P., Lippa, E. A., Royer, J., Bron, A., Greve, E., Gunning, F., Flament, J., Benichou, C., Buntinx, A., and Brunner-Ferber, F., 1989, *Invest. Ophthalmol. Visual Sci.* **30**(Suppl.):98.
19. Gloster, J., and Perkins, E. S., 1955, *Br. J. Ophthal.* **39**:647–658.
20. Graham, S. L., Shepard, K. L., Anderson, P. S., Baldwin, J. J., Best, D. B., Christy, M. E., Freedman, M. B., Gautheron, P., Habecker, C. N., Hoffman, J. M., Lyle, P. A., Michelson, S. R.,

Ponticello, G. S., Robb, C. M., Schwam, H., Smith, A. M., Smith, R. L., Sondey, J. M., Strohmaier, K. M., Sugrue, M. F., and Varga, S. L., 1989, *J. Med. Chem.* **32**:2548–2554.
21. Green, H., Bocher, C. A., Calnan, A. F., and Leopold, I. H., 1955, *Arch. Ophthalmol.* **53**: 463–471.
22. Greve, E., personal communication.
23. Hennekes, R., Pfeiffer, N., Lippa, E. A., Garus, H., Grehn, F., and Jaeger, A., 1988, *Invest. Ophthalmol. Visual Sci.* **29**(Suppl.):82.
24. Higginbotham, E., 1988, Glaucoma Management in the 1990's, Chicago, November 4, 1988.
25. Higginbotham, E., Kao, S. F., Kass, M., Weinreb, R., Lippa, E. A., Skuta, G., Reiss, G., Shaw, B., Batenhorst, R., Wilensky, J., Lichter, P., 1989, *Invest. Ophthalmol. Visual Sci.* **30**(Suppl.):23.
26. Kalina, P. H., Shetlar, D. J., Lewis, R. A., Kullerstrand, L. J., and Brubaker, R. F., 1988, *Ophthalmology* **95**:772–777.
27. Kass, M. A., 1988, American Academy of Ophthalmology Annual Meeting, Las Vegas, October 11, 1988.
28. Kass, M. A., personal communication.
29. Lewis, R. A., Schoenwald, R. D., and Barfknecht, C. F., 1988, *J. Ocular Pharmacol.* **4**:215–219.
30. Lewis, R. A., Schoenwald, R. D., Eller, M. G., Barfknecht, C. F., and Phelps, C. D., 1984, *Arch. Ophthalmol.* **102**:1821–1824.
31. Lewis, R. A., Schoenwald, R. D., Barfknecht, C. F., and Phelps, C. D., 1986, *Arch. Ophthalmol.* **104**:842–844.
32. Lindskog, S., 1969, in: *CO_2: Chemical, Biochemical, and Physiological Aspects*, National Aeronautics and Space Administration, Special Publication SP-188,157, Washington, D.C.
33. Lippa, E. A., Hofmann, H. M., Feicht, B., Bron, A., Royer, J., Brunner-Ferber, F., and von Denffer, H., 1988, Third Congress of the European Glaucoma Society, Estoril, Portugal, May 24, 1988.
34. Lippa, E. A., von Denffer, H. A., Hofmann, H. M., and Brunner-Ferber, F. L., 1988, *Arch. Ophthalmol.* **106**:1694–1696.
35. Lotti, V. J., Schmitt, C. J., and Gautheron, P. D., 1984, *Graefes Arch. Clin. Exp. Ophthalmol.* **222**:13–19.
36. Mallorga, P., Reiss, E. R., Ponticello, G. S., Baldwin, J. J., and Sugrue, M. F., 1989, *Invest. Ophthalmol. Visual Sci.* **30**(Suppl.):445.
37. Maren, T. H., 1963, *Pharmacol. Exp. Ther.* **139**:140–153.
38. Maren, T. H., 1967, *Physiol. Rev.* **47**:595–781.
39. Maren, T. H., 1987, *Drug Dev. Res.* **10**:255–276.
40. Maren, T. H., and Bar-Ilan, A., 1987, *Invest. Ophthalmol. Visual Sci.* **28**(Suppl.):268.
41. Maren, T. H., and Bar-Ilan, A., 1988, *Invest. Ophthalmol. Visual Sci.* **29**(Suppl.):16.
42. Maren, T. H., Haywood, J. R., Chapman, S. K., and Zimmerman, T. J., 1977, *Invest. Ophthalmol. Visual Sci.* **16**:730–742.
43. Maren, T. H., Jankowska, L., Sanyal, G., and Edelhauser, H. F., 1983, *Exp. Eye Res.* **36**:457–480.
44. Maren, T. H., Bar-Ilan, A., Caster, K. C., and Katritzky, A. R., 1987, *J. Pharmacol. Exp. Ther.* **241**:56–63.
45. Michelson, S. R., Schwam, H., Baldwin, J. J., Mallorga, P., Ponticello, G. S., Smith, R. L., and Sugrue, M. F., 1989, *Invest. Ophthalmol. Visual Sci.* **30**(Suppl.):24.
46. Mills, K. B., personal communication.
47. Nagasubramanian, S., Bloom, J., Poinoosawmy, D., and Hitchings, R. A., 1987, *Glaucoma Update III* (G. K. Krieglstein, ed.), Springer-Verlag, Berlin, pp. 255–259.
48. Ponticello, G. S., Freedman, M. B., Habecker, C. N., Lyle, P. A., Schwam, H., Varga, S. L., Christy, M. E., Randall, W. C., and Baldwin, J. J., 1987, *J. Med. Chem.* **30**:591–597.
49. Ponticello, G. S., Baldwin, J. J., Freedman, M. B., Habecker, C. N., Christy, M. E., Sugrue, M. F., Mallorga, P. J., and Schwam, H., 1988, abstract 134, Medicinal Chemistry Section, 3rd Chemical Congress of North American, Toronto, Canada, June 5, 1988.

50. Putnam, M. L., Schoenwald, R. D., Duffel, M. W., Barfknecht, C. F., Segarra, T. M., and Campbell, D. A., 1987, *Invest. Ophthalmol. Visual Sci.* **28**:1373–1382.
51. Randall, W. C., and Schwam, H., personal communication.
52. Reynolds, P., personal communication.
53. Quint, M.-P., Durr, M., Plazonnet, B., and Grove, J., 1989, *Invest. Ophthalmol. Visual Sci.* **30**(Suppl.):248.
54. Schwam, H., Michelson, S. R., Sondey, J. M., and Smith, R. L., 1984, *Invest. Ophthalmol. Visual Sci.* **25**(Suppl.):180.
55. Schwam, H., Michelson, S. R., deSolms, S. J., Duprat, P., Gautheron, P., Shephard, K. L., Smith, R. L., and Sugrue, M. F., 1988, 8th International Congress of Eye Research, San Francisco, Calif., September 1988, p. 116.
56. Shepard, K. L., Anderson, P. S., Graham, S. L., Schwam, H., Smith, R. L., and Sugrue, M. F., 1986, abstr. MEDI 50, 192nd American Chemical Society National Meeting, Anaheim, Calif., September 1986.
57. Stein, A., Pinke, R., Krupin, T., Glabb, E., Podos, S. M., Serle, J., and Maren, T. H., 1983, *Am. J. Ophthalmol.* **95**:222–228.
58. Sugrue, M. F., Gautheron, P., Schmitt, C., Viader, M. P., Conquet, P., Smith, R. L., Share, N. N., and Stone, C. A., 1985, *J. Pharmacol. Exp. Ther.* **232**:534–540.
59. Sugrue, M. F., Gautheron, P., Grove, J., Mallorga, P., Schwam, H., Viader, P., Baldwin, J. J., and Ponticello, G. S., 1988, *Invest. Ophthalmol. Visual Sci.* **29**(Suppl.):81.
60. Sugrue, M. F., Gautheron, P., Mallorga, P., Graham, S. L., Schwam, H., Shepard, K. L., and Smith, R. L., 1989, *Br. J. Pharmacol.* **96**:18P.
61. Sugrue, M. F., Mallorga, P., Schwam, H., Baldwin, J. J., and Ponticello, G. S., 1989, *Invest. Ophthalmol. Visual Sci.* **30**(Suppl.):99.
62. Tuulonen, A., Høvding, G., Gustad, L., Dithmer, O., Mönestam, E., Lippa, E. A., Krogh, E., Alm, A., Calissendorff, B., Aasved, H., Bertelsen, T., and Airaksinen, P. J., 1989, *Invest. Ophthalmol. Visual Sci.* **30**(Suppl.):24.
63. Wang, R.-F., Serle, J. B., Podos, S. M., and Sugrue, M. F., 1988, *Invest. Ophthalmol. Visual Sci.* **29**(Suppl.):16.
64. Wang, R.-F., Serle, J. B., Podos, S. M., Severin, C. H., and Sugrue, M. F., 1989, *Invest. Ophthalmol. Visual Sci.* **30**(Suppl.):99.
65. Werner, E. B., Gerber, D. S., and Yoder, Y. J., 1987, *Can. J. Ophthalmol.* **22**:316–319.
66. Wistrand, P. J., 1974, in: *Pharmacology and Pharmacokinetics* (T. Teorell, R. L. Dedrick, and P. G. Comcliff, eds.), Plenum Press, New York, pp. 191–194.
67. Woltersdorf, O. W., Jr., Schwam, H., Bicking, J. B., Brown, S. L., deSolms, S. J., Fishman, D. R., Graham, S. L., Gautheron, P. D., Hoffman, J. M., Larson, R. D., Lee, W. S., Michelson, S. R., Robb, C. M., Share, N. N., Shepard, K. L., Smith, A. M., Smith, R. L., Sondey, J. M., Strohmaier, K. M., Sugrue, M. F., and Viader, M. P., 1989, *J. Med. Chem.* **32**:2486–2492.

CHAPTER 15

Carbonic Anhydrase II Deficiency Syndrome

Clinical Delineation, Interpretation, and Implications

WILLIAM S. SLY

Since osteopetrosis (marble bone disease) was first described by Albers-Schonberg in 1904, over 300 cases have been reported.[5] An autosomal dominant form, the adult, benign form, has a relatively benign course and is compatible with a normal life span. The clinically more severe, autosomal recessive, "malignant, lethal" form has its onset in infancy and produces anemia, leukopenia, hepatomegaly, failure to thrive, cranial nerve symptoms, and early death. Intermediate forms have also been described.

The association of renal tubular acidosis (RTA) with osteopetrosis was first recognized in 1972.[4,17] The initial pedigrees suggested autosomal recessive inheritance, but the clinical course was much milder than that associated with the recessive lethal form; the hematologic abnormalities associated with the recessive lethal form of osteopetrosis were absent.

The additional finding of cerebral calcification in four children with osteopetrosis and RTA from Saudi Arabia was reported in 1980,[10] as was calcification of the basal ganglia in the original American kindred.[19]

In 1983, it was reported from this laboratory that the three sisters from the original American kindred with this syndrome lacked carbonic anhydrase (CA) II in their erythrocytes.[12] Erythrocyte lysates from their normal-appearing parents had half-normal activities of CA II.[3] These observations led to the proposal that

WILLIAM S. SLY • Department of Biochemistry and Molecular Biology, St. Louis University School of Medicine, St. Louis, Missouri 63104.

CA II deficiency is the primary defect in this newly recognized metabolic disorder of bone, kidney, and brain. This hypothesis was consistent with CA II being the only known soluble isozyme of CA in kidney and brain cytoplasm.

Eighteen additional patients in 11 unrelated families of different geographical and ethnic origins were then studied.[14] Subsequently four additional Saudi Arabian patients were reported, including the first affected neonate with the clinical features of 21 reported patients.[8] One more case plus a comprehensive review of the clinical findings on the 30 patients reported to date[2] was recently reported. Every patient identified with this syndrome lacked erythrocyte CA II.

1. Clinical Manifestations

1.1. Osteopetrosis

Defective bone resorption leads to a generalized accumulation of bone mass that prevents normal development of marrow cavities, normal tubulation of long bones, and the enlargement of osseous foramina. Anemia is rarely significant in patients with CA II deficiency; however, two Algerian patients needed to be referred to Paris for bone marrow transplantation.[2]

The radiologic findings in patients with CA II deficiency syndrome resemble those seen in patients with other forms of osteopetrosis. These findings include increased bone density, abnormal modeling, delay or failure of normal tubulation of long bones, transverse banding of metaphyses, fractures, and "bone-in-bone" appearance. The changes vary considerably with age. In the only neonate studied to date, the radiologic features were too subtle to even permit the diagnosis at 23 days of age, even though the hyperchloremic metabolic acidosis and alkaline urine were prominent. This observation suggests that the osteopetrosis is usually a postpartum developmental abnormality. One patient had no osteopetrosis at 4 months; typical findings evolved and progressed over the first 3 years of life.[2] In some patients, the radiologic features of osteopetrosis, prominent in childhood, improved substantially after puberty.[19] Some patients report 15–30 fractures by midadolescence, with the frequency of bone fractures decreasing after puberty. Fractures were the most prominent symptoms in the American patients[19] and the Belgian patient in whom mental retardation was not present[17] but were not seen at all in Guibaud's patients.[2,4]

Symptoms of cranial nerve compression secondary to osteopetrosis have been reported to occur in 60% of patients to date. However, the symptoms are milder than those seen in the recessive, lethal form. Strabismus is also common, as is hearing impairment. Facial weakness has been noted in two reports. Frank optic atrophy is infrequent, but optic pallor is common. Optic atrophy has been found in patients in whom the optic foramina were of normal size. The mechanism of optic atrophy in these patients is unclear.[7]

1.2. Renal Tubular Acidosis

The RTA varies in type and severity in different pedigrees. It is present at birth. Although one of the first patients reported had only proximal RTA, evidenced by low bicarbonate threshold, and had normal distal acidification,[4] most of the patients have a combination of proximal and distal RTA.[2] In most patients, hypochloremia, a normal anion gap, and inappropriately alkaline urine pH (>6.0) provide evidence of distal RTA. Symptomatic hyperkalemia was reported in four patients. However, unlike in other conditions with distal RTA, there is neither hypercalciuria nor nephrocalcinosis. Glomerular filtration rate is not reduced, and serum creatinine and blood urea nitrogen are normal.

Most patients also have a proximal RTA evidenced by a reduced transport maximum for bicarbonate. Although they have no bicarbonaturia when acidotic, they lose bicarbonate when plasma bicarbonate levels are raised to normal levels by loading. They have no aminoaciduria, glycosuria, or any other manifestations of Fanconi's syndrome.

1.3. Mental Retardation

The mental retardation was not appreciated initially because affected patients in two of the first four families recognized with this syndrome were not retarded.[17,19] Over 90% of the patients reported subsequently had mild to moderate mental retardation. Affected patients in most families do not receive education in regular schools.

1.4. Cerebral Calcification

Cerebral calcifications were first demonstrated by computed tomography (CT) scans[10] and by X rays.[19] The cerebral calcifications are much easier to document on CT scan than on X rays because of the increased bone density. They are not present at birth but appear some time during the first decade (in one case, by 18 months). Calcifications involve the caudate nucleus, putamen, and globus pallidus, and they also appear peripherally in the periventricular and subcortical white matter.

1.5. Growth Retardation

Almost all reported patients had short stature, and many were underweight. Bone age was retarded and corresponded to height age. Genu valgum is a common finding in older patients. At least part of the growth retardation probably results from chronic metabolic acidosis. Guibaud *et al.* reported acceleration of growth after correction of the acidosis[4] but later noted that growth retardation persisted, even after treatment,[2] and final height attained was 2–4 standard deviations below normal.

1.6. Dental Malocclusion

Delayed dentition and dental malocclusion were prominent findings in affected patients, complicating dental hygiene. Dental caries may be severe. Enamel hypoplasia has also been noted.

1.7. Other Clinical Features

Features include craniofacial disproportion. The forehead is prominent, and the cranial vault is large relative to the size of the face. A characteristic facies has been reported in affected Saudi Arabian patients.[10] The mouth is small, and there is micrognathia. The nose is narrow but prominent. The philtrum is short, the upper lip is thin, and the lower lip is thick. Squint is common and contributes to the unusual facies.

Restrictive lung disease has been observed in two patients.[9] Chest films showed no signs of parenchymal lung disease, but the rib cages were very dense, restricting lung expansion and respiratory exchange.

2. Bone Pathology

Histologic features in bone biopsies from iliac crest were typical of osteopetrosis.[19] The cortical bone showed small Haversian systems widely separated by dense bone. Trabeculae were broad and irregular. Osteoid and normal-appearing osteoblasts were seen lining trabecular bone in several areas. On routine microscopy, osteoclast morphology was unremarkable. A minute sample of femoral cortex was obtained during open reduction of a femoral fracture. Osteoclasts were normal in appearance on light microscopy. Four osteoclasts were identified on electron microscopy and showed a normal rim of cytoplasm adjacent to the bone surface. This clear zone was free of organelles. The osteoclasts appeared normal, although no ruffled borders were seen.

3. Pathogenesis

In 1983, the three affected sisters described[19] were shown to have no detectable CA II in their erythrocytes.[12] No immunoreactivity was detectable with antibody specific to CA II. CA I was present in near-normal concentration. The obligate heterozygote parents had half-normal CA activity in their erythrocytes.[3] These findings were subsequently extended to 18 similarly affected patients from 11 unrelated families of different geographic and ethnic origins.[2,9] In this group, and in every patient with osteopetrosis and RTA since tested, CA II was nondetectable.[2,9]

CA II deficiency has not yet been demonstrated directly in bone and kidney in humans with the CA II deficiency syndrome. However, the deficiency in these tissues can be inferred, since CA II is the major soluble isozyme present in these tissues, and the observed metabolic abnormalities can be explained by the CA II deficiency in these two organs.

Although CA II activity and immunoreactivity have been nondetectable in erythrocytes, there could be residual activity in cells that, unlike erythrocytes, continue to synthesize protein (such as osteoclasts in bone and cells in the proximal and distal tubules of the kidney). Some of the clinical heterogeneity in this syndrome may be explained by differences in residual CA II activity in bone and kidney in patients with different mutations.

4. Pathophysiology

4.1. Osteopetrosis

All forms of osteopetrosis result from failure to resorb bone. Studies showing inhibition of parathyroid hormone-induced release of Ca^{2+} from bone by CA inhibitors suggested a role for CA in bone resorption. This finding was supported by the demonstration of CA histochemically in chick and hen osteoclasts and of CA II immunohistochemically in rat and human osteoclasts. The finding of osteopetrosis in the CA II deficiency syndrome provided genetic evidence for a role for CA in bone resorption to confirm the earlier pharmacologic and immunohistochemical evidence, and it specifically implicated the CA II isozyme.

We have suggested that the role of CA II in acidifying the bone-resorbing compartment is an indirect one.[12] It had been suggested that CA aids the resorptive process by mediating the secretion of H^+ directly. However, recent evidence indicates that the acidification of the bone-resorbing compartment is mediated by a proton-translocating ATPase, which secretes protons into the lumen.[1] This reaction would simultaneously generate an OH^- ion in the cytoplasm for each H^+ translocated to the lumen. We have postulated that CA II is required for titration of the OH^- ions produced in the cytosol and that formation of HCO_3^- from OH^- and CO_2 is essential for the proton-translocating ATPase to maintain the pH gradient (7.0–4.5) between the cytosol of the osteoclast and the bone-resorbing compartment.[14] This would explain[1] the pharmacologic evidence for a requirement for CA in bone resorption and[2] the osseous manifestations of CA II deficiency.

4.2. Renal Tubular Acidosis

Three things need explanation. First, most CA II-deficient patients have both a proximal and a distal component to the RTA. Second, involvement of the two sites varies in different pedigrees. Some patients have predominantly proximal RTA, whereas others have predominantly distal RTA. Third, CA II-deficient patients have a nearly normal bicarbonaturia after ingestion or infusion of CA inhibitors.

We explained[13] these findings with a model in which the functions of CA II in the proximal and distal tubules are physiologically and biochemically distinct, and a major role for CA in bicarbonate reclamation is assigned to CA IV, the luminal CA in the brush border of the proximal tubule. CA IV is biochemically and immunologically distinct from CA II, and it appears to be normal in CA II-deficient patients.[13]

What causes proximal RTA? Bicarbonate reclamation takes place predominantly in the proximal tubule and is known to be blocked by inhibitors of CA. It is now clear that two distinct CAs, CA II and CA IV, participate in different ways in bicarbonate reclamation by the proximal tubule.

CA IV acts in the lumen to promote production of CO_2 from bicarbonate. Bicarbonate reclamation is driven by H^+ secretion, which is mediated by Na^+–H^+ exchange in the proximal tubule. The H^+ secreted into the lumen is titrated by the HCO_3^- in the glomerular filtrate to produce H_2CO_3, which is in contact with the membrane-bound CA IV. The luminal CA IV catalyzes the dehydration of H_2CO_3 to H_2O and CO_2. The bicarbonaturia seen in already acidotic CA-deficient patients in response to infused acetazolamide we attribute to inhibition of this luminal CA IV.[13]

CA II acts in the cytosol at the next step in bicarbonate reclamation. The CO_2 produced by the CA IV-catalyzed reaction in the lumen diffuses freely into the cytosol of the proximal tubule. Here in the cytoplasm CO_2 encounters CA II, which hydrates the CO_2 to H_2CO_3, which in turn dissociates spontaneously to HCO_3^- and H^+. The HCO_3^- generated from CO_2 in the cytosol is transported from the cytosol to the interstitial fluid or peritubular capillary, by still incompletely understood mechanisms, and this step completes the reclamation of the filtered bicarbonate. The H^+ regenerated in the cytosol by the CA II-catalyzed reaction can be secreted in exchange for Na^+ to initiate another round of HCO_3^- reclamation.

When acidotic, CA II-deficient patients do not spill HCO_3^- in the urine. This finding suggests that CA II is not required for HCO_3^- reclamation when patients have low bicarbonate loads, i.e., when they are acidotic. However, CA II-deficient patients do have a lowered transport maximum for bicarbonate and lose bicarbonate when the filtered load is increased by bicarbonate infusion or ingestion. This finding demonstrates that CA II is required to regenerate H^+ for bicarbonate reclamation under normal bicarbonate loads. This requirement explains the proximal component of the RTA in CA II-deficient patients.[13]

There is a prominent distal component to the RTA in most CA II-deficient patients, evidenced by inappropriately high urine pH values when patients are acidotic. This nearly constant finding indicates a need for CA II for distal acidification. This finding is consistent with immunohistochemical evidence of a much more intense reaction for CA II in the distal tubule and the intercalated cells of the collecting ducts than in the proximal tubules. The CA II-rich cells in the distal human nephron and collection duct appear to be analogous to the distal nephron and collecting system in the amphibian. Here, CA-rich cells are specialized cells that secrete H^+ and are capable of generating a steep pH gradient. However, the acidification of the lumen by these cells is sensitive to inhibition by acetazolamide. This has been explained by the requirement for CA to titrate the OH^- produced in the cytosol by the proton-translocating Mg^{2+} ATPase. Unless the OH^- is titrated by CO_2, the proton-translocating ATPase cannot generate a pH gradient and acidify the lumen. The absence of CA II for this reaction in CA II-deficient patients can explain their defect in distal tubular acidification. Note the similarity in the explanation for the distal defect to the explanation for the defect in acidifying the bone-resorbing compartment.

The heterogeneity in the renal lesion in patients with CA II deficiency requires explanation. Why is there variability in prominence of the proximal and distal lesions in different pedigrees? The explanation for this is unknown, since the mutational basis for CA II deficiency has not been demonstrated, even in a single patient. However, if one assumes that different structural gene mutations produce CA II deficiency in the different pedigrees of patients with the CA II deficiency syndrome, it is plausible that different mutations could lead to this heterogeneity. First, different mutations could affect the rate of enzyme turnover in proximal and distal tubular cells differentially, resulting in different levels of residual enzyme activity in the two locations. Second, different structural gene mutations might affect the two different enzymatic activities in the two locations differentially. Hydration of CO_2 to produce H^+ and HCO_3^- in the proximal tubule and the condensation of OH^- and CO_2 to produce HCO_3^- in the distal tubule might be differentially affected by different mutations in the CA II gene. Delineation of the mutations in different CA II-deficient patients should soon be feasible and allow one to test this hypothesis.

4.3. Brain Calcification and Cerebral Function

The mechanism of the cerebral calcification is unclear. CA II is primarily a glial enzyme that occurs predominantly in oligodendrocytes. It is the major soluble CA in brain homogenates. As much as 50% of the total CA II activity occurs in a membrane-bound or myelin-associated form. Its function in the brain is not known. Whether the cerebral calcification in CA II deficiency is a direct effect of the deficiency of CA II in the brain or an indirect effect (for example, of CA II deficiency in erythrocytes or of chronic systemic acidosis) is not clear.

4.4. Growth Retardation

Growth failure appears to result from the combined effects of (1) the osteopetrosis on bone elongation and (2) the chronic metabolic acidosis on general health. Correction of the acidosis has been followed by a growth spurt in one patient, but the dramatic reduction in final height achieved makes it clear that the growth retardation is not due to the acidosis alone.

5. Inheritance

The CA II deficiency syndrome is inherited as an autosomal recessive trait. Affected patients are offspring of normal-appearing heterozygote carrier parents who have half-normal levels of CA II in their erythrocyte lysates. Heterozygotes have no symptoms and no signs of the disorder. Males and females are affected with equal frequency and severity. Consanguinity is very common (87%) in parents of affected offspring.

More than half of the known cases of this syndrome were observed in predominantly Bedouin families from Kuwait, Saudi Arabia, and North Africa, where consanguineous marriages are acceptable.

6. Diagnosis

CA II deficiency should be suspected in any newborn infant with metabolic acidosis and failure to thrive, especially if the urine pH is alkaline. Enzymatic confirmation can be made by measuring CA activity at 4°C in erythrocyte lysates. CA I activity is virtually abolished by inclusion of 8 mM sodium iodide in the assay.[15] To quantitate the two isozymes, one simply measures the total activity (CA I + CA II) and also the activity seen in the presence of 8 mM sodium iodide (CA II). Patients with CA II deficiency have no iodide-resistant enzyme (i.e., no CA II). Obligate heterozygotes have about half-normal iodide-resistant activity.[15]

7. Genetic Counseling

The appropriate counseling for an autosomal recessive trait is indicated. First-degree relatives can be tested for heterozygosity.[15] The osteopetrosis does not appear prenatally, and prenatal diagnosis is currently not available. CA concentrations in erythrocytes are normally extremely low at birth, and it is not clear that CA II deficiency could be diagnosed by measuring CA II activity in samples of fetal blood. No disease-associated DNA markers are yet available. Although the cDNA probe for CA II is available, the disease has not yet been linked to the structural gene for CA II, and it is not yet certain that the underlying CA II deficiency involves a structural gene locus.

8. Treatment

No specific treatment for CA II deficiency is available. Treatment for the metabolic acidosis is recommended, at least until after adolescence. It appears that the RTA may stabilize at a milder level after puberty. Frequent fractures require conventional orthopedic management. Bone healing is usually normal. Most patients require special education because of serious mental retardation. There is no specific treatment for the cranial nerve abnormalities, which may lead to impaired vision, hearing deficits, and facial nerve weakness. Attention to dental hygiene is important because of the susceptibility to caries.

In the initial American family,[19] treatment with bicarbonate was withheld for fear that the acidosis may be compensating for the osteopetrosis and that treatment of the acidosis might aggravate the osteopetrosis, with further loss of vision and hearing. However, prolonged treatment of several patients by Guibaud and colleagues appeared to have a beneficial effect on general health without any marked progression of the osteopetrosis and with no aggravation of cranial nerve symptoms.[2] It is not clear whether the development of cerebral calcification is influenced favorably, unfavorably, or not at all by correction of the acidosis.

Bone marrow transplantation is probably not indicated, since the hematologic manifestations for which this is usually considered appropriate in the infantile, recessive lethal form of osteopetrosis are not present in the CA II deficiency syndrome. Although the bone manifestations might improve after bone marrow transplantation, since CA II-containing osteoclasts would be provided by stem cells from the donor marrow, the renal insufficiency will not improve.

We had the opportunity to replace the CA II-deficient red cells with CA II-replete blood cells after severe uterine hemorrhage in one of the patients we followed.[18] Raising the circulating erythrocyte levels of CA II to the heterozygote range by transfusion with replete erythrocytes had no effect on plasma pH or urine pH. These observations supported the proposal that the metabolic acidosis is due to the renal CA II deficiency, not a secondary consequence of CA II deficiency in erythrocytes.

9. Implications for Osteoporosis

Although osteopetrosis in this syndrome results from a defect in bone resorption, there are other metabolic disorders in which the reverse is true, and accelerated bone loss is the problem. In organ culture, Ca^{2+} release from bones was shown to be hormone responsive (parathormone and dibutyryl cyclic AMP) and sensitive to inhibition by acetazolamide and other inhibitors of CA.[11] Animal studies have suggested that bone loss associated with disuse can be partially prevented by CA inhibitors.[6] These observations raise hopes that CA inhibitors might have a role in treating common causes of bone loss such as osteoporosis. One problem, however, is that chronic administration of currently available agents produces a systemic acidosis as a result of their actions on the kidney, and systemic acidosis itself can lead to calcium mobilization from bone. For this reason, it has been suggested that effective use of CA inhibitors for metabolic bone disease may require development of agents that act selectively on CA II in bone or that can be selectively targeted to bone-resorbing osteoclasts to avoid inhibition of CA II in kidney and other sites.[11]

10. Other Implications and Future Prospects

Several laboratories are attempting to identify the molecular defect in the CA II deficiency syndrome; success could make prenatal diagnosis possible. A mouse with CA II deficiency[8] has been developed. The mutation was produced by exposing mice that were heterozygotes for electrophoretically distinguishable CA II gene products to a powerful mutagen and screening progeny electrophoretically for loss of one of the alleles. A null mutation was found, and a breeding colony was established. The affected mouse has severe acidosis but does not appear to have osteopetrosis or cerebral calcification.

Finally, the remarkable utility of this human disease in shedding light on the physiological roles of the various CAs should stimulate clinical research aimed at identifying disorders due to deficiencies of other members of the CA gene family.[16] An inherited deficiency of CA I has already been found and proved to have no clinical consequences.[16] Presumably, this reflects the fact that CA I is expressed primarily in the erythrocytes and CA II, which is also expressed in erythrocytes, is present in normal levels in CA I-deficient patients. CA II could more than handle the requirements for CA activity in the erythrocytes.[20] It seems likely that deficiencies for CA III, CA IV, or CA V would produce significant clinical abnormalities. Such experiments of nature probably exist, and once they are identified, they will likely add greatly to our understanding of why we have evolved so many isozymes to catalyze a reaction as simple as the reversible hydration of CO_2.

References

1. Baron, R., Neff, L., Louvard, D., and Courtoy, P. J., 1985, *J. Cell Biol.* **101**:2210–2222.
2. Cochat, P., Loras-Duclaux, I., and Guibaud, P., 1987, *Pediatrie* **42**:121–128.
3. Dodgson, S. J., Forster, R. E., II, Sly, W. S., and Tashian, R. E., 1988, *J. Appl. Physiol.* **65**:1472–1480.
4. Guibaud, P., Larbre, F., Freycon, M. T., and Genoud, J., 1972, *Arch. Fr. Pediatr.* **29**:269–286.
5. Johnston, C. C., Jr., Lavy, N., Lord, T., Vellios, F., Merritt, A. D., and Deiss, W. P., Jr., 1968, *Medicine* (Baltimore) **47**:149–167.
6. Kenny, A. D., 1985, *Calcif. Tissue Int.* **37**:126–133.
7. Krupin, T., Sly, W. S., Whyte, M. P., and Dodgson, S. J., 1985, *Am. J. Ophthalmol.* **99**:396–399.
8. Lewis, S. E., Erickson, R. P., Barnett, L. B., Venta, P. J., and Tashian, R. E., 1988, *Proc. Natl. Acad. Sci. USA* **85**:1962–1966.
9. Ohlsson, A., Cumming, W. A., Paul, A., and Sly, W. S., 1986, *Pediatrics* **77**:371–381.
10. Ohlsson, A., Stark, G., and Sakati, N., 1980, *Dev. Med. Child. Neurol.* **22**:72–84.
11. Raisz, L. G., Simmons, H. A., Thompson, W. J., Shepard, K. L., Anderson, P. S., and Rodan, G. A., 1988, *Endocrinology* **122**:1083–1086.

12. Sly, W. S., Hewett-Emmett, D., Whyte, M. P., Yu, Y.-S. L., and Tashian, R. E., 1983, *Proc. Natl. Acad. Sci. USA* **80**:2752–2756.
13. Sly, W. S., Whyte, M. P., Krupin, T., and Sundaram, V., 1985, *Pediatr. Res.* **19**:1033–1036.
14. Sly, W. S., Whyte, M. P., Sundaram, V., Tashian, R. E., Hewett-Emmett, D., Guibaud, P., Vainsel, M., Baluarte, H. J., Gruskin, A., Al-Mosawi, M., Sakati, N., and Ohlsson, A., 1985, *N. Engl. J. Med.* **313**:139–144.
15. Sundaram, V., Rumbolo, P., Grubb, J., Strisciuglio, P., and Sly, W. S., 1986, *Am. J. Hum. Genet.* **38**:125–136.
16. Tashian, R. E., Hewett-Emmett, D., Dodgson, S. J., Forster, R. E., and Sly, W. S., 1984, *Ann. N.Y. Acad. Sci.* **429**:262–275.
17. Vainsel, M., Fondu, P., Cadranel, S., Rocmans, C., and Gepts, W., 1972, *Acta Paediatr. Scand.* **61**:429–434.
18. Whyte, M. P., Hamm, L. L., and Sly, W. S., 1988, *J. Bone Mineral Res.* **3**:385–388.
19. Whyte, M. P., Murphy, W. A., Fallon, M. D., Sly, W. S., Teitelbaum, S. L., MacAlister, W. H., and Avioli, L. V., 1980, *Am. J. Med.* **69**:64–74.
20. Wistrand, P. J., 1981, *Acta Physiol. Scand.* **113**:417–426.

PART IV

Genetic Regulation of the Carbonic Anhydrase Isozymes

CHAPTER 16

The Structure and Regulation of the Human Carbonic Anhydrase I Gene

PETER H. W. BUTTERWORTH, JONATHAN H. BARLOW, HUGH J. M. BRADY, MINA EDWARDS, NICHOLAS LOWE, and JANE C. SOWDEN

1. Introduction

It has long been known that the genes coding for the closely related carbonic anhydrase (CA) I and II isozymes are tightly linked,[19] and recent data from the analysis of somatic cell hybrids using cloned molecular probes have located not only CA I and CA II but also CA III to the long arm of chromosome 8.[4,14] In collaboration with Yvonne Edwards' group, we have used pulse-field electrophoresis of large fragments of human DNA to show that the three genes lie within 200 kb of each other (unpublished data). Because these three genes have quite different patterns of tissue-specific expression (reviewed in reference 17), their proximity on chromosome 8 poses interesting questions concerning the molecular events responsible for differential gene activity. In the first instance, we need to define the organization of each gene and the characteristics of each transcription unit. The CA III gene, which is expressed in muscle and the liver of male rats, and the more generally expressed CA II gene have been cloned by groups in London[12] and Ann Arbor,[18] respectively; we have cloned the entire region containing the human CA I transcription unit, which is activated late in fetal development and expressed at high levels in erythroid tissues. It is now known that there is a second promoter within this transcription unit which is functional in colon in mice[7] and humans (our unpublished data). Presented below is an outline of our current

PETER H. W. BUTTERWORTH • University of Surrey, Guilford, Surrey GU2 5XH, United Kingdom. **JONATHAN H. BARLOW, HUGH J. M. BRADY, MINA EDWARDS, NICHOLAS LOWE, and JANE C. SOWDEN** • Department of Biochemistry, University College London, London WC1E 6BT, United Kingdom.

progress in identifying the different levels at which expression of the CA I gene is regulated in erythroid cells.

2. Methods

A human erythroleukemic cell line (HEL), a human embryonic erythroid cell line (K562), and HeLa cells were grown in Dulbecco modified eagle medium supplemented with 10% fetal calf serum (FCS) in the presence of 100 μg/ml penicillin, 100 μg/ml streptomycin, and 2 μg/ml amphotericin. A mouse erythroleukemic cell line (MEL C88) was grown in α-minimal essential medium supplemented with 10% FCS and 50 μg/ml diaminopurine. HUT-78 (a human lymphoid cell line) was grown in RPMI 1640 supplemented with FCS and antibiotics as described above.

Preparation of nuclear proteins from cell lines in log phase was carried out by a modification of a previous technique.[21] Analysis involved gel retardation assays,[9] DNase I footprinting,[8] isolation of total RNA from cells and tissues,[3] Northern analysis,[11] fusions between MEL and K562 cells,[1] and subsequent RNase protection analysis[13] of the heterokaryon RNA, which was carried out by using probes for human CA I (HCA I) and mouse embryonic globin (ϵ-globin) generated from sequences subcloned into SP6 vectors.[22]

3. Results

3.1. CA I Gene Structure and Organization

We have cloned the CA I gene on a series of six overlapping recombinants selected from a human lambda 2001 genomic library (Fig. 1). Approximately 50 kb contains the entire protein-coding sequence, transcriptional start site for reticulocyte CA I mRNA (defined by S1 mapping and primer extension[2]; our unpublished results) and 5′ and 3′ untranslated sequences, the latter containing at least two discrete polyadenylation sites. The protein-coding sequence is divided into seven exons as with the CA II and CA III genes.[12,18] The location of the intervening sequences in the protein-coding region is conserved between CA I, CA II, and CA III, although the sizes of the introns are different. We have described elsewhere[2] that there is considerable complexity in the 5′ untranslated region of the reticulocyte CA I transcription unit, which alone covers more than 35 kb. Comparison of several cDNA sequences with data from genomic clones indicates intervening sequence within the 5′ leader and alternative splicing events that can give rise to at least two populations of mRNAs with different 5′ ends. For consistency with designations for other CA genes, the exons making up the reticulocyte CA I 5′ leader are indicated as exons 1a (68 bp beginning with the transcription start site at

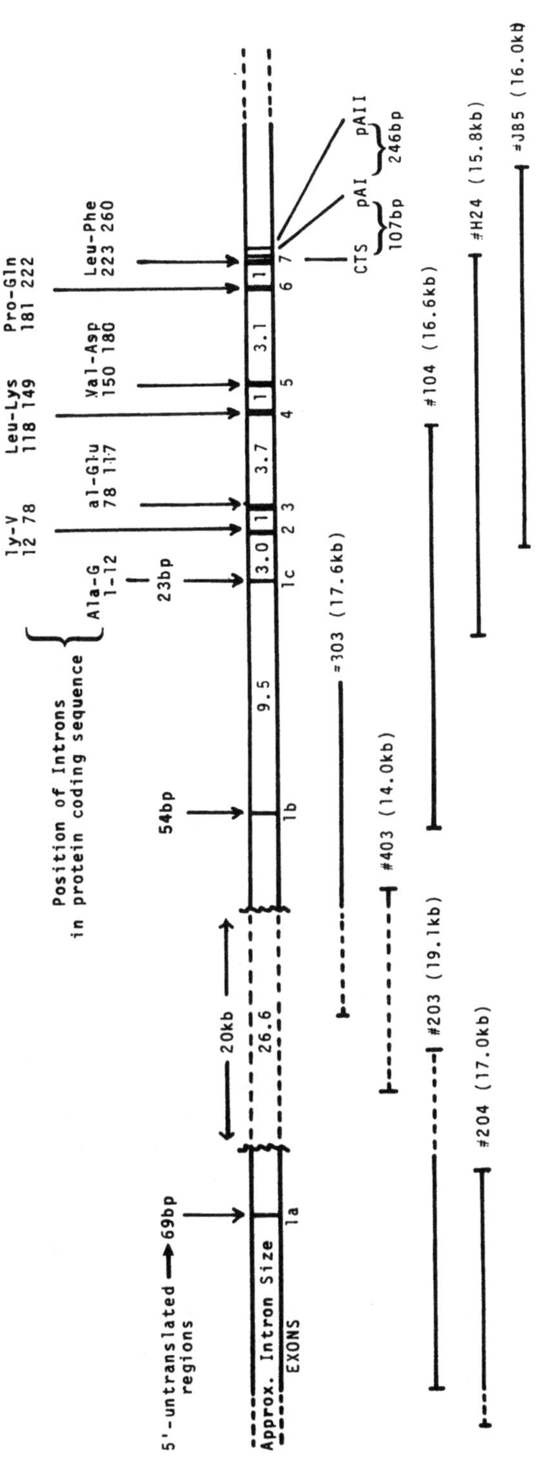

Figure 1. Structure and organization of the human CA I gene. The gene has been cloned on six overlapping recombinants shown at the bottom, one of which (H24) was kindly provided by P. J. Venta and R. E. Tashian.

+1), 1b (54 bp), and 1c (23 bp prior to the start of the protein-coding sequence). From Northern analysis of total reticulocyte RNA and studies of cDNA clones, it would appear that exons 1a and 1c are invariant components of CA I mRNA, whereas exon 1b is only an occasional component. The physiological significance of alternative splicing in this 5′ leader is not known. However, recent discoveries of an alternative promoter, upstream of the mouse equivalent of exon 1c, which is apparently functional in colon,[7] focuses attention on the potential for regulatory events operating on the formation of the 5′ end of transcripts of this gene.

3.2. Expression of the CA I Gene in Erythroid Cells

Extensive work from several laboratories on humans,[5,20] mouse,[16] and chick[6,15] globin genes has highlighted conserved short-sequence cassettes that occur 5′ and 3′ to a gene and are thought to be signature sequences required for expression of specific genes in erythroid cells. The consensus for this putative erythroid-specific sequence element has been suggested to be

$$5'\text{-}\substack{A\\C}\text{-Py-}\substack{T\\A}\text{-A-T-C-}\substack{A\\T}\text{T-Py-}3' \quad \text{(according to reference 5)}$$

$$3' \text{ -G-A-A-T-A-G-} \qquad 5' \quad \text{(according to reference 16)}$$

or the reverse complement thereof, which we will refer to as GATAAG elements.

Six GATAAG-like elements occur, three within 300 bp 5′ to the transcription start site (+1) and three within 600 bp of the 3′ end of the protein-coding sequence. We will focus here on only one of these, that located at position −190. A 23-mer oligonucleotide from −177 to −199 (Hcgl), containing this GATAAG element, has been synthesized; it is illustrated in Fig. 2, together with restriction fragments of the CA I gene containing this −190 motif that have been used in these experiments.

Nuclear proteins have been extracted from human and mouse erythroleukemic cell lines (HEL and MEL, respectively), which express the CA I gene, and from the human embryonic cell line K562, which, although it is of erythroid lineage, does not express CA I (see below). We have also used HeLa and HUT-78 (lymphoid) cells as nonerythroid controls. Nuclear proteins from HEL and K562 cells contain one or more factors that bind to the GATAAG-containing oligonucleotide probe, resulting in a shift in the mobility of the radioactively labeled probe on native gel electrophoresis (Fig. 3, lanes 5 and 7, respectively). This effect is eliminated by the inclusion of a GATAAG-containing, nonradioactive competitor in the binding reaction (lanes 6 and 8). The nonerythroid cell extracts do not appear to contain GATAAG-binding factors (lanes 1–4). The difference in the proportion of the two bands in band shift between K562 and HEL cell extracts is probably a procedural artifact, the smaller species being derived from the higher-molecular-weight complex.

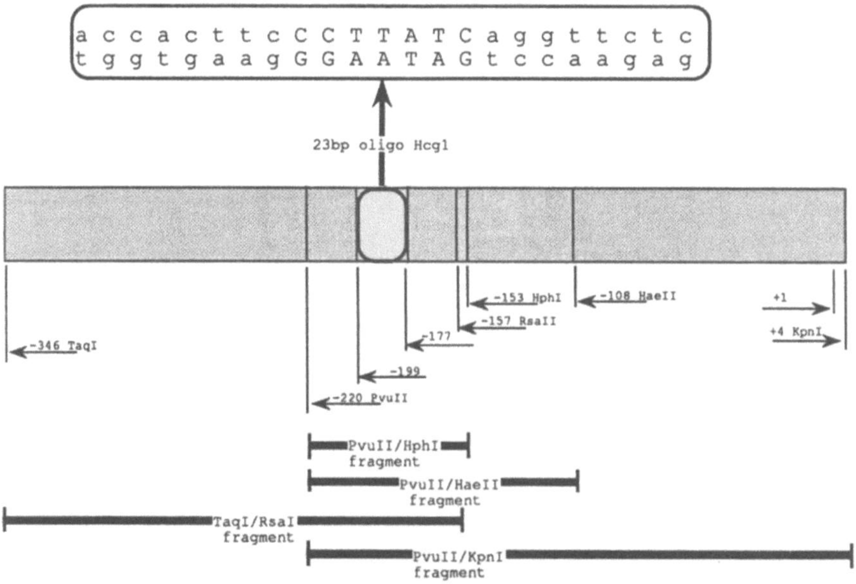

Figure 2. Detailed diagram of the 5′-flanking region of the CA I gene containing the GATAAG element at position −190. The sequence of oligonucleotide Hcgl is shown at the top, together with diagrams of the various restriction fragments used.

That erythroid-specific factors bind at or close to the putative erythroid-specific motif in the CA I gene is shown in the DNase I footprint in Fig. 4. In this experiment, extracts from HEL cells (erythroid) show DNase I protection in the region of the GATAAG sequence and the generation of at least one strong hypersensitive site immediately adjacent to the region of protection (lanes 3 and 4). These effects are competed out by an oligonucleotide containing GATAAG (Fig. 4, lane 5) and not by one in which this sequence motif is absent (lane 6); extracts from nonerythroid HUT-78 cells do not produce any evidence of factor binding at this site (lanes 7 and 8).

Therefore, *in vitro*, the evidence is that the GATAAG sequence motif in the 5′-flanking region of the CA I gene acts in *cis* by binding factors acting in *trans*, factors found in abundance only in erythroid cells. Is this sequence element required for erythroid-specific gene expression *in vivo*? In preliminary experiments, we have placed the CA I 5′ *Taq*/*Rsa* fragment (−346 to −157; see Fig. 2) containing the GATAAG sequence upstream of the reporter gene encoding chloramphenicol acetyltransferase fused to the thymidine kinase gene promoter. This construct and a similar control construct lacking the CA I sequence have been transfected into MEL (erythroid) cells. Inclusion of the CA I GATAAG-containing sequence in the construct stimulated the activity of this nonerythroid promoter in

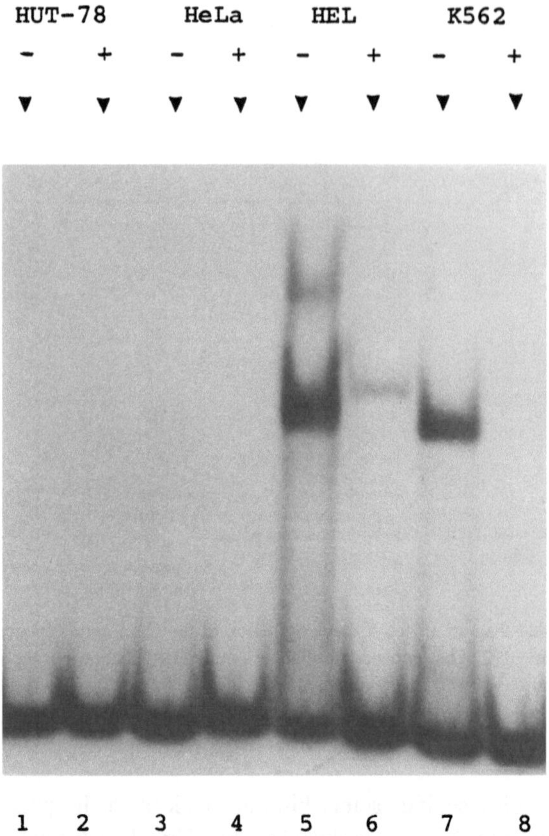

Figure 3. Binding of nuclear proteins to the HCA I (−190) GATAAG-containing oligonucleotide (Hcgl), explored by using extracts from erythroid and nonerythroid cells. Nuclear proteins from HEL (erythroid) and from HUT-78 and HeLa cells (nonerythroid) were incubated with radioactively labeled double-stranded oligonucleotide Hcgl in the absence (−) or presence (+) of an unlabeled, GATAAG-containing competitor oligonucleotide. Complexes were resolved by native gel electrophoresis and were visualized by autoradiography. The band produced by HEL cell nuclear proteins in the presence of competitor is an artifact produced by the probe.

erythroid cells four- to fivefold (data not shown). Careful deletion experiments are now being carried out to prove definitively the involvement of the GATAAG element in erythroid tissue-specific expression of the CA I gene.

3.3. Activation of the CA I Gene in Development

In erythroid cells, the human CA I gene is activated late in fetal development, at about the time of the switch from fetal to adult globin gene expression. This is

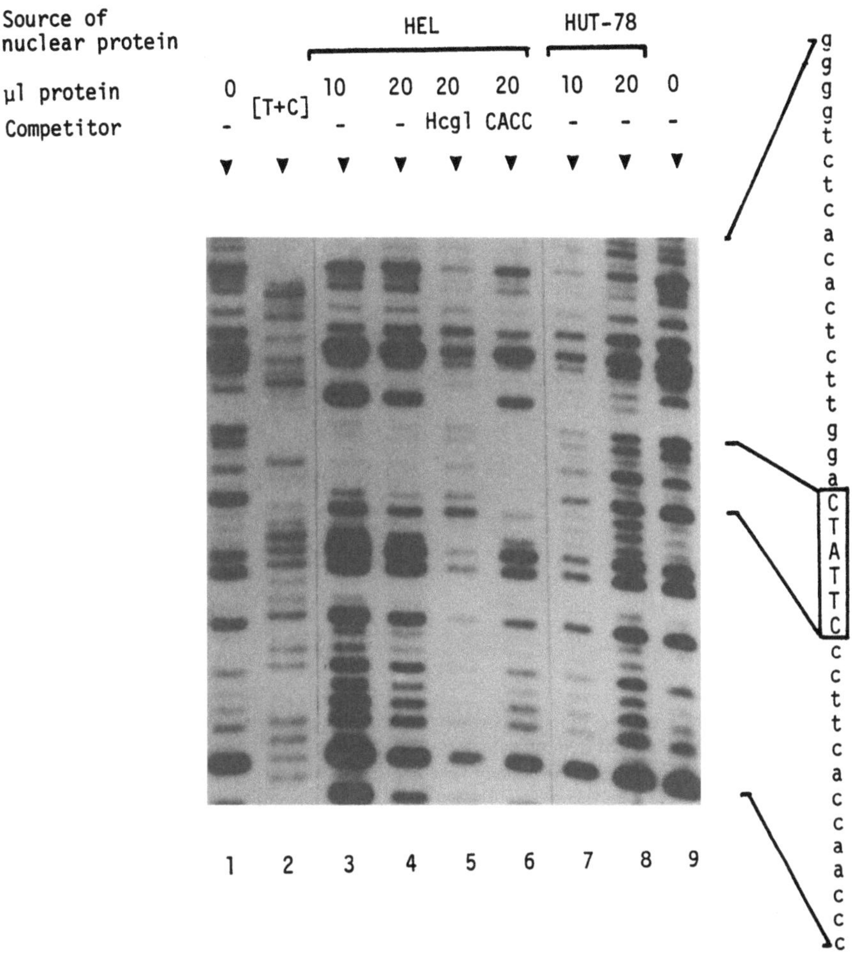

Figure 4. Binding of nuclear proteins to 5′-flanking regions of HCA I, examined by DNase I footprinting. A ^{32}P-end-labeed *Pvu*II–*Hae*II fragment (see Fig. 2) was incubated with various amounts of nuclear protein in the presence of 1 µg of poly(dI-dC)·poly(dI-dC) and, where indicated, 400 of ng unlabeled competitor oligonucleotide [either GATAAG-containing Hcgl (see Fig. 2) or the CACCC oligonucleotide of Plumb *et al.*,[16] which does not contain the GATAAG motif]. After partial DNase I digestion, the fragments were resolved on an 8% polyacrylamide gel.

illustrated simply in Fig. 5, which compares CA I, γ-globin, and β-globin mRNA levels in total reticulocyte RNA at different stages of development (18-week fetus, at birth, and adult).

In a number of instances, cell fusion techniques have been used to show that *trans*-acting factors are probably involved in developmental stage-specific activation of specific genes (actin genes[10] and globin genes[1]). To explore this question

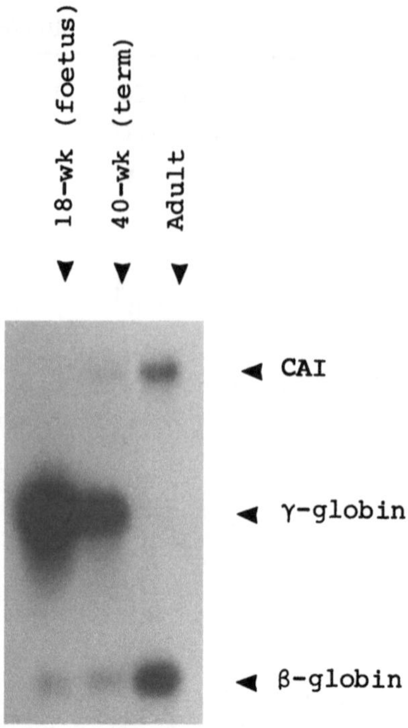

Figure 5. Northern analysis of total RNA from reticulocytes at different stages of human development. Total RNA was isolated from reticulocytes, fractionated by agarose gel electrophoresis, blotted onto nitrocellulose, and probed sequentially for CA I mRNA and for human γ-globin and β-globin mRNAs, using nick-translated sequences. For comparative purposes, the results of the three analyses are superimposed.

with respect to CA I, we have fused mouse MEL cells, which express the CA I gene, with human embryonic erythroid K562 cells, in which the CA I gene is silent. In the resulting transient heterokaryons, RNase protection analysis shows that the human CA I gene has been activated (Fig. 6B) and that there is a parallel activation of the mouse embryonic globin gene (Fig. 6A). In transient heterokaryons, nuclei stay separate; thus, we can infer that diffusible factors from mouse and human cell nuclei have *trans*-activated the human CA I and mouse embryonic globin genes, respectively.

4. Discussion

We have initiated a study of how expression of the human CA I gene is controlled. At least two regulatory processes have been identified. Each involves *trans*-acting factors.

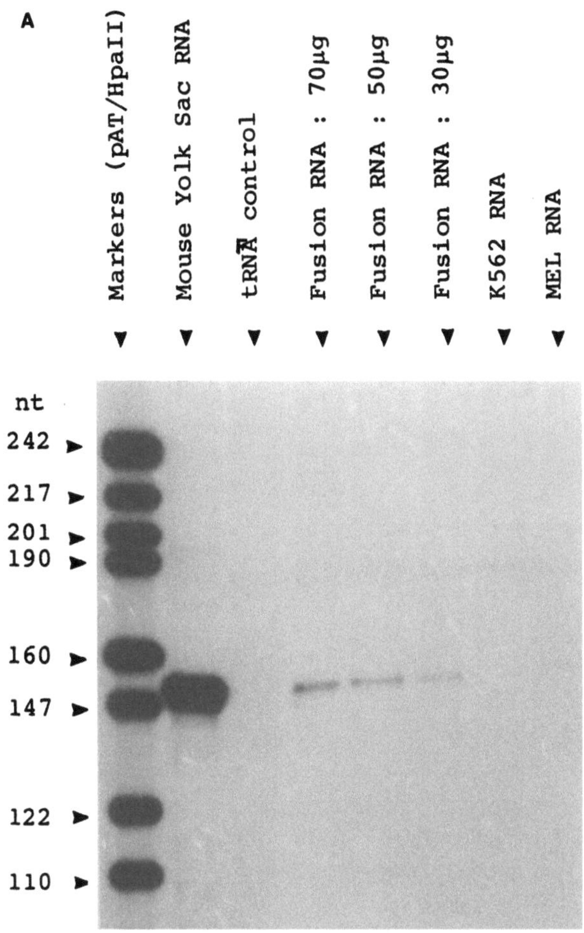

Figure 6. RNase protection analyses of RNA from heterokaryons formed between mouse MEL cells and human embryonic erythroid (K562) cells. Cell fusions were carried out according to the protocol identified in Section 2. (A) Activation of the mouse ε− globin gene is indicated by protection of 145 nucleotides (nt) of the ϵ-globin probe; (B) activation of the human CA I gene is indicated by protection of 404 nt of the CA I probe, i.e., to the 5′ end of exon lc. Exon lb is only rarely included in mature mRNA (as discussed in Section 3.1), although the probe was derived from cDNA containing exons la, lb, and lc (see Fig. 1).

One system involves the binding of a factor(s) to the *cis*-acting GATAAG sequence motif that is (invariably?) found in regions immediately flanking genes which are expressed selectively in erythroid cells. In this respect, CA I is probably typical of other genes specifically expressed in erythroid cells. These sequence cassettes are found both upstream and downstream of the CA I gene, and we have

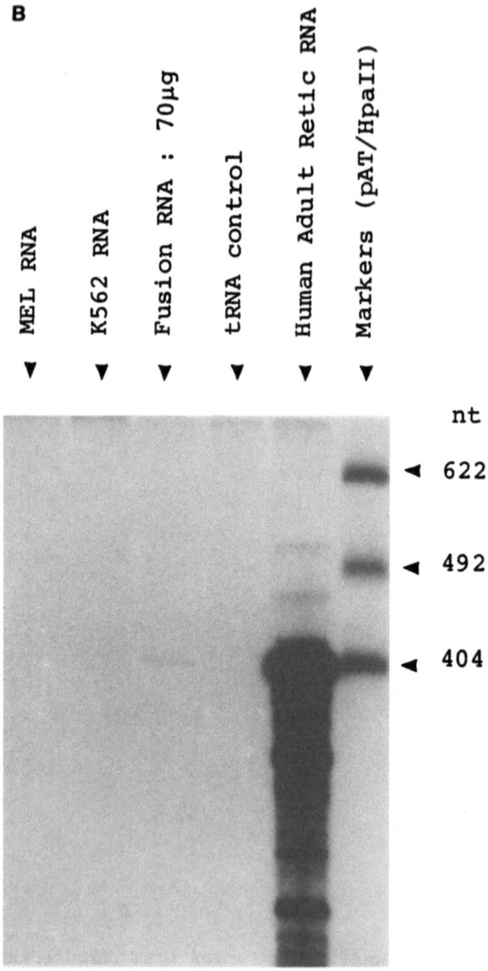

Figure 6. (*Continued*)

yet to resolve whether all, some, or only one of these elements is required to ensure expression of CA I in an erythroid specific manner.

The second system of regulation by *trans*-acting factors operates in a developmental stage-specific manner. Expression of CA I is classically of the adult phenotype. Activation of a silent CA I gene in an embryonic erythroid cell (K562) results from fusion with a CA I-expressing cell, which suggests that the latter contains a factor or factors that are capable of effecting a major change in the developmental program of gene expression. Nothing is known about the genomic sequences in a CA I-associated domain with which such factors interact, nor do we know whether such factors are related to (or are even the same as) factors involved in developmental switching with respect to other erythroid genes, particularly the globins.

Note added in proof: Detailed experimental procedures and interpretations of the data given in Fig. 6 have been published (Brady, J. M., Edwards, M., Linch, D. C., Knott, L., Barlow, J. J., and Butterworth, P. H. W., 1990, *Brit. J. Haematol.* **76**:135–142).

ACKNOWLEDGMENTS. This work was generously supported by a grant from the Wellcome Trust.

References

1. Baron, M. H., and Maniatis, T., 1986, *Cell* **46**:591–602.
2. Brady, H. J. M., Lowe, N., Sowden, J. C., Barlow, J. H., and Butterworth, P. H. W., 1989, *Biochem. Soc. Trans.* **17**:184–185.
3. Chirgwin, J. M., Przybyla, A. E., MacDonald, R. J., and Rutter, W. J., 1979, *Biochemistry* **18**:5294–5299.
4. Davis, M. B., West, L. F., Barlow, J. H., Butterworth, P. H. W., Lloyd, J. C., and Edwards, Y. H., 1988, *Somat. Cell Mol. Genet.* **13**:173–178.
5. deBoer, E., Antoniou, M., Mignotte, V., Wall, L., and Grosveld, G., 1988, *EMBO J.* **7**:4203–4212.
6. Evans, T., Treitman, M., and Felsenfeld, G., 1988, *Proc. Natl. Acad. Sci. USA* **85**:5976–5980.
7. Fraser, P., Cummings, P., and Curtis, P., 1989, *Mol. Cell. Biol.* **9**:3308–3313.
8. Fried, M., and Crothers, D. M., 1981, *Nucleic Acids Res.* **9**:6505–6525.
9. Galas, D. J., and Schmitz, A., 1978, *Nucleic Acids Res.* **5**:3157–3170.
10. Hardeman, E. C., Chiu, C.-P., Minty, A., and Blau, H., 1986, *Cell* **47**:123–130.
11. Konialis, C. P., Barlow, J. H., and Butterworth, P. H. W., 1985, *Proc. Natl. Acad. Sci. USA* **82**:663–667.
12. Lloyd, J., Brownson, C., Tweedie, S., Charlton, J., and Edwards, Y. H., 1987, *Genes Dev.* **1**:594–602.
13. Melton, D. A., Krieg, P. A., Sebagliati, M. R., Maniatis, T., Zinn, K., and Green, M. R., 1984, *Nucleic Acids Res.* **12**:7035–7056.
14. Nakai, H., Byers, M. G., Venta, P. J., Tashian, R. E., and Shows, T. B., 1987, *Cytogenet. Cell Genet.* **44**:234–235.
15. Perkins, N. D., Nicolas, R. H., and Goodwin, G. H., 1989, *Nucleic Acids Res.* **17**:1299–1314.
16. Plumb, M., Frampton, J., Wainwright, H., Walker, M., Macleod, K., Goodwin, G., and Harrison, P., 1989, *Nucleic Acids Res.* **17**:72–92.
17. Tashian, R. E., Hewett-Emmett, D., and Goodman, M., 1983, in: *Curr. Top. Biol. Med. Res.* **7**:79–100.
18. Venta, P. J., Montgomery, J. C., Hewett-Emmett, D., Wiebauer, K., and Tashian, R. E., 1985, *J. Biol. Chem.* **260**:12130–12135.
19. Venta, P. J., Shows, T. B., Curtis, P. J., and Tashian, R. E., 1983, *Proc. Natl. Acad. Sci. USA* **80**:4437–4440.
20. Wall, L., DeBoer, E., and Grosveld, F., 1988, *Genes Dev.* **2**:1089–1100.
21. Wildeman, A. G., Sassone-Corsi, P., Grundstrom, T., Zenke, M., and Chambon, P., 1984, *EMBO J.* **3**:3129–3133.
22. Zinn, K., Di Maio, D., and Maniatis, T., 1983, *Cell* **34**:865–879.

CHAPTER 17

Expression of Carbonic Anhydrases I and II in Mouse Erythropoiesis

WILLIAM THIERFELDER, PATRICK CUMMINGS, PETER FRASER, and PETER J. CURTIS

1. Introduction

Carbonic anhydrase (CA) is the second most abundant protein in red blood cells, where it exists in two isozymic forms, CA I and CA II. CA I is found in a limited number of tissues besides red blood cells, including intestinal epithelium, vascular endothelium, corneal epithelium, and lens of the eye. CA II, though most abundant in red blood cells, is found in a wide variety of cells and tissues. The ratio of these two isozymes in red blood cells varies from species to species; the ratios of CA I to CA II are 6:1 in humans, 27:1 in orangutans, 2.6:1 in macaques, and 1:2 in mice. However, CA II is considerably more active than CA I; thus, the physiological significance of CA I is uncertain, since there exists CA I deficiency in red blood cells of macaques and humans with no observable effect.

2. Expression of the CA I and CA II Genes

CA accumulates during the later stages of erythropoiesis as the cells complete their final divisions. CA expression has been studied mainly in mouse erythroleukemia (MEL) Friend cells.[7] These cells, derived from spleen tumors of mice infected with the Friend virus complex, have many similarities to proerythroblasts and, when treated with a variety of agents (dimethyl sulfoxide is commonly used), accumulate hemoglobin as well as many other proteins characteristic of red blood cells, including CA.[8,10] In fact, under the correct tissue culture conditions, the cells

WILLIAM THIERFELDER, PATRICK CUMMINGS, PETER FRASER, and PETER J. CURTIS • The Wistar Institute of Anatomy and Biology, Philadelphia, Pennsylvania, 19104-4268.

undergo morphological changes similar to those of normal mouse orthochromatophilic erythroblasts and reticulocytes, including enucleation.[14] During induction of MEL cells, total CA activity increases.[9] Using specific antibodies, Stern *et al.*[11] showed that CA I protein remained more or less constant whereas CA II protein increased significantly and that the ratio of CA I to CA II approaches the *in vivo* state. Further analysis of the regulated expression of the CA genes has been possible with the cloned mouse CA I[4] and CA II[2,3] cDNAs. Analysis of total RNA isolated from uninduced MEL cells and MEL cells induced for 1, 2, and 3 days showed that the steady-state level of CA I mRNA was highest in the uninduced cells and decreased dramatically upon induction, implying that the CA I protein must be metabolically stable in the induced cells.[5] The steady-state level of CA II mRNA did not show any significant change during induction. Measurements of the transcriptional rate of the CA genes by the labeling of newly synthesized RNA in isolated nuclei showed that the CA I gene was actively transcribed in the uninduced cells but was shut down upon induction, whereas transcription of the CA II gene increased five- to sevenfold after 2 days of induction, the increase occurring in parallel with transcription of β-globin, 5′-aminolevulinic acid synthase, and band 3.[5]

The results presented above demonstrate that the regulation of CA expression occurs primarily at the level of transcription and that expression of the CA II gene is coordinated at this level with that of β-globin. The decreased expression of the CA I gene indicates that CA I is expressed initially at an early stage of erythroid differentiation. Villeval *et al.*[13] demonstrated the presence of CA I in human cells phenotypically similar to BFU-Es/CFU-Es, which did not contain hemoglobin.

3. Genomic Structure of the CA Genes

The differential expression of the CA genes during erythroid differentiation is particularly interesting, since they are closely linked in the mouse and human genomes and thus could share the same structural domain of chromatin. To understand how expression of these genes is regulated, it is first necessary to clone and characterize the genomic DNA, identify the *cis*-acting DNA elements essential for erythroid-specific expression, and then identify the *trans*-acting DNA-binding proteins that recognize the *cis* elements.

The mouse CA II gene was initially cloned as a 38-kb segment in a cosmid.[12] The gene contains seven exons and six introns spanning about 16 kb. Upstream of the first exon, there is a TATA box and a possible CCAAT box (CCACT) in locations similar to those of β-globin. Except for CA I, the other CA genes isolated so far have essentially the same structure.

The mouse CA I gene has been cloned in three recombinant phages.[6] The CA I gene is distinct from the CA II gene since it contains an additional exon, which encodes 63 bp of the 5′ untranslated region of the CA I mRNA and is located more than 10 kb upstream of the remaining seven exons. A somewhat similar structure has been found for the human CA I gene.[1] The region preceding the mouse first exon contains a TATA box at −30, a CCAAT box at −70, and a CCACACCC box at −212, features found in the β-globin gene. Analysis of mouse colon RNA by primer extension analysis revealed a 5′ sequence different from that of the erythroid CA I mRNA. The colon CA I mRNA 5′ sequence matched the genomic sequence preceding exon 2. Thus, the mouse CA I gene has at least two tissue-specific promoters.

Pulsed-field gel electrophoresis of MEL cell DNA digested with restriction enzymes *Sal*I, *Sac*II, *Cla*I, and *Apa*I, which cut eukaryotic DNA rarely, followed by Southern blot analysis demonstrated that the erythroid exon 2 was between 10 and 250 kb upstream of the rest of the CA I gene. Hybridization of the same Southern blot with the CA II cDNA indicated that the CA II gene is no more than 50 kb 3′ of the CA I gene.

4. Mapping of DNase I Hypersensitive Sites

An impressive correlation between the location of transcriptional regulatory elements and the positions of DNase I hypersensitive sites has been documented for a number of genes. Hence, to identify potential regulatory sequences and *trans*-acting factors that are important for the erythroid-specific expression of the CA I and CA II genes, we have mapped DNase I hypersensitive sites throughout the entire gene and its flanking regions in both uninduced and induced MEL cells.

Nuclei were isolated from MEL cells and incubated with DNase I for increasing lengths of time, and DNA was extracted. DNA was subsequently digested with restriction enzymes and analyzed by the Southern blot method, using labeled DNA fragments from CA I and CA II cDNA or genomic DNA (Fig. 1). In the CA I gene (Fig. 2), four hypersensitive sites were detected; the 5′ site mapped to 1 kb upstream of the erythroid promoter/exon 1, and the second site was located close to the erythroid promoter. The third and fourth sites were located in a similar way with respect to the colon promoter just upstream of the erythroid exon 2. The pattern of appearance and intensity of each site was essentially unchanged in uninduced and induced MEL cells. In the CA II gene, four hypersensitive sites were also detected (Fig. 3). Site 1, which was clearly more intense than the other three sites, mapped to the promoter region, sites 2 and 3 mapped to intron 1, about 1 kb 3′ of the promoter, and site 4 was located just 5′ of exon 3 in intron 2. The promoter site (site 1) was somewhat stronger in the induced cells.

5. Summary

During the final stages of erythropoiesis, the CA I and CA II genes are differentially regulated at the level of transcription; the rate for the CA I gene decreases dramatically, whereas that for the CA II gene increases five- to sevenfold.

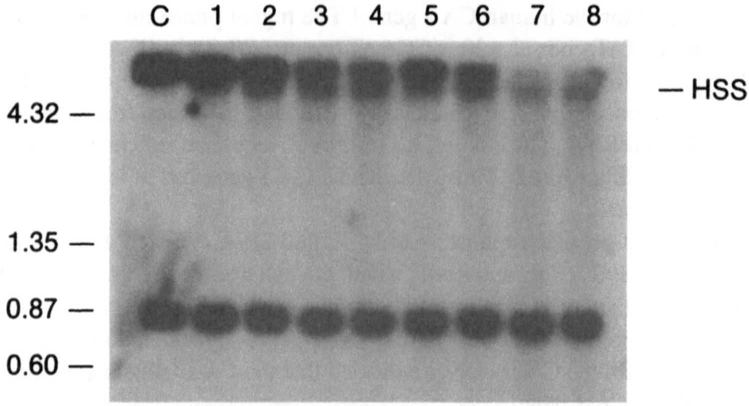

Figure 1. DNase hypersensitive site at the CA II promoter. Nuclei were treated with DNase I for 10–45 min; DNA was purified, digested with *Hin*dIII, resolved on 1.5% agarose by electrophoresis, and transferred to a nylon filter. The filter was hybridized with the first three exons of the CA II gene. Lanes: C, DNA isolated directly from cells with no DNase I treatment; 1–4, DNA from uninduced cells; 5–8, DNA from induced cells. Marker DNA sizes are shown at the left in kilobase pairs.

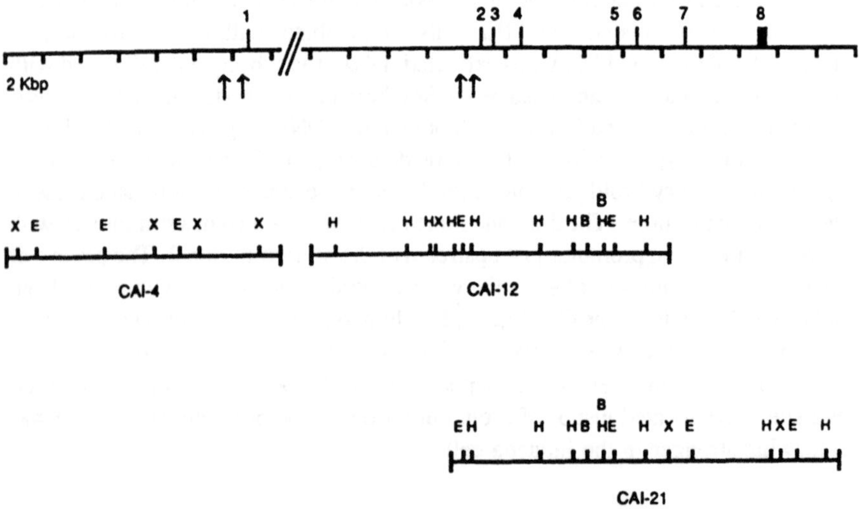

Figure 2. DNase I hypersensitive sites of the mouse CA I gene. Locations of the mouse CA I gene exons 1–8 are indicated by the numbered black boxes marked on the upper line. The discontinuity in the gene indicates the undetermined size of intron 1. The DNase I hypersensitive sites are indicated by arrows. Below are shown the restriction maps of λCAI-4, -12, and -21, which contain all of the exons. Restriction sites: B, *Bam*HI; E, *Eco*RI; H, *Hin*dIII; X, XbaI.

Figure 3. DNase I hypersensitive sites of the mouse CA II gene. Locations of the mouse CA II gene exons 1–7 are indicated by the numbered boxes marked on the upper line. The DNase hypersensitive sites are marked by arrows. Below are shown the restriction maps of λCAII-6 and -13, which contain all of the exons. Restriction sites: B, *Bam*HI; E, *Eco*RI; H, *Hin*dIII.

To determine the events that regulate these genes, the two genes have been cloned and characterized. DNase I hypersensitive sites have been mapped, since it is likely that one or more of these sites plays a role in regulation of gene expression. The next important step is to identify which sites or regions of CA I and CA II are essential for erythroid-specific expression.

ACKNOWLEDGMENTS. This work was supported in part by NIH grant CA10815. P. Cummings, P. Fraser, and W. Thierfelder were supported by NIH training grant CA09171 to the Wistar Institute.

References

1. Brady, H. J. M., Lowe, N., Sowden, J. C., Barlow, J. H., and Butterworth, P. W., 1989, *Biochem. Soc. Trans.* **17**:184–185.
2. Curtis, P. J., 1983, *J. Biol. Chem.* **258**:4459–4463.
3. Curtis, P. J., Withers, E., Demuth, D., Watt, R., Venta, P. J., and Tashian, R. E., 1983, *Gene* **25**:325–332.
4. Fraser, P. J., and Curtis, P. J., 1986, *J. Mol. Evol.* **23**:294–299.
5. Fraser, P. J., and Curtis, P. J., 1987, *Genes Dev.* **1**:855–886.
6. Fraser, P., Cummings, P., and Curtis, P., 1989, *Mol. Cell. Biol.* **9**:3308–3313.
7. Friend, C., 1957, *J. Exp. Med.* **105**:307–318.
8. Friend, C., Scher, W., Holland, J. G., and Sato, T., 1971, *Proc. Natl. Acad. Sci. USA* **68**:378–382.
9. Kabat, D., Sherton, C. C., Evans, L. H., Bigley, R., and Koler, R. D., 1975, *Cell* **5**:331–338.
10. Marks, P. A., and Rifkind, R. A., 1978, *Annu. Rev. Biochem.* **47**:419–448.
11. Stern, R. H., Boyer, S. H., Conscience, J. I., Friend, G., Margalet, I., Tashian, R. F., and Ruddle, F. H., 1977, *Proc. Soc. Exp. Biol. Med.* **156**:52–55.
12. Venta, P. J., Montgomery, J. C., Hewett-Emmett, D., Wiebauer, K., and Tashian, R. E., 1985, *J. Biol. Chem.* **260**:12130–12135.
13. Villeval, J. L., Testa, U., Vinci, G., Tonthat, H., Bettelab, A., Titeux, M., Cramer, P., Edelman, L., Rochant, M., Breton Gorius, J., and Veinchenker, W., 1985, *Blood* **66**:1162–1170.
14. Volloch, V., and Housman, D., 1982, *J. Cell Biol.* **93**:390–394.

CHAPTER 18

Structure and Expression of the Carbonic Anhydrase III Gene

YVONNE H. EDWARDS

1. Introduction

Carbonic anhydrase (CA) has evolved into at least six separate forms, CA I to CA VI (for reviews, see references 36 and 37). Structural comparisons at both the protein and DNA levels show that the CA gene family has expanded from a single ancestral gene by the process of duplication; this origin is further evidenced by chromosomal synteny of at least three of the CA genes. The close linkage of mouse *CA1* and *CA2* near the centromere of chromosome 3 was reported more than a decade ago.[12] More recently, we have shown by linkage analysis and *in situ* hybridization that mouse *CA3* maps to the same region.[1] Similarly, the human *CA1*, *CA2*, and *CA3* genes have been shown to be clustered on the long arm of chromosome 8 at q22.[7,9,10,28] The probable presence of CA I, CA II, and CA III isoforms in birds and reptiles suggests that the duplicated CA genes have maintained linkage for more than 300 million years. It is not clear why this linkage has been retained and whether the chromosomal proximity of the CA genes plays any part in their widely diverse patterns of expression.

Three of the mammalian CA genes are each expressed predominantly in one cell type, cytoplasmic CA I and CA III in erythrocytes and skeletal muscle cells, respectively, and the secreted form CA VI in parotid salivary glands. These CA genes show a greater degree of tissue specificity than do the more ubiquitously expressed cytoplasmic CA II or the membrane-bound CA IV genes, and the mitochondrially associated CA V gene. It will be of considerable interest to understand the molecular mechanisms that control expression of these various CA genes and the timing of regulatory events during embryogenesis and cell type

YVONNE H. EDWARDS • MRC Human Biochemical Genetics Unit, The Galton Laboratory, University College London, London NW1 2HE, United Kingdom.

differentiation. Toward these objectives, we have cloned and characterized the mRNA and the gene for CA III.[24] This chapter describes some recent observations on the regulation of this gene.

2. Expression and Regulation of the CA III Gene

CA III is a major component of the sarcoplasm, accounting for 7 mg/g (wet weight) of muscle tissue in human limb muscle.[18] The properties of this isoform are reviewed elsewhere in this volume.[13,15] The exact role of CA III in muscle is still uncertain but an understanding of those factors which determine how and when this gene is expressed may throw some light on its physiological function.

2.1. Cell Type Specificity and CA III Expression during Myogenesis

There are three phases of myogenesis, which follow on from the commitment of the mesodermal stem cell to the muscle cell lineage. In the first phase, the mononucleate myoblasts fuse to form multinucleate myotubes. In the second phase, the contractile apparatus is assembled and morphogenesis takes place to give rise to recognizable muscles. In the last period, the individual muscle fibers differentiate into the major adult fiber types: type 1 slow, oxidative fibers; type 2A fast, intermediate, glycolytic–oxidative fibers; and type 2B fast, glycolytic fibers. This ultimate step is influenced directly by neuronal and hormonal signals. The expression of CA III in the very earliest phase of myogenesis has been examined by using the rodent myogenic cell line G8. Both CA III mRNA and protein are found in significant amounts (0.05–1% of the adult skeletal muscle level) in the undifferentiated myoblast, and these levels are not apparently induced by myotube formation.[38] This finding was unexpected, since myoblast fusion is marked by the coordinate induction of many other muscle-specific genes, including those of the contractile apparatus (such as myosin, actin, and tropomyosin), myoglobin, and creatine kinase (for reviews, see references 5 and 33). Furthermore, these gene products are either absent or barely detectable in mononucleate myoblasts. Thus, it appears that CA III activity is established upon commitment of the mesodermal stem cell to the muscle lineage and prior to the contractile apparatus. This may imply that CA III activity and contractile activity are unrelated.

During the next phase of muscle development, when individual muscles are formed and become active, CA III activity is slow to rise. For example, in human fetal muscle at 24 weeks of gestation, CA III is only about 20% of adult amounts even though muscle morphogenesis is complete and significant limb movement is occurring. Amounts of CA III rise more rapidly during the last trimester, to reach 50–60% of adult levels at birth.[18,23] This late developmental increase appears to coincide with the appearance of differentiated adult muscle fiber types, in particular the type 1 slow-twitch fibers. Histological analyses of rat muscle using labeled CA III-specific antibodies[41] and hybridization studies using a cDNA probe to mRNAs from rabbit muscles[4] have established that CA III mRNA and protein are largely confined to the type 1, slow-twitch fibers (Fig. 1).

Muscle fiber type differentiation, and thus the expression of genes that characterize a particular fiber type, is in part dependent on the pattern of electrical stimulation imposed on the muscle cell by the motor nerve. This influence can be monitored experimentally by using implanted electrical stimulators that artificially alter the contractile activity of muscle. When a fast-twitch muscle, such as rabbit anterior tibialis or extensor digitorum longus, is stimulated in a pattern characteristic of a slow-twitch muscle, a generalized fast-to-slow transformation occurs (for reviews, see references 30 and 31). This transition is marked, *inter alia*, by an 11-fold increase of both CA III mRNA and protein after 4–10 days of stimulation (Fig. 2). A maximal level, close to that found in an untreated type 1 muscle, is reached between 14 and 21 days, and preliminary data suggest that higher levels of CA III protein are retained up to 42 days after the electrostimulation has ceased.[4]

These observations indicate that the repression of CA III gene activity in the

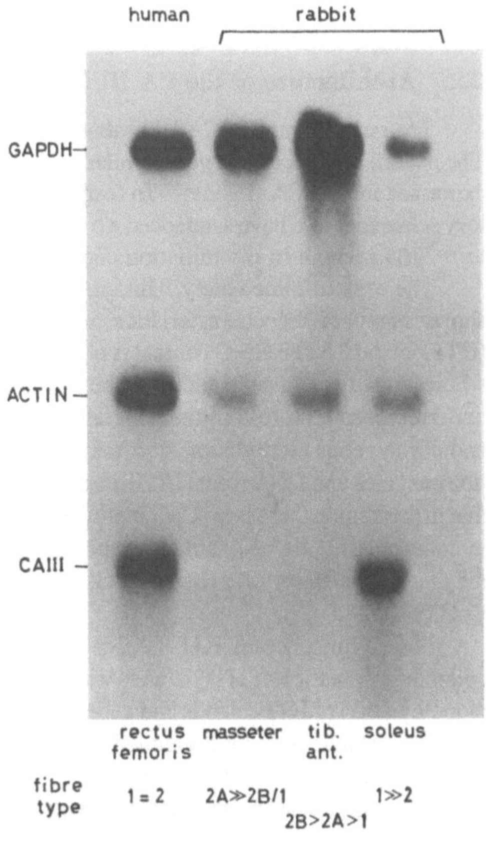

Figure 1. Northern analyses showing the distribution of CA III, α skeletal muscle actin, and glyceraldehyde-3-phosphate dehydrogenase (GAPDH) mRNAs in rabbit muscles of different fiber type composition; 1, slow-twitch oxidative fibers; 2A, fast-twitch intermediate fibers; and 2B, fast-twitch glycolytic fibers. Three identical filters were probed with ^{32}P-labeled cDNA probes. tib. ant., Anterior tibialis.

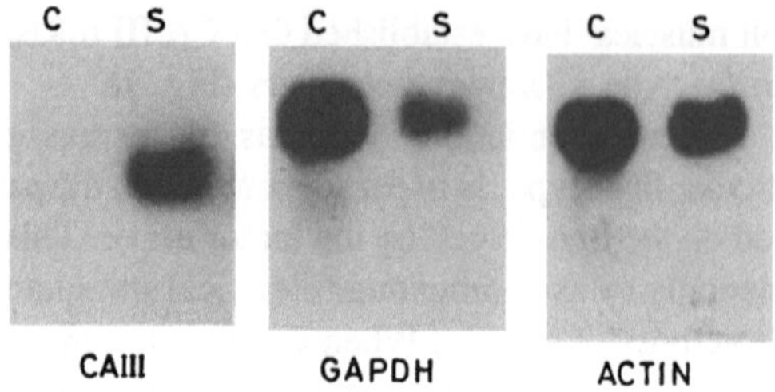

Figure 2. Northern analyses of CA III, GAPDH, and actin mRNAs in 21-day stimulated anterior tibialis muscle (S) and contralateral unstimulated control (C).

type 2, fast-twitch muscle fibers is reversible and is responsive to changing external influences. In this context, it is interesting that a study of CA III DNA methylation patterns in human tissues suggests that type 1 and type 2 muscle fibers show the same pattern of undermethylation that is characteristic of CA III-expressing cells, in contrast to the overmethylation found in the nonexpressing liver DNA.[24]

2.2. Architecture of the CA III Gene

The gene for human CA III is about 10.3 kb long and comprises seven exons. The overall architecture is very similar to that described for the more ubiquitously expressed mouse CA II gene.[39] To facilitate a detailed search for potential regulatory sequences, we have sequenced a 5.2-kb region of DNA from positions −2945 to +2264 relative to the initiation codon.[24]

The region immediately flanking the 5′ end of the CA III gene and exon I shows many of the characteristics of a methylation-free *Hpa*II tiny-fragment (HTF) island.[2,3] The G+C content is 60% in this region, compared with 40% in the whole human genome. In addition, the incidence of CpG dinucleotides, which are underrepresented 5-fold in the general genome, increases about 10-fold at the 5′ end of the gene. Methylation studies using various methylation-sensitive restriction enzymes and DNA from CA III-expressing and -nonexpressing tissues confirm that in this region there is no CpG methylation either in CA III-expressing tissues or in nonexpressing tissues, including sperm.[11] Thus, human CA III appears to be in the unusual category of a tissue-specific gene associated with a methylation-free CpG cluster.

The significance of HTF islands in tissue-specific genes is uncertain. The common association of HTF islands with the 5′ ends of genes and the concurrence of methylation of HTF islands and gene inactivation on the X chromosome suggest that HTF islands may be prerequisite for gene transcription. This function may involve binding with transcription factors, and such binding might occur preferentially at nonmethylated sites. However, since such methylation-free zones occur in

DNA from all tissue, regardless of gene expression, it seems unlikely that they play a role in tissue-specific expression. In the case of CA III, the HTF island may represent an evolutionary remnant from an ancestral gene with a more ubiquitous pattern of expression.[8]

The upstream sequences of the CA III gene exhibit other features that are characteristic of housekeeping genes and thought to define constitutive expression. These are illustrated in Fig. 3. There is fairly good alignment (48%) with a 25-bp consensus sequence found downstream of the TATA box in a number of housekeeping genes.[25] Overlapping this region is a 41-bp sequence spanning the TATA box which shows close homology (90%) between human CA II and CA III. In addition, the Sp1 transcription factor-binding hexanucleotide[20] occurs twice within 170 bp of the cap site. At one of these positions, −173 bp in CA III, the Sp1 site is flanked by the repeated element CACCTC. This motif is conserved in the CA II gene at −110 bp[34] and has been shown to contribute to transcriptional efficiency.

About 1.5 kb upstream of the CA III transcription initiation site is a stretch of sequence with striking homology with the simian virus 40 (SV40) early enhancer. This enhancer is encoded in a 125-bp sequence and comprises several distinct motifs that have been identified as transcriptionally important.[32] All of the SV40 motifs are present, in either single or double copies, within a 283-bp region of the CA III gene and point to this region as being a transcription enhancer (Fig. 4). It seems unlikely that these motifs confer muscle specificity, since the SV40 enhancer is active in all cell types and allows ubiquitous expression. However, SV40 motifs have been found in the immediate area of tissue-specific enhancers in other genes, in particular near the muscle-specific creatine kinase enhancer.[17]

Since fast-to-slow muscle fiber transitions and concomitant increases in muscle CA III are induced by thyroid hormone (T_3; 3,3,′5-triiodo-L-thyonine) treatment,[14] a search for possible thyroid hormone receptor binding elements (TRE) in the CA III sequence was made. The most likely candidate was found at −854 bp relative to the cap site and shows homology across a 20-bp region with the TRE in the rat growth hormone (GH) gene[22]:

AAAGGTA-AGATCAGGGACGTG rat GH TRE

AGAGGTACAGCTCAGGGA-GTG human CA III TRE

The T_3 receptor is capable of binding to a large family of related sequences including, for example, estrogen receptor elements and glucocortocoid receptor elements (GRE).[16] A possible GRE, similar to that of the rat metallothionein II gene, occurs at −2100 bp in the CA III gene, and this site is immediately adjacent to a progesterone receptor element-like sequence.[27]

Direct experiments using *in vitro* and *in vivo* gene expression systems will establish which of these candidate regulatory sequences are transcriptionally active.

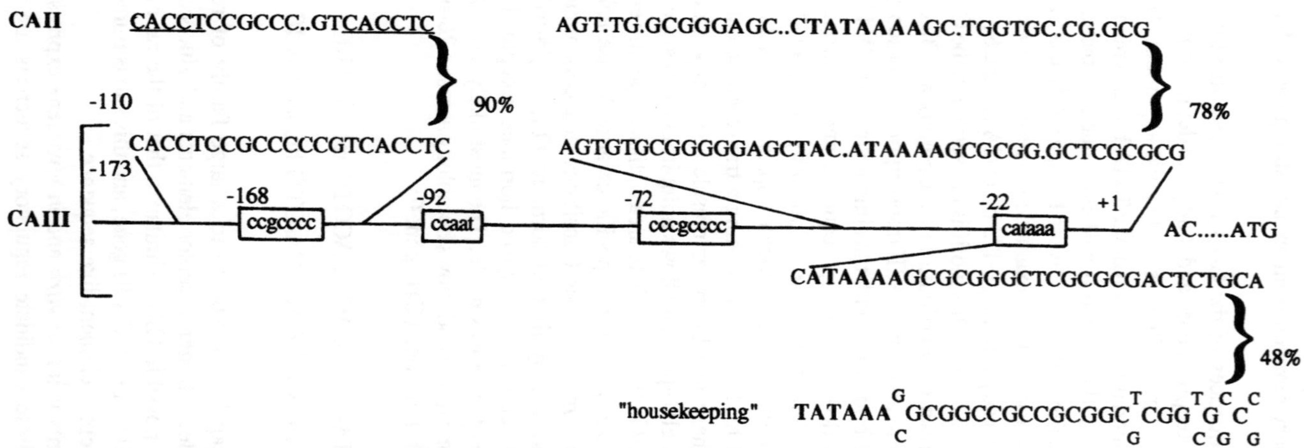

Figure 3. 5′-Flanking sequence of the human CA III gene showing the disposition of sequences associated with transcription by RNA polymerase II and indicating homologies with the human CA II sequence[34] and housekeeping genes.[25]

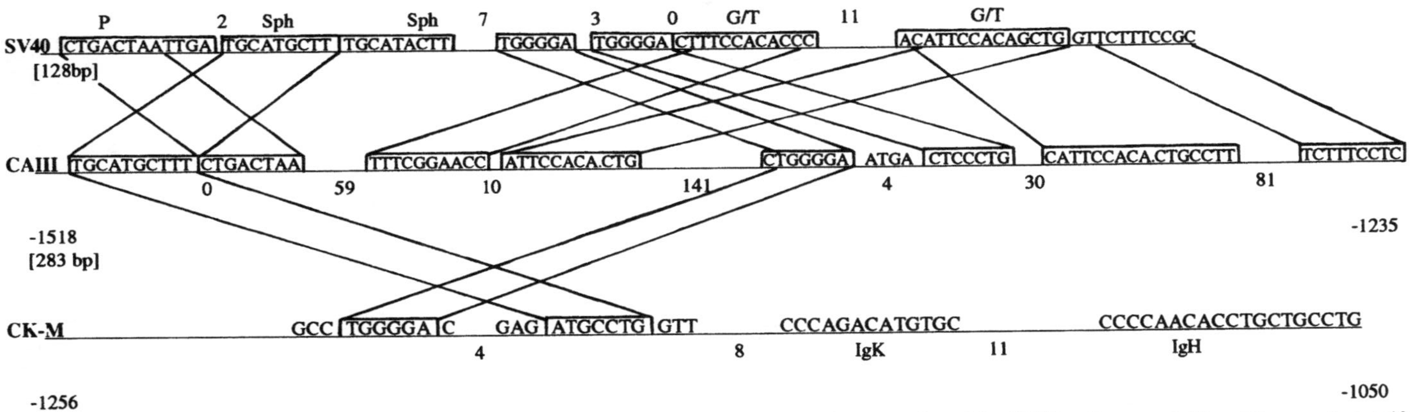

Figure 4. Comparisons of a putative enhancer region in the human CA II gene with the enhancer region of the SV40 early promoter[32] and the muscle-specific enhancer of creatine kinase M (CK-M).[17] The relative positions upstream and the distances between elements are indicated.

2.3. CA III Expression in Rodent Liver

In humans and other large mammals, CA III activity is essentially confined to skeletal muscle, whereas in rats and mice, CA III is also expressed in liver. Data from peptide analysis,[6] isoelectric focusing (Fig. 5), and RNase protection assays[38] show that the liver and muscle tissue forms are determined by the same gene.

Rat liver exhibits a sexual dimorphism, with males showing hepatic CA III levels 10- to 20-fold those of females.[35] This difference appears to be associated with the pattern of secretion of growth hormone[19,26] but appears to be peculiar to the rat. No or only slight differences between the levels of male and female mouse liver CA III protein and mRNA were found with use of immunodetection after isoelectric focusing and RNase protection assays using the mouse cDNA as probe[38] (Fig. 5). This result was unexpected, since the patterns of growth hormone secretion are very similar in rats and mice; indeed, the growth hormone-related sexual differentiation of the major urinary protein[29] could readily be detected in the mouse livers. This observation may indicate that the intermediary steps between GH secretion and CA III regulation are complex and show interspecies variation.

In preliminary gene structure studies, we have used the rat CA III cDNA[21] to isolate genomic clones. Analysis of one clone, λCA3RG.1, which contains the entire rat gene in a 15-kb insert, has provided interesting data. A 5.5-kb *Eco*RI fragment of the clone, which hybridizes to the 5′ end of the CA III cDNA, has been mapped and partially sequenced (J. Charlton and Y. H. Edwards, unpublished data). At position +132 relative to the putative cap site and in the 5′ untranslated

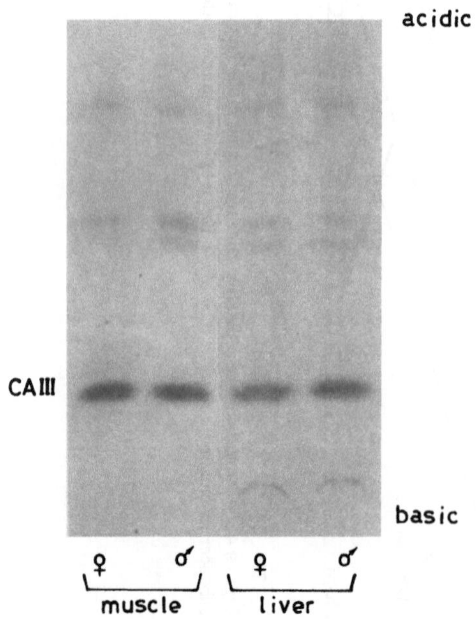

Figure 5. CA III protein in aqueous extracts prepared from male and female mouse muscle and liver, detected by using a specific antibody after isoelectric focusing and Western blotting.

region, the first exon is apparently interrupted by an intron. However, no intron is found at this position in the human gene, and the splice site at the exon/intron junction AGCA/tgaa does not conform to the consensus sequence for donor splice sites. At least two other explanations are possible. First, λCA3RG.1 may contain sequences that have become rearranged during the cloning process; other clones are being analyzed to test this idea. Second, the cDNA that has been cloned may represent an exceptional mRNA with an unusual 5′ untranslated region. To test this possibility, CA III mRNA from rat muscle was examined by using an RNase protection assay. The probe was an *Eco*RI–*Pst*I cDNA fragment containing the entire 5′ untranslated region and 135 bp of coding sequence. This analysis showed that rat muscle appears to contain a mixed population of mRNA. One class gives a full-length protected fragment and must contain sequences identical to the cloned cDNA. The other class apparently contains sequence which is not fully protected by the cDNA probe so that at least two RNA fragments are generated.

3. Summary

The detailed characterization of the human and rat CA III genomic sequences provides the necessary tools with which to confront some of the important questions relating to CA III gene expression:

1. What are the factors which determine the initial expression of CA III activity in cells committed to the muscle cell lineage and the subsequent induction of CA III to high levels in adult differentiated type 1 fibers? A full answer on these points will require some understanding of the component pathway whereby neural impulse is translated into gene activation or repression.
2. How is tissue-specific expression imposed and maintained during development and after tissue differentiation?
3. What species-specific differences, either in the structure of the CA III gene or in availability of regulatory factors, leads to the activation of CA III in rodent liver and absence in human liver? How are these features modulated to give rise to the sexually dimorphic pattern of expression characteristic of the rat liver?

References

1. Beechey, C., Tweedie, S., Spurr, N., Peters, J., and Edwards, Y., 1990, *Genomics* **4**:672–696.
2. Bird, A. P., 1986, *Nature* (London) **321**:209–213.
3. Bird, A., Taggart, M., Nicholls, R., and Higgs, D., 1987, *EMBO J.* **6**:999–1004.
4. Brownson, C., Isenberg, H., Brown, W., Salmons, S., and Edwards, Y., 1988, *Muscle Nerve* **11**:1183–1189.
5. Caravatti, M., Minty, A., Robert, B., Montarras, D., Weydert, A., Cohen, A., Daubas, P., and Buckingham, M., 1982, *J. Mol. Biol.* **160**:59–76.

6. Carter, N., Hewett-Emmett, D., Jeffery, S., and Tashian, R., 1981, *FEBS Lett.* **128**:114–118.
7. Davis, M., West, L., Barlow, J., Butterworth, P. H. W., Lloyd, J., and Edwards, Y. H., 1987, *Somat. Cell Mol. Genet.* **13**:173–178.
8. Edwards, Y. H., 1990, *Phil. Trans. R. Soc.* **326**:207–215.
9. Edwards, Y. H., Barlow, J. M., Konialias, C. P., Povey, S., and Butterworth, P. H., 1986, *Ann. Hum. Genet.* **50**:123–129.
10. Edwards, Y. H., Lloyd, J., Parkar, M., and Povey, S., 1986, *Ann. Hum. Genet.* **50**:41–47.
11. Edwards, Y. H., Charlton, J., and Brownson, S., 1988, *Gene* **71**:473–481.
12. Eicher, E. M., Stern, R. H., Womack, J. E., Davisson, M. T., Roderick, T. H., and Reynolds, S. C., 1976, *Biochem. Genet.* **14**:651–660.
13. Eriksson, A. E., and Liljas, A., this volume, Chapter 3.
14. Fremont, P., Charcot, P., Cote, C., and Rogers, P., 1988, *J. Histochem. Cytochem.* **36**:775–782.
15. Geers, C., and Gros, G., this volume, Chapter 19.
16. Glass, C., Holloway, J., Devary, O., and Rosenfeld, M., 1988, *Cell* **54**:313–323.
17. Jaynes, J., Johnson, J., Buskin, J., Gartside, C., and Hauschka, S., 1988, *Mol. Cell. Biol.* **8**:62–70.
18. Jeffery, S., Edwards, Y., and Carter, N., 1980, *Biochem. Genet.* **18**:843–849.
19. Jeffery, S., Wilson, C. A., Mode, A., Gustafsson, J.-A., and Carter, N., 1986, *J. Endocrinol.* **110**:123–126.
20. Kadonaga, J., Jones, K., and Tjian, R., 1986, *Trends Biochem. Soc.* **II**:20–23.
21. Kelly, C., Carter, N., Jeffery, S., and Edwards, Y., 1988, *Biosci. Rep.* **3**:401–406.
22. Koenig, R., Brent, G., Warne, R., Larsen, P., and Moore, D., 1984, *Proc. Natl. Acad. Sci. USA* **84**:5670–5674.
23. Lloyd, J., McMillan, S., Hopkinson, D. A., and Edwards, Y., 1986, *Gene* **41**:233–239.
24. Lloyd, J., Brownson, C., Tweedie, S., Charlton, J., and Edwards, Y. H., 1987, *Genes Dev.* **1**:594–602.
25. Martini, G., Toniola, D., Vulliamy, T., Luzzato, L., Dono, R., Viglietto, G., Paonessa, G., Durso, M., and Perisco, M., 1986, *EMBO J.* **5**:1849–1855.
26. Mode, A., Gustafsson, J.-A., Jansson, J.-O., Eden, S., and Isaksson, O., 1982, *Endocrinology* **111**:1692–1697.
27. Mulvihill, E., LePenneo, J.-P., and Chambon, P., 1982, *Cell* **28**:621–632.
28. Nakai, H., Byers, M. G., Venta, P. J., Tashian, R. E., and Shows, T. B., 1987, *Cytogenet. Cell Genet.* **44**:234–235.
29. Norstedt, G., and Palmiter, R., 1984, *Cell* **36**:805–812.
30. Pette, D., 1984, *Med. Sci. Sports* **16**:517–528.
31. Salmons, S., and Henriksson, J., 1981, *Muscle Nerve* **4**:94–105.
32. Senke, M., Grundstrom, T., Matthes, H., Witzerith, M., Schatz, C., Wildeman, A., and Chambon, P., 1986, *EMBO J.* **5**:387–397.
33. Shani, M., Zevin-Sonkin, D., Saxel, O., Carmon, Y., Katcoff, D., Nudel, U., and Yaffe, D., 1981, *Dev. Biol.* **86**:483–492.
34. Shapiro, L., Venta, P., and Tashian, R., 1987, *Mol. Cell. Biol.* **7**:4589–4593.
35. Shiels, A., Jeffery, S., Phillips, I., Shephard, E., Wilson, C., and Carter, N., 1983, *Biochem. Biophys. Acta* **760**:335–342.
36. Tashian, R. E., 1989, *BioEssays* **10**:186–192.
37. Tashian, R. E., and Hewett-Emmett, D. (eds.), 1984, *Ann. N.Y. Acad. Sci.* **249**.
38. Tweedie, S., and Edwards, Y., 1989, *Biochem. Genet.* **27**:17–30.
39. Venta, P., Montgomery, J., Hewett-Emmett, D., Wiebauer, K., and Tashian, R., 1985, *J. Biol. Chem.* **260**:12130–12135.
40. Venta, P., Shows, T., Curtis, P., and Tashian, R., 1983, *Proc. Natl. Acad. Sci. USA* **80**:4437–4440.
41. Wistrand, P., Carter, N., and Askmark, H., 1987, *Comp. Biochem. Physiol.* **86A**:177–184.

PART V

Muscle Carbonic Anhydrases

CHAPTER 19

Muscle Carbonic Anhydrases

Function in Muscle Contraction and in the Homeostasis of Muscle pH and pCO_2

CORNELIA GEERS and GEROLF GROS

1. Localization in Muscle Cells

In striated muscles, at least three types of carbonic anhydrase (CA) have been demonstrated: (1) a sulfonamide-resistant isozyme, CA III, which appears in the cytosol of red skeletal muscles,[7,33,38] (2) a sulfonamide-sensitive cytosolic isozyme, most likely CA II,[51] and (3) a membrane-bound form present in the sarcolemma[16,23,57] and the sarcoplasmic reticulum (SR).[5] Depending on the muscle type, these CA isozymes are variably present at these different sites in the muscle cell.

1.1. Cytoplasm

In muscle, CA III is present almost exclusively in the cytoplasm of slow-oxidative fibers[35,50,56] and is largely absent in all other skeletal muscle fibers and in heart.[9,49] This isozyme is also found in the liver of adult male rats.[8] Immunoelectron microscopic studies showed a uniform distribution within the sarcomere of cells[56] (Frémont *et al.*, this volume). In the rat soleus, CA III is about 50% of the soluble protein,[49] about 13 g/kg wet weight.[9] In the cytosol from white muscles of the rabbit [containing fast-twitch glycolytic and fast-twitch oxidative–glycolytic (FOG) fibers], a sulfonamide-sensitive CA identical to CA II was found with biochemical methods.[51] Histochemical studies confirmed the presence of a sulfonamide-sensitive CA in the cytoplasm FOG fibers.[43] There has been difficulty in

CORNELIA GEERS and GEROLF GROS • Zentrum Physiologie, Medizinische Hochschule Hannover, 3000 Hannover 61, Federal Republic of Germany.

localizing this enzyme. Because of its high specific activity, the activity exhibited by this CA is comparable to that of CA III in red skeletal muscles. In the cytosol from rabbit heart, several studies have confirmed the absence of any CA.[9,30,33,41]

1.2. Sarcolemma

From measurements of the distribution space of $H^{14}CO_3^-$ in perfused hindlimbs of various mammalian species, the presence of an extracellular CA in skeletal muscle was proposed. It was observed that this distribution space is greatly reduced by CA inhibitors.[18,60] pH stop-flow studies[42] indicated that H^+–HCO_3^-–CO_2 is near equilibrium in a CA-free perfusate leaving a rabbit hindlimb but that a pH disequilibrium develops when a CA inhibitor is added to the perfusate; thus, an extracellular CA was predicted in skeletal muscle. Most investigators speculated that this extracellular CA is on the capillary endothelial cell wall. Measurement of the $H^{14}CO_3^-$ distribution space in the presence of CA inhibitors of different molecular sizes has indicated that the extracellular CA of skeletal muscle is in the interstitial space, probably bound to the sarcolemma.[23] This has been confirmed histochemically.[11,35,43] Further indirect evidence for an extracellular sarcolemmal CA came from muscle surface pH measurements[15]; direct evidence provided by studies on isolated sarcolemmal vesicles[57,58] demonstrated for sarcolemma of white and red muscles from the rabbit that there is a membrane-bound CA. On the basis of its inhibition constants toward sulfonamides and anions, the latter authors concluded that this enzyme is different from the cytosolic enzymes CA I, CA II, and CA III.

1.3. Sarcoplasmic Reticulum

Preparations of SR from red and white skeletal muscles of the rabbit possess a membrane-bound CA[5] with acetazolamide sensitivity between that of human CA I and CA II. Histochemical studies with dansylsulfonamide showed a pattern of intracellular staining in red and white muscles that indicated sarcoplasmic reticulum staining. An immunocytochemical study[35] revealed a similar intracellular pattern.

1.4. Heart

The pattern of CA distribution in heart differs from that in skeletal muscle. Heart seems to be (at least in the rabbit) the only striated muscle without cytosolic CA but with a membrane-bound Triton-extractable CA (our unpublished observation). Such a Triton-extractable CA was also found in all rabbit skeletal muscles, although to a lesser degree, and was there determined to be CA from sarcolemma and SR. In heart muscle, the activity of this CA is three to eight times higher than in skeletal muscle, but the origin of this membrane-bound CA is not yet clear. In histochemical studies (unpublished observations), a pattern similar to that of skeletal muscle is observed, i.e., stained "spots" within the muscle cells. This finding indicates that in heart, CA is associated with the SR. There are conflicting physiological results on the question of whether a sarcolemmal CA is also present. Results from indicator dilution studies[61] in which, in contrast to the results with skeletal muscle, no change in the distribution space of $H^{14}CO_3^-$ was observed upon addition of CA inhibitors to the perfusate seem to suggest that no sarcolemmal CA is present in heart. From surface pH measurements, however, de Hemptinne *et al.*[15] concluded that an extracellular CA associated with heart sarcolemma should exist.

2. Known Functions of Muscle Carbonic Anhydrases

For about 10 years it has been known that the presence of CA in muscle is useful for the transfer of CO_2 from the site of production within muscle cells to the blood. Two separate mechanisms improving this transfer from the muscle cell to the blood can be discriminated: facilitation of CO_2 diffusion within the muscle cell and extracellular catalysis of CO_2 transport. In addition, it has been recognized during the last few years that muscle CA plays a role in H^+ buffering in muscle.

2.1. Facilitated CO_2 Diffusion

It has been shown that CA activity and a sufficient concentration of mobile buffers lead to a CO_2 diffusion facilitated by simultaneous HCO_3^- and H^+ transport.[28,29] The known presence of CA, and of mobile buffers as well, would then lead one to expect a facilitation of CO_2 diffusion in striated muscle. That a facilitation does occur has been shown for the red abdominal muscle of the rat by Kawashiro and Scheid,[36] and it was demonstrated by Gros *et al.*[31] (see also reference 26) that in this muscle facilitation is due to CA III. The latter authors demonstrated that (1) facilitated CO_2 diffusion in this muscle can be suppressed by incubating the muscle in acetazolamide and (2) the inhibitor concentration necessary to reduce facilitation by 50% ($IC_{50} = 3 \cdot 10^{-3}$ M) indicates the involvement of the relatively sulfonamide-resistant CA III.

Measurements of CO_2 diffusion across thin sections from various rabbit skeletal muscles and the heart (G. Gros, F. Romanowski, and J. Schierenbeck, unpublished results) reveal that facilitated diffusion occurs not only in skeletal muscles containing CA III but in all muscles of the rabbit that have been investigated. Table I shows the results of these studies. To eliminate uncertainties about the geometry of the muscle sections studied (with a rather inhomogeneous thickness of ~1 mm), the diffusion of CO_2 was compared with the diffusion of a gas that shows no facilitation, acetylene, and the ratio of diffusion constants, $K_{CO_2}/K_{acetylene}$, was determined rather than the CO_2 diffusion constant, K_{CO_2}.

The ratio $K_{CO_2}/K_{acetylene}$ was 0.9 in all muscles studied when the muscle sections had been incubated in 0.15 M lactate (pH 4) before the diffusion measure-

TABLE I
Facilitated CO_2 Diffusion in Various Rabbit Muscles[a]

Muscle	$K_{CO_2}/K_{acetylene} \pm$ SD Control	Lactate pH 4	Acetazolamide 1–6 mM	Facilitation factor	CA isozymes present
Heart	2.4 ± 0.4	0.9 ± 0.1	1.3 ± 0.2	2.7	Membrane bound
Soleus	3.1 ± 0.4	0.9 ± 0.1	2.0 ± 0.3	3.4	CA III; membrane bound
EDL	1.9 ± 0.3	0.9 ± 0.02	1.0 ± 0.1	2.1	CA II; membrane bound
Masseter	1.9 ± 0.3	0.9 ± 0.1	1.3 ± 0.2	2.1	CA II(?); membrane bound

[a]Measurements were performed on 1-mm-thick sections cut from the muscles parallel to the longitudinal fiber axes. The sections were mounted between two chambers that were filled with 95% O_2–5% CO_2. Into one of the chambers, small amounts of $^{14}CO_2$ and acetylene were introduced; their appearance in the other chamber was monitored over time by taking small samples that were analyzed for ^{14}C and acetylene. Control, Muscle sections preincubated in Ringer solution, pH 7.4; lactate pH 4, sections preincubated in 0.15 M sodium lactate, pH 4.0; acetazolamide 1–6 mM, sections preincubated in Ringer solution with between 1 and 6 mM acetazolamide. Given are the ratios of the diffusion constants for CO_2 and acetylene. The facilitation factor is calculated as ratio of the figure in column 1 over that in column 2. Temperature, 22°C. Diffusion data are from Gros, Romanowski, and Schierenbeck (unpublished); isozyme distribution data are from Geers *et al.* (cited in reference 26).

ments were performed (Table I). After the ensuing acidification of the tissue, any facilitated diffusion of CO_2 is expected to be entirely suppressed. Thus, 0.9 is the figure expected when no facilitation of CO_2 diffusion occurs; this is roughly identical to the ratio of the CO_2 and acetylene diffusion constants in water. Any increase of $K_{CO_2}/K_{acetylene}$ above 0.9 indicates that the diffusion of CO_2 is facilitated. In all untreated muscle sections studied, $K_{CO_2}/K_{acetylene}$ was significantly greater than 0.9 (Table I). Thus, facilitated diffusion of CO_2 occurs in all of these muscles. The facilitation factor indicates the factor by which CO_2 diffusion is accelerated as a result of facilitation ($=K_{CO_2}/K_{acetylene}$ in untreated control sections divided by $K_{CO_2}/K_{acetylene}$ in sections incubated at pH 4). It varies between 3.4 for soleus and 2.1 for extensor digitorum longus (EDL) and masseter. The third column of Table I shows that preincubation of the muscle sections in acetazolamide leads to a marked decrease in $K_{CO_2}/K_{acetylene}$, which demonstrates that facilitation of CO_2 diffusion in all muscles depends on the presence of an active CA.

Which CA isozyme is responsible for the facilitation of CO_2 diffusion? The most clear-cut case is heart, where cytoplasmic CAs are absent and only the membrane-bound form occurs. Thus, membrane-bound CAs, either sarcolemmal or SR bound or both, can produce facilitated CO_2 diffusion. Since we have previously shown that in the red abdominal muscle of the rat it is mainly CA III that causes facilitated diffusion, the same is likely to hold for the red soleus, which has a very high activity of CA III and relatively little membrane-bound CA.[26] In the case of EDL and masseter, either the cytoplasmic CA II that is known to be present in EDL or the membrane-bound forms that are present in both muscles, but not CA III, which is hardly present in either muscle, can be responsible for facilitated CO_2 diffusion.

The acceleration of CO_2 diffusion apparent from these data two- to fourfold, is much greater than that observed for the red cell[27] and that found previously for rat abdominal muscle.[26,36] It may be noted that the degree of facilitation is greatest in the two muscles with the highest rate of resting oxygen consumption, heart and soleus, and smaller in the two muscles that have a high glycolytic capacity, EDL and masseter.

2.2. Extracellular Catalysis of CO_2 Transport

A CA available in the interstitial space of skeletal muscle is thought to accelerate the uptake of CO_2 by the blood.[23] Such an interstitial CA activity will result in the (partial) hydration of metabolically produced CO_2 right after it has left the cells (Fig. 1). About 50% of the CO_2 taken up by the blood is transported in form of plasma HCO_3^-. The formation of part of this plasma HCO_3^- in the extracellular space would thus circumvent the formation of HCO_3^- within red cells, with subsequent HCO_3^-–Cl^-, exchange and allow the rapid utilization of the buffer capacity of the plasma for H^+. From prior studies,[4,12,19,32] such a

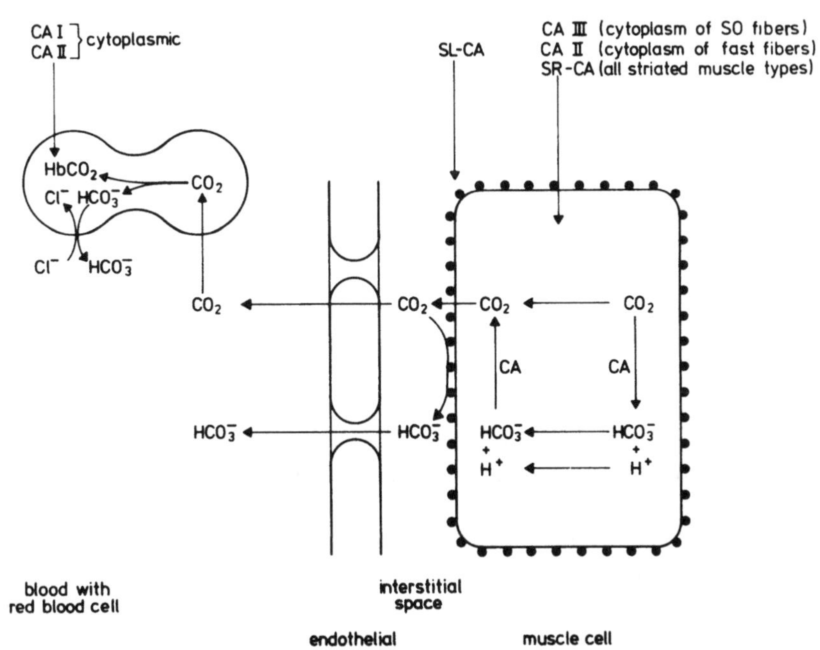

Figure 1. Schematic diagram of the localization of CAs inside striated muscles. Their functions for the transfer of CO_2 from the site of production within muscle cells to the blood are (1) facilitation of CO_2 diffusion within the muscle cell and (2) extracellular catalysis of CO_2 transport in a muscle capillary. SL-CA, Extracellular sarcolemmal CA; SR-CA, sarcoplasmic reticulum CA.

capability for the rapid formation of HCO_3^- and H^+ in the extracellular space is expected to accelerate the CO_2 uptake by the blood and increase the amount of CO_2 taken up during capillary passage of blood in skeletal muscle. However, direct experimental evidence demonstrating the functional importance of this mechanism in resting and exercising muscle is lacking.

2.3. Intra- and Extracellular H^+ Buffering

In skeletal muscle, cytoplasmic and extracellular CA may function in the kinetics of H^+ buffering. It has been postulated that the CO_2–HCO_3^- system contributes about 40% to the intracellular buffer capacity.[1,44] However, without CA this system may be too slow when H^+ appear in the cytoplasm rapidly. Intracellular lactic acid probably rises to 20–30 mM within less than 1 min, whereas the hydration–dehydration reaction requires $\geqslant 1$ min for near completion. If the system CO_2–HCO_3^- is too slow to participate in H^+ buffering, the loss in immediate intracellular buffering power can be illustrated as follows: assuming an intracellular "nonbicarbonate" buffer factor of 40 mM/pH, $pCO_2 = 45$ mm Hg, intracellular pH (pH_i) = 7.20, and $[HCO_3^-]_i = 18.4$ mM. When the concentration of lactic acid rises to 30 mM and the system CO_2–HCO_3^- reacts fast enough while pCO_2 remains constant, pH_i will fall to 6.75; if the CO_2–HCO_2^--system does not participate in buffering, pH_i will transiently fall to 6.45. Thus, almost half of the intracellular buffering power under these conditions is due to HCO_3^-.

In a snail neuron, the importance of CA for the intracellular buffering by CO_2–HCO_3^- has been demonstrated.[55] The pH_i was recorded with microelectrodes, and the change of pH_i upon intracellular injections of defined amounts of HCl was observed. Each injection was followed by a considerable fall in pH_i when the neuron was superfused with Ringer solution equilibrated with 0.5% CO_2 ($pH_0 = 7.5$). When the superfusion solution was equilibrated with 4% CO_2 (pH_0 constant at elevated $[HCO_3^-]_0$), the fall in pH_i after HCl injection was considerably reduced because the concentration of intracellular HCO_3^- and thus the intracellular buffer power had increased. When acetazolamide was added to the latter superfusate, the fall in pH_i upon HCl injection increased twofold, indicating that in the absence of CA activity the intracellular buffering power of the system CO_2–HCO_3^- could not be fully utilized. Similar results are expected for skeletal muscle, where intracellular CA activity is known present, and possibly also in heart muscle. There is no direct evidence in the case of muscle.

A function for the extracellular sarcolemmal CA in extracellular buffering was demonstrated.[15] Using pH microelectrodes, pH_i and surface pH (pH_s) were recorded from skeletal and heart muscle. Sudden addition and withdrawal of propionic acid caused transient changes of pH_s, which were increased in the presence of 0.1 mM acetazolamide. Since CO_2–HCO_3^- is almost the only buffer system in the interstitial space, an important role of the sarcolemmal CA may be to prevent drastic acidification of the interstitium when lactic acid moves rapidly out of muscle fibers. Since fast white muscles produce more lactic acid than do slow red muscles, a higher specific CA activity in white than in red muscle sarcolemma[57] is in keeping with such a function of extracellular sarcolemmal CA.

3. Functional Consequences of Muscle Carbonic Anhydrase Inhibition

To investigate the involvement of one or more CA in muscle tissues in excitation–contraction coupling, we have studied isolated (unperfused) muscles in the presence and absence of CA inhibitors.[20–22]

3.1. Contractile Parameters

Several parameters of isometric contraction of isolated bathed (20–22°C, 95% O_2–5% CO_2, pH 7.4) skeletal muscles have been found to be altered by CA inhibition.

3.1.1. Isometric Force

Barclay[3] studied the isometric tetanic force developed by isolated, directly stimulated mouse soleus in the absence and in the presence of the CA inhibitors acetazolamide and cyanate (CNO^-). Exposure to 10^{-5} M acetazolamide for 25 min did not affect isometric tension, unsurprising because of the low permeability for acetazolamide across cell membranes. Using 10^{-5} M CNO^-, Barclay observed a decrease in isometric force by 25%, but only after pCO_2 had been increased by equilibrating the solution with 10% CO_2 instead of 5%. However, we read from his Fig. 1 a decrease in isometric force by 10% with $1 \cdot 10^{-5}$ M CNO^- compared with control muscle even in normocapnic solution (5% CO_2). He concluded that the effect of CNO^- on force was due to inhibition of CA III, which he believed might be important in regulating intracellular pH, especially at high pCO_2. Similar experiments with isolated, directly stimulated rat skeletal muscle led us to different conclusions.[20] With chlorzolamide (CLZ) at a concentration of $5 \cdot 10^{-4}$ M and with incubation times of at least 3 hr, we observed in normocapnic conditions qualitatively the same changes in isometric contractions as with CNO^-, which at a concentration of 10^{-2} M and with an incubation time of 1 hr produced drastic changes in contractile parameters. Table II shows that isometric tetanic force is significantly reduced by both types of CA inhibitor in EDL, which is almost free of CA III, as well as in the CA III-rich soleus. Isometric force of single twitches is decreased in EDL by CNO^- as well as by CLZ, whereas in soleus only CNO^- reduces the twitch force.

3.1.2. Relaxation Time

Half-relaxation times of twitches and tetani were markedly prolonged by both CA inhibitors in EDL as well as in soleus (Table II).[20–22]

TABLE II
Effects of CA Inhibition of Contractile Parameters[a]

Muscle	Force: Twitch (Mean ± SE)	Force: Tetanus (Mean ± SE)	Half-relaxation time: Twitch (Mean ± SE)	Half-relaxation time: Tetanus (Mean ± SE)	Twitch time to peak (Mean ± SE)
Soleus					
CLZ, 6 h	0.98 ± 0.10 n.s.	0.82 ± 0.05 *	2.00 ± 0.16 **	1.10 ± 0.01 **	1.24 ± 0.05 **
NaCNO, 4 h	0.49 ± 0.05 **	0.51 ± 0.03 **	1.32 ± 0.05 **	1.38 ± 0.09 **	1.16 ± 0.06 n.s.
EDL					
CLZ, 6 h	0.71 ± 0.09 *	0.71 ± 0.07 *	1.49 ± 0.15 *	1.25 ± 0.04 **	1.24 ± 0.12 **
NaCNO, 3 h	0.32 ± 0.05 **	0.24 ± 0.06 **	2.53 ± 0.39 **	2.21 ± 0.16 **	1.00 ± 0.05 n.s.

[a]All values are given as fractions of the corresponding control values obtained in the absence of inhibitors. Temperature, 22°C. Levels of significance of differences between control and corresponding inhibitor values: *, $P < 0.05$; **, $P < 0.01$; n.s., not significant. Values are for between 6 and 16 animals. CLZ concentration was $5 \cdot 10^{-4}$ M; CNO^- concentration was 10^{-2} M; times given after name of inhibitor are incubation times. All muscles were from rats. Absolute values under control conditions: for soleus, twitch force 0.2 N, tetanic force 1.2 N, twitch half-relaxation time 200 msec, tetanus half-relaxation time 1000 msec, twitch time to peak 140 msec; for EDL, twitch force 0.25 N, tetanic force 1.3 N, twitch half-relaxation time 40 msec, tetanus half-relaxation time 100 msec, twitch time to peak 45 msec.

3.1.3. Time to Peak

Time to peak of twitches was increased in EDL and soleus muscle after incubation in CLZ ($5 \cdot 10^{-4}$ M), whereas CNO^- showed no significant effect on this parameter (Table II).[20–22]

3.1.4. Possible Mechanisms by which CA Inhibition Produces Changes in Different Contractile Parameters

Isometric force may be determined by the number of cross-bridges in the force-generating state.[13] Processes that are likely to determine relaxation time or time to peak are thought to be cross-bridge cycling rate and calcium uptake[17] or release rate of the SR.[24,45] Possibly one or more of the CAs in skeletal muscle are involved in one of these mechanisms. The changes in contractile parameters seen with CA inhibition are similar to those seen with muscular fatigue, which is known to correlate with pH_i and concentrations of energy-rich phosphates.[14]

3.2. Intracellular pH

Incubation of isolated rat skeletal muscle with CA inhibitors decreases the pH_i of the muscles. The pH_i values in soleus and EDL, measured by 5,5-dimethyl-oxazolidine-2,4-dione (DMO) distribution, are 7.16 and 7.10,[20-22] respectively. These pH_i values decrease by ~0.1–7.05 for soleus and 6.97 for EDL when CLZ ($5 \cdot 10^{-4}$ M) is present and more drastically, to 6.75 and 6.28, respectively, when sodium cyanate (10^{-2} M) is added.

The inhibitor concentrations used here and their incubation times, as discussed above, are sufficient to inhibit facilitation of CO_2 diffusion. When this is suppressed after CA inhibition, the mean muscle pCO_2 should increase. For the experiments with rat muscles, the mean pCO_2 of the muscles can be estimated in analogy to the equations for calculating the mean O_2 tension in a cylinder consuming oxygen as derived by Thews.[54] By using an estimated mean radius of 1.5 mm for the isolated muscle, a CO_2 production rate of 0.95 ml/100 g per min, and a CO_2 diffusion constant of $13.6 \cdot 10^{-4}$ cm²/min per atm in the presence of facilitated CO_2 diffusion as it occurs in soleus (see above), it is estimated that the mean pCO_2 within the muscle should be approximately 13 mm Hg higher than in the surrounding medium. After complete suppression of facilitated CO_2 diffusion, the CO_2 diffusion constant is $4 \cdot 10^{-4}$ cm²/min per atm,[26] and the estimated muscle pCO_2 then is 50 mm Hg higher than that of the surrounding medium. Assuming a buffer capacity of 45 mmol/liter per ΔpH[1] an increase in pCO_2 from 48 to 85 mm Hg should decrease the intracellular pH by 0.1 pH unit. Thus, the effect of CLZ on intracellular pH can be explained by suppression of facilitated diffusion. On the basis of the facilitated diffusion occurring in EDL (Table I), a similar effect of CA inhibition on pH_i of EDL is expected and, with CLZ, indeed is found. Cyanate, which decreased pH_i by 0.4 units in soleus and by 0.8 units in EDL, must, however, have another effect besides inhibition of facilitated CO_2 diffusion. It may be noted that since the muscles were not perfused, any contribution of sarcolemmal CA to CO_2 transfer from muscle cell to the blood, as discussed above, should not be important in these studies.

3.3. Energy-Rich Phosphates

In rat soleus muscles that have been incubated with chlorzolamide ($5 \cdot 10^{-4}$ M) for 6 hr, there is an increase in creatine and P_i (Table III). Incubation in CNO^- (10^{-2} M) for 4 hr leads to an even higher increase in creatine and P_i of rat soleus and EDL muscles and also decreases the concentrations of ATP and phosphocreatine (PCr) in both muscles (Table III).[21,22]

It is possible that these two CA inhibitors act by different mechanisms, since one is believed to be a specific CA inhibitor (CLZ) and the other (CNO^-) is known to have a number of additional effects.

TABLE III

Effects of CA Inhibition on Energy-Rich Phosphates and Their Products[a]

Muscle	Mean ± SE				n
	ATP	PCr	Creatine	P_i	
Soleus					
CLZ	1.00 ± 0.04 n.s.	0.93 ± 0.05 n.s.	1.54 ± 0.09 *	1.30 ± 0.04 **	7
NaCNO	0.48 ± 0.03 *	0.41 ± 0.06 *	2.27 ± 0.34 *	2.70 ± 0.33 *	4
EDL					
NaCNO	0.82 ± 0.05 *	0.27 ± 0.03 **	2.48 ± 0.30 **	3.22 ± 0.24 **	5

[a]All values are given as fractions of the corresponding control values obtained in the absence of inhibitors. Muscles from rats were incubated in $5 \cdot 10^{-4}$ CLZ for 6 hr and in 10^{-2} M NaCNO for 4 hr. Levels of significance of differences beween control and corresponding inhibition values: *, $P < 0.05$; **, $P < 0.01$; n.s., not significant. n, Number of animals. Absolute values under control conditions are (in millimoles per kilogram wet weight): for soleus, ATP 3.3, PCr 12, creatine 4.0, P_i 3.5; for EDL, ATP 5.5, PCr 18.5, creatine 7.5, P_i 4.3.

3.3.1. CLZ

Since inhibition of muscle CA results in increased pCO_2, with an accordingly decreased pH_i, the effect of CA inhibition on phosphate concentrations can be compared with results of experiments on which the effects of increased CO_2 concentrations have been investigated. In isolated rat EDL muscle, Sahlin *et al.*[46] observed a decrease of PCr to 44% of control values when the CO_2 concentration of the equilibration gas was increased from 6.5 to 30%. This finding agrees with our own observations showing a decrease of PCr in isolated rat EDL muscle to 42% and in soleus to 70% when the CO_2 concentration was raised from 5 to 30%. An increase in the concentration of P_i and a decrease in PCr upon elevation of CO_2 concentration from 5 to 45% have also been observed in frog skeletal muscle.[37] Spriet *et al.*,[53] upon a moderate increase in pCO_2 of the perfusate from 38 to 63 torrs, which is closer to what is expected from inhibition of intracellular facilitated diffusion of CO_2, observed no change in ATP/ADP and PCr/ATP ratios in various perfused skeletal muscles of the rat. It is concluded that CA inhibition may change the concentration of phosphates by way of an increased tissue partial pressure of CO_2 but that this does not explain all changes.

3.3.2. CNO^-

Several reports have linked application of CNO^- to animals *in vivo* with reduced high-energy phosphate levels in heart[2] and brain[10,39] although in no study was the intracellular concentration of CNO^- known. Assuming a distribution of CNO^- in the body,[11] a concentration of 10^{-2} M CNO^- can be calculated for heart tissue. The effect of CNO^- on phosphates has been explained by the well-known increase in O_2 affinity of the blood, i.e., by an impaired O_2 delivery to tissues. Since this does not explain the effects seen in isolated muscle experiments, it is probably also incorrect for the cited *in vivo* experiments. Part of the effects observed in these *in vivo* experiments may thus be due to CA inhibition.

However, besides its reversible inhibition of CA, CNO^- has other inhibitory effects at high concentrations (>0.1 M) that are mostly due to irreversible carbamylation of proteins.[25,34,48] Another possible explanation for the CNO^- effects on phosphates comes from a report[6] that high concentrations of CNO^- ($>1\text{–}4\cdot10^{-2}$ M) uncouple the oxidative phosphorylation in preparations of rat liver mitochondria. The occurrence of this effect in skeletal and heart muscle could explain the marked decreases in the concentrations of ATP and PCr. Thus, it is crucial to know the intracellular concentration of CNO^- in the isolated muscle experiments in order to distinguish effects due to CA inhibition from this uncoupling effect.

From measurements of ^{14}C-cyanate uptake in rat skeletal muscle, we calculate that the intracellular concentration of CNO^- is $1.7\cdot10^{-3}$ M in soleus and $0.8\cdot10^{-3}$ M in EDL muscle when the muscles have been incubated for 4 hr in 10^{-2} M CNO^-. These intracellular concentrations are small compared with the concentrations necessary for the uncoupling effect on oxidative phosphorylation and the other unspecific effects.

Another argument in favor of a specific CA inhibition effect of CNO^- in the isolated muscle experiments, and in disfavor of an effect due to irreversible carbamylation, is that the effect of CNO^- on contraction parameters is readily reversible, as is CA inhibition.[22] That this reversibility also applies to the changes in energy-rich phosphate appears likely from the rapid restoration of muscle function but has not been demonstrated directly.

3.4. Can the Effect of CA Inhibition on Isometric Contraction be Explained by Low pH_i or Changed Phosphates?

It seems conceivable that the decrease in isometric force and the increase in relaxation time of tetani observed under CA inhibition are caused by the combined effects of a decrease in pH_i and an increase in P_i, both of which are produced by CA inhibition. The results of experiments with isolated muscles bathed in acidifying solutions provide evidence for this. Solutions with high CO_2 (30% in equilibrating gas), which reduce muscle pH_i as well as increase muscle phosphate concentration, decrease isometric force,[46] and solutions with low HCO_3^- (2 mM) but normal CO_2 have similar but slightly less pronounced effects.

The increase in relaxation time of twitches and the increase in time to peak can be explained by neither the decrease in pH_i nor the increase in P_i. Experiments in which a low pH_i and high P_i concentration were induced by high CO_2 concentrations showed no change in half-relaxation time of twitches, no change in time to peak for soleus, and even a decrease in time to peak for EDL[46] (our observations).

The effects of CA inhibition on muscle contraction were seen in the CA III-rich soleus as well as in the almost CA III-free EDL muscle of the rat; thus, an isozyme present in both muscles, not CA III, is probably responsible. It cannot be excluded that CA III in soleus and another isoenzyme in EDL muscle have identical functions, as in the facilitation of CO_2 diffusion.

4. A Role of the Carbonic Anhydrase of the Sarcoplasmic Reticulum in Ca^{2+} Mobilization?

What causes the prolongation of time to peak and relaxation of twitches seen during CA inhibition in EDL and soleus? As just discussed, neither altered pH_i nor phosphate metabolities are likely to cause these effects. We have previously proposed that the CA of the SR might be involved in the process of Ca^{2+} movements across the SR membrane and that its inhibition might cause such changes in contractile parameters.[5,20,22]

It has been shown[52] that only about one-half of the calcium released from the SR during a tetanus is electrically compensated for by movements of Mg^{2+} and K^+ in the opposite direction. Possible additional counterions are protons, and there is considerable evidence for this possibility.[40] The system CO_2–HCO_3^- could constitute a proton sink within the SR for this exchange. However, during onset of contraction, calcium release is very fast and time to peak is only 100 msec, whereas in the absence of CA, the CO_2 hydration–dehydration reaction has a half-time of ~10 sec. This would be too slow for such an exchange. A possible role of the CA associated with the SR could be to accelerate the CO_2 hydration–dehydration within SR so that CO_2 and HCO_3^- react fast enough to provide a rapid source and sink for protons leaving and entering the SR in exchange for Ca^{2+}. Inhibition of SR-localized CA would interfere with the proton exchange for calcium and thus could slow down the calcium release and uptake rate of the SR. This provides a reasonable explanation for the observed increases in time to peak and relaxation time that cannot be related to altered pH_i and intracellular phosphate. More direct experimental evidence for this hypothesis has recently been provided by our laboratory.[59]

5. Carbonic Anhydrase in Neuromuscular Transmission

Scheid and Siffert[47] compared the effect of CA inhibitors on the isometric force developed by directly and indirectly stimulated frog gastrocnemius. After incubating the muscles for 2 hr in solutions containing the inhibitors, they found that with indirect stimulation, isometric force decreased to 50% in $3 \cdot 10^{-9}$ M ethoxzolamide, 10^{-4} M acetazolamide, or 10^{-4} M methazolamide, whereas these same inhibitor concentrations had no effect on the force development of directly stimulated muscles. Because CA inhibitors did not affect the conduction velocity of the nerve and the amplitude of the compound nerve action potential, a CA was concluded to be involved in neuromuscular transmission.

ACKNOWLEDGMENT. We are greatly indebted to Lederle Laboratories, Pearl River, N.Y., for generous gifts of chlorzolamide.

References

1. Aickin, C. C., and Thomas, R. C., 1977, *J. Physiol.* **267**:791–810.
2. Allison, T. B., Pieper, G. M., Clayton, F. C., and Eliot, R. S., 1976, *Am. J. Physiol.* **230**(6):1751–1754.
3. Barclay, J. K., 1987, *Can. J. Physiol. Pharmacol.* **65**:100–104.
4. Bidani, A., Crandall, E. D., and Forster, R. E., 1978, *J. Appl. Physiol.* **44**(5):770–781.
5. Bruns, W., Dermietzel, R., and Gros, G., 1986, *J. Physiol.* **371**:351–364.
6. Cammer, W., 1982, *Biochem. Biophys. Acta* **679**:343–346.
7. Carter, N., Shiels, A., and Tashian, R., 1978, *Biochem. Soc. Trans.* 574th Meeting, Bath, pp. 552–553.
8. Carter, N. D., Hewett-Emmett, D., Jeffery, S., and Tashian, R. E., 1981, *FEBS Lett.* **128**:114–118.
9. Carter, N., Jeffery, S., and Shiels, A., 1982, *FEBS Lett.* **139**:265–266.
10. Cassel, J., Kogure, K., Busto, R., Kim, C. Y., and Harkness, D. R., 1973, *Adv. Exp. Med. Biol.* **37A**:319–324.
11. Cerami, A., Allen, T. A., Graziano, J. H., De Furia, F. G., Manning, J. M., and Gillette, P. N., 1973, *J. Pharmacol. Exp. Ther.* **185**:653–666.
12. Crandall, E. D., Bidani, A., and Forster, R. E., 1977, *J. Appl. Physiol.* **43**(4):582–590.
13. Dawson, M. J., Gadian, D. G., and Wilkie, D. R., 1978, *Nature* (London) **274**:861–866.
14. Dawson, M. J., Gadian, D. G., and Wilkie, D. R., 1980, *J. Physiol.* **299**:465–484.
15. De Hemptinne, A., Marrannes, R., and Vanheel, B., 1987, *Can. J. Physiol. Pharmacol.* **65**: 970–977.
16. Dermietzel, R., Leibstein, A., Siffert, W., Zamboglou, N., and Gros, G., 1985, *J. Histochem. Cytochem.* **33**(2):93–98.
17. Edwards, R. H. T., Hill, D. K., and Jones, D. A., 1975, *J. Physiol.* **251**:287–301.
18. Effros, R. M., and Weissman, M. L., 1979, *J. Appl. Physiol.* **47**(5):1090–1098.
19. Forster, R. E., and Crandall, E. D., 1975, *J. Appl. Physiol.* **38**(4):710–718.
20. Geers, C., and Gros, G., 1988, *Life Sci.* **42**:37–45.
21. Geers, C., and Gros, G., 1988, *Pfluegers Arch.* **411**:R189.
22. Geers, C., and Gros, G., 1990, *J. Physiol.* **423**:279–297.
23. Geers, C., Gros, G., and Gärtner, A., 1985, *J. Appl. Physiol.* **59**(2):548–558.
24. Gillis, J. M., 1985, *Biochem. Biophys. Acta* **811**:97–145.
25. Graceffa, P., and Lehrer, S. S., 1986, *Biochem. Biophys. Res. Commun.* **134**(1):64–70.
26. Gros, G., and Dodgson, S. J., 1988, *Annu. Rev. Physiol.* **50**:669–694.
27. Gros, G., and Moll, W., 1971, *Pfluegers Arch.* **324**:249–266.
28. Gros, G., and Moll, W., 1974, *J. Gen. Physiol.* **64**:356–371.
29. Gros, G., Moll, W., Hoppe, H., and Gros, H., 1976, *J. Gen. Physiol.* **67**:773–790.
30. Gros, G., Siffert, W., and Schmid, A., 1980, in: *Biophysics and Physiology of Carbon Dioxide* (C. Bauer, G. Gros, and H. Bartels, eds.), Springer-Verlag, Berlin, pp. 409–416.
31. Gros, G., Ganghoff, F., Scheid, P., Siffert, W., Teske, W., and Krüger, D., 1984, *Pfluegers Arch.* **400**:R58.

32. Hill, E. P., Power, G. G., and Gilbert, R. D., 1977, *J. Appl. Physiol.* **42**(6):928–934.
33. Holmes, R. S., 1977, *Eur. J. Biochem.* **78**:511–520.
34. Huang, C.-C., and Madsen, N. B., 1966, *Biochemistry* **5**(1):116–125.
35. Jeffery, S., Carter, N. D., and Smith, A., 1986, *J. Histochem. Cytochem.* **34**:513–515.
36. Kawashiro, T., and Scheid, P., 1976, *Pfluegers Arch.* **362**:127–133.
37. Kitano, T., 1988, *J. Physiol.* **397**:643–655.
38. Koester, M. K., Register, A. M., and Noltmann, E. A., 1977, *Biochem. Biophys. Res. Commun.* **76**(1):196–204.
39. Kogure, K., Busto, R., Cassel, J., Kim, C. Y., and Harkness, D. R., 1975, *Pharmacology* **13**: 391–400.
40. Meissner, G., 1981, *J. Biol. Chem.* **256**(2):636–643.
41. Moynihan, J. B., 1977, *Biochem. J.* **168**:567–569.
42. O'Brasky, J. E., and Crandall, E. D., 1980, *J. Appl. Physiol.* **49**(2):211–217.
43. Riley, D. A., Ellis, S., and Bain, J., 1982, *J. Histochem. Cytochem.* **30**(12):1275–1288.
44. Roos, A., and Boron, W. F., 1981, *Physiol. Rev.* **61**:296–434.
45. Rüegg, J. C., 1986, *Calcium in Muscle Activation*, Springer-Verlag, Berlin.
46. Sahlin, K., Edström, L., and Sjöholm, H., 1983, *Am. J. Physiol.* **245**:C15–C20.
47. Scheid, P., and Siffert, W., 1985, *J. Physiol.* **361**:91–101.
48. Shen, W.-C., and Colman, R. F., 1975, *J. Biol. Chem.* **250**(8):2973–2978.
49. Shiels, A., Jeffery, S., Wilson, C., and Carter, N., 1984, *Biochem. J.* **218**:281–284.
50. Shima, K., Tashiro, K., Hibi, N., Tsukada, Y., and Hirai, H., 1983, *Acta Neuropathol.* **59**: 237–239.
51. Siffert, W., and Gros, G., 1982, *Biochem.* **205**:559–566.
52. Somlyo, A. V., Gonzalez-Serratos, H., Shuman, H., McClellan, G., and Somlyo, A. P., 1981, *J. Cell Biol.* **90**:577–594.
53. Spriet, L. L., Matsos, C. G., Peters, S. J., Heigenhauser, G. J. F., and Jones, N. L., 1985, *Am. J. Physiol.* **248**:C337–C347.
54. Thews, G., 1953, *Acta Biotheor.* **10**:105–138.
55. Thomas, R. C., 1984, *J. Physiol.* **354**:3P–22P.
56. Väänänen, H. K., Takala, T., and Morris, D. C., 1986, *Histochemistry* **86**:175–179.
57. Wetzel, P., and Gros, G., 1988, *Pfluegers Arch.* **412**:R80.
58. Wetzel, P., and Gros, G., 1990, *Arch. Biochem. Biophys.* **279**:345–354.
59. Wetzel, P., Liebner, T., and Gros, G., 1990, *FEBS Lett.* **267**:66–70.
60. Zborowska-Sluis, D. T. L-Abbate, A., amd Klassen, G. A., 1974, *Respir. Physiol.* **21**:341–350.
61. Zborowska-Sluis, D. T., L'Abbate, A., Mildenberger, R. R., and Klassen, G. A., 1975, *Respir. Physiol.* **23**:311–316.

CHAPTER 20

Distribution and Ultrastructural Localization of Carbonic Anhydrase III in Different Skeletal Muscle Fiber Types

PIERRE FRÉMONT, PIERRE M. CHAREST, CLAUDE CÔTÉ, and PETER A. ROGERS

1. Introduction

Prior to the discovery of carbonic anhydrase (CA) III two groups reported the existence of a protein of approximately 30 kDa in the soluble enzyme fraction of skeletal muscle. This unidentified protein was designated "protein F"[38] and basic muscle protein.[5] Several years later, a number of studies[18,19,27,35] clearly established that this 30-kDa protein, which is found predominantly in skeletal muscle, was a third CA isozyme different from CA I and CA II. This CA isozyme, first called CA A,[18] is now designated CA III.

The biochemical properties of several CA III isozymes have been documented (reviewed in references 9 and 16). Like the other CA isozymes, it is a metalloenzyme that contains one zinc atom per molecule linked to three His residues at position 94, 96, and 119 in the active site. An important distinctive characteristic of CA III is its low specific activity for the CO_2 hydration–dehydration reaction when compared with the CA I and CA II isozymes. Studies of bovine CA II and CA III[36] strongly support the hypothesis that this lower specific activity is due to the presence of two basic residues in the active site of CA III (Lys-64 and Arg-67 instead of His-64 and Asn-67 for CA II). It is believed that the His-64 residue of CA II is implicated in proton transfer out of the active site and that substitution with a

PIERRE FRÉMONT, PIERRE M. CHAREST, CLAUDE CÔTÉ, and PETER A. ROGERS • Muscle Biology Research Group, Laval University, Ste-Foy, Quebec, Canada G1V 4G2.

more basic residue (such as Lys in the case of ox,[44] rat,[25] and human[29] or Arg in the case of horse[49] interferes with the water molecule protolysis step of the CO_2 hydration reaction. It is estimated that the high concentration of CA III present in muscles containing predominantly type I fibers, such as the rabbit soleus (SOL) muscle, can accelerate the rate of CO_2 hydration by several hundred-fold.[16]

It has been shown in cat[24,45] and bovine[11,46] muscles that CO_2 hydration by CA III is not affected by variations in pH ranging from 6.0 to 8.0. This is very different from results for CA I and CA II, for which an optimal pH for CO_2 hydration activity is observed between 7 and 8.[26] The His-64 residue may be partially responsible for the pH dependency of the human CA II isozyme.[42] The characteristic of CA III may be physiologically relevant since skeletal muscle contraction may significantly reduce pH (from 7.05 to 6.5 within 3 min.[8]) without affecting CA III activity.

2. Diversity and Plasticity of Skeletal Muscle Fibers

Using myofibrillar actomyosin ATPase activity as the principal criterion, mammalian skeletal muscle fibers are generally classified into three types that differ in their structural, contractile, and metabolic characteristics. These are referred to as type I (slow-twitch oxidative), type IIa (fast-twitch oxidative and glycolytic), and type IIb (fast-twitch glycolytic) fibers. The characteristics of these different muscle fiber types have been well studied, and a number of reviews on this subject have appeared.[7,10,17] Fiber types do not have consistent metabolic and contractile characteristics.[10,34] There are also significant interspecies differences in the metabolic characteristics of the different fiber types.[17]

Aside from the fact that muscle fibers display a remarkable degree of variation, they also have an impressive capacity to adapt to different external stimuli. Some well-characterized experimental models such as chronic electrical stimulation, denervation, and thyroid hormone treatment have been used to induce important changes of the muscle fiber phenotype (reviewed in reference 43). As for fiber type characteristics, interspecies differences have been reported as regards the adaptive response of skeletal muscle to an altered functional demand.[41]

The diversity of muscle fibers and their adaptability ensure the retention of a close structure–function relationship in the muscle tissue. In this regard, our major goal was to clearly establish the distribution of CA III in the different types of rat skeletal muscle fibers in order to identify possible associations between the presence of CA III and certain characteristics of the different muscle phenotypes. Assuming that CA III is no exception to the prominent structure-function relationship of muscle, any such association could represent important information in our attempt to elucidate its physiological role.

Muscles that contain "almost exclusively" or "predominantly" one type of muscle fibers are used to investigate the structural, metabolic, and contractile characteristics of these fibers,[10] permitting correlation of biochemical parameters with the contractile properties of a muscle.

The fiber type composition of the muscles used in our investigations is shown in Table I. In rats, the superficial portion of the vastus lateralis (SVL) muscle is almost exclusively composed of type IIb fibers and is devoid of type I fiber, whereas the SOL muscle contains a large majority of type I fibers and no type IIb fibers. Finally, the deep portion of the vastus lateralis (DVL) and tibialis anterior (AT) contains approximately 66% type IIa, 33% type IIb, and very few type I fibers. Rat muscles were used in part because most studies dealing with the distribution of CA III were also performed on rat muscles. Moreover, the metabolic characteristics of the rat SOL, DVL, and SVL muscles are very well documented.[3,4,30]

3. Levels and Localization of CA III in Skeletal Muscle Fiber Types

A high concentration of CA III is present in type I fibers.[13,14,23,37,48] Until recently, it was not clear whether CA III was expressed in type II fibers and, if so, whether type IIa and IIb fibers differ in CA III concentration.

The first observation of a nonuniform distribution of CA III in muscles of varying composition was reported by Holmes.[18,19] Starch gel electrophoresis followed by bromothymol blue staining indicated that CA III could be detected only in extracts from "red" leg muscles and not in the breast muscles of the chicken. Similar results using rat skeletal muscles were then reported[32]: the extravascular CA activity of different muscles was found to be "approximately" correlated with the proportion of dark (red) fibers in these muscles. Since these initial studies, it is generally agreed that higher concentrations of CA III are present in muscles of high oxidative potential (red or dark muscles) than in those of low oxidative potential (white muscles). However, both type I and IIa fibers can be classified as red, and both possess a high oxidative capacity. Whether significant amounts of CA III were present in these two fiber types remained unclear.

TABLE I
Fiber Type Population in Rat Skeletal Muscles of Different Composition

	Fiber population[a]		
Muscle	Type I	Type IIa	Type IIb
SOL	85	15	0 (28, 2)
AT	2	66	32 (1)
DVL	6	58	36 (28, 2)
SVL	0	4	96 (28, 2)

[a]Numbers in parentheses indicate number of reference. When more than one study was available, the mean value is presented. Values are percentage of fiber population in the muscle.

More detailed information concerning the distribution of CA III became available with the quantification of the enzyme content in muscles for which the exact fiber type composition had been established. A number of studies using immunoassay techniques consistently reported values for CA III concentration approximately 13-fold higher in the rat SOL muscle than in the AT muscle.[6,22,39] Considering that the AT muscle contains only 2% type I fibers as compared to 85% for the SOL (see Table I), these results strongly suggested that in rat muscles CA III was not restricted to type I fibers. In another report,[14] it was estimated, with CA III activity present in the SOL muscle as a reference, that type I fibers account for less than 30% of the CA III activity found in the DVL muscle. Together, these studies provided indirect evidence suggesting that CA III is present in more than just type I fibers.

More direct evidence concerning the distribution of the enzyme came from analysis of CA III activity and content in individual fibers classified on the basis of their actomyosin ATPase activity. Using histochemical staining for sulfonamide-resistant CA activity, it was found that CA III activity was the highest in type I fibers, that type IIa fibers exhibited an intermediate level of activity, and that type IIb fibers contained no demonstrable activity.[37] In an immunohistochemical study with rat muscle sections, CA III was found only in type I fibers.[21] In a subsequent report with a modified antibody incubation medium, the same authors were able to demonstrate the presence of CA III in some type IIa fibers of the rat AT muscle.[23]

Type II human fibers were not stained clearly compared with controls in an immunohistochemical study.[40] Results of immunofluorescent studies were equivocal.[21,31] More recently, CA III was found only in type I fibers,[47] and another reported that "CA III was mainly localized in type I fibers."[20] Finally, in equine muscle, besides the high concentration of CA III found in type I fibers, small amounts of the enzyme were reported to be present in type IIa fibers.[33] Although these histochemical studies of CA III distribution in rat, equine, and human muscles are not in entire agreement, they clearly established that if CA III is present in type II fibers, its concentration is much lower than in type I fibers. They also consistently showed that the enzyme is not detectable in type IIb fibers. It could not be definitely established whether an intermediate concentration of CA III was present in type IIa fibers.

A number of factors could explain these conflicting results. First, as a consequence of the highly soluble nature of this enzyme, insufficient fixation could have resulted in a loss of the enzyme and failure to demonstrate its presence in tissues containing a low concentration. In the studies described previously, the fixation procedures varied considerably, and none included perfusion *in situ* with the fixating agent. Second, denaturation of CA isozymes has been shown to affect the specificity of CA antiserum,[15] and cross-reactions of CA antiserum with different isozymes are common.[12] These observations indicate that histochemical studies using a CA antiserum should include a demonstration of antibody specificity under the precise conditions used for immunohistochemistry. In most of the studies described, the antibody specificity was demonstrated by immunodiffusion, radioimmunoassay, or both but never under the specific conditions used for immunohistochemistry.

These observations constituted two major concerns in our own investigation of the enzyme distribution in the rat muscle fibers.[14] The tissue was fixed *in situ* by perfusion with 3% glutaraldehyde and then *in vitro* after dissection of the muscle tissue. The antibody specificity for CA III was confirmed by the absence of labeling in the cytosol of an erythrocyte that had not been removed from the SOL by the perfusion procedure. The results revealed a uniform intracellular distribution of CA III and, in agreement with results of others,[48] confirmed the truly cytosolic nature of the enzyme. Quantification of the labeling on electron micrographs from type IIa fibers of the DVL muscle demonstrated the presence of a low but detectable concentration of CA III. The results supported our previous conclusion that the intermediate CA III content and activity found in the rat AT and DVL muscles cannot be explained solely by their content of type I fibers. We therefore concluded that the relative CA III content in the three principal types of rat muscle fibers is type I > IIa > IIb ~ 0.

As a consequence of the heterogeneity observed within fiber types and between different species, the conclusion reached concerning the distribution of CA III in the fiber types of the rat SOL, DVL, and SVL muscles does not necessarily apply to fibers from other muscles and to muscles from other species.

In rat skeletal muscle, despite some previous controversial results, the most recent studies seem to finally confirm that the relative CA III content in the three principal fiber types is type I > IIa > IIb ~ 0. In other species, although it is well established that type I fibers contain a high concentration of CA III, the presence of the enzyme in type IIa fibers has not yet been consistently demonstrated.

ACKNOWLEDGMENT. The work presented here is dedicated to the memory of Dr. Graham W. Mainwood, Professor of Physiology, University of Ottawa.

References

1. Ariano, M. A., Armstrong, R. B., and Edgerton, V. R., 1973, *J. Histochem. Cytochem.* **21**:51–55.
2. Armstrong, R. B., and Phelps, R. O., 1984, *Am. J. Anat.* **171**:259–272.
3. Baldwin, K. M., Klinkerfuss, G. H., Terjung, R. L., Molé, P. A., and Holloszy, J. O., 1972, *Am. J. Physiol.* **222**:373–378.
4. Baldwin, K. M., Winder, W. W., Terjung, R. L., and Holloszy, J. O., 1973, *Am. J. Physiol.* **225**:962–966.
5. Blackburn, M. N., Chirgwin, J. M., James, G. T., Kempe, T. D., Parsons, T. F., Register, A. M., Schnackerz, K. D., and Noltmann, E. A., 1972, *J. Biol. Chem.* **247**:1170–1179.
6. Carter, N., Jeffery, S., and Shiels, A., 1982, *FEBS Lett.* **139**:265–266.
7. Close, R. I., 1972, *Physiol. Rev.* **52**:129–197.
8. Connett, R. J., 1987, *J. Appl. Physiol.* **63**:2360–2365.
9. Deutsch, H. F., 1987, *Int. J. Biochem.* **19**:101–113.
10. Eisenberg, B. R., 1983, in: *Handbook of Physiology*, Section 10 (L. D. Peachey, R. H. Adrian, and S. R. Geiger, eds.), American Physiological Society, Bethesda, M.D., pp. 73–112.

11. Engberg, P., Millqvist, E., Pohl, G., and Lindskog, S., 1985, *Arch. Biochem. Biophys.* **241**:628–638.
12. Erickson, R. P., Kay, G., Hewett-Emmett, D., Tashian, R. E., and Claflin, J. L., 1982, *Biochem. Genet.* **20**:809–819.
13. Frémont, P., Boudriau, S., Tremblay, R. R., Côté, C., and Rogers, P. A., 1989, *Int. J. Biochem.* **21**:143–147.
14. Frémont, P., Charest, P. M., Côté, C., and Rogers, P. A., 1988, *J. Histochem. Cytochem.* **36**:775–782.
15. Frémont, P., Lazure, C., Tremblay, R. R., Chrétien, M., and Rogers, P. A., 1987, *Biochem Cell Biol.* **65**:790–797.
16. Gros, G., and Dodgson, S. J., 1988, *Annu. Rev. Physiol.* **50**:669–694.
17. Holloszy, J. O., and Booth, F. W., 1976, *Annu. Rev. Physiol.* **38**:273–291.
18. Holmes, R. S., 1976, *J. Exp. Zool.* **197**:289–295.
19. Holmes, R. S., 1977, *Eur. J. Biochem.* **78**:511–520.
20. Ibi, T., Haimoto, H., Nagura, H., Sahashi, K., Kato, K., Mokuno, K., Svgimura, K., and Matsuoka, Y., 1985, *Acta Neuropathol.* **68**:74–76.
21. Jeffery, S., Carter, N. D., and Smith, A., 1986, *J. Histochem. Cytochem.* **34**:513–516.
22. Jeffery, D., Edwards, Y. H., Jackson, M. J., Jeffery, S., and Carter, N. D., 1982, *Comp. Biochem. Physiol.* **73B**:971–975.
23. Jeffery, S., Smith, A., and Carter, N., 1987, *J. Histochem. Cytochem.* **35**:663–668.
24. Kararli, T., and Silverman, D. H., 1985, *J. Biol. Chem.* **260**:3484–3489.
25. Kelly, C. D., Carter, N. D., Jeffrey , S., and Edwards, Y. H., 1988, *Biosci. Rep.* **8**:401–406.
26. Khalifah, R. G., 1971, *J. Biol. Chem.* **246**:2561–2573.
27. Koester, M. K., Register, A. M., and Noltmann, E. A., 1977, *Biochem. Biophys. Res. Commun.* **76**:196–204.
28. Laughlin, M. H., and Armstrong, R. B., 1983, *Am. J. Physiol.* **244**:H814–H824.
29. Lloyd, J., McMillan, S., Hopkinson, D., and Edwards, Y. H., 1986, *Gene* **41**:233–239.
30. Miller, W. C., Palmer, W. K., Arnall, D. A., and Oscai, L. B., 1987, *Can. J. Physiol. Pharmacol.* **65**:317–322.
31. Moyle, S., Jeffery, S., and Carter, N. D., 1984, *J. Histochem. Cytochem.* **32**:1262–1264.
32. Moynihan, J. B., 1977, *Biochem. J.* **168**:567–569.
33. Nishita, T., and Matsushita, H., 1987, *Equine Vet. J.* **19**:509–513.
34. Pette, D., and Spamer, C., 1986, *Fed. Proc.* **45**:2910–2914.
35. Register, A. M., Koester, M. K., and Noltmann, E. A., 1978, *J. Biol. Chem.* **253**:4143–4152.
36. Ren, X., Sandström, A., and Lindskog, S., 1988, *Eur. J. Biochem.* **173**:73–78.
37. Riley, D. A., Ellis, S., and Bain, J., 1982, *J. Histochem. Cytochem.* **30**:1275–1288.
38. Scopes, R. K., 1966, *J. Biol. Chem.* **98**:193–197.
39. Shiels, A., Jeffery, S., Wilson, C., and Carter, N., 1984, *Biochem. J.* **218**:281–284.
40. Shima, K., Tashiro, K., Hibi, N., Tsukada, Y., and Hirai, H., 1983, *Acta Neuropathol.* **59**:237–239.
41. Simoneau, J.-A., and Pette, D., 1988, *Pfluegers Arch.* **412**:86–92.
42. Steiner, H., Jonsson, B.-H., and Lindskog, S., 1975, *Eur. J. Biochem.* **59**:253–259.
43. Swynghedauw, B., 1986, *Physiol. Rev.* **66**:710–771.
44. Tashian, R. E., Hewett-Emmett, D., Stroup, S. K., Goodman, M., and Yu, Y.-S. L., 1980, *Biophysics and Physiology of Carbon Dioxide* (C. Bauer, G. Gros, and H. Bartels, eds.), Springer-Verlag, New York, pp. 165–176.
45. Tu, C., Sanyal, G., Wynns, G. C., and Silverman, D. N., 1983, *J. Biol. Chem.* **258**:8867–8871.
46. Tu, C., Thomas, H. G., Wynns, G. C., and Silverman, D. N., 1986, *J. Biol. Chem.* **261**:10100–10103.
47. Väänänen, H. K., Paloniemi, M., and Vuori, J., 1985, *Histochemistry* **83**:231–235.
48. Väänänen, H. K., Takala, T., and Morris, D. C., 1986, *Histochemistry* **86**:175–179.
49. Wendorff, K. M., Nishita, T., Jabusch, J. R., and Deutsch, H. F., 1985, *J. Biol. Chem.* **260**:6129–6132.

CHAPTER 21

Hormonal and Neuronal Control of Carbonic Anhydrase III Gene Expression in Skeletal Muscle

NICHOLAS D. CARTER

Muscle development has been extensively investigated in terms of the pre- and postnatal changes that occur in the expression of myofibrillar proteins and isozymes. DNA probes now enable the defining of changes in mRNA levels for the various proteins and hence the profiles of gene expression.

Carbonic anhydrase (CA) III is a cytosolic isozyme that is thought to facilitate the transport of CO_2 in skeletal muscle by catalyzing the reversible hydration of CO_2; other functions are unknown. This anhydrase activity is involved in the maintenance of ionic balance and acid–base homoeostasis within the muscle tissue. The concentrations of CA III are 2% of weight in adult red muscle such as rat soleus[9,17,21] but are at trace levels in adult white muscle, such as rat anterior tibialis (AT) and chicken pectoralis muscle[8,9] These patterns of enzyme have been also defined in different muscle fiber types by immunocytochemistry studies.[11,21,25] The concentration of CA III is found to be highest in type 1 (slow) fibers, variable to low in type 2A (fast) fibers, and low or absent in type 2B (fast) fibers. The development of the specific fiber types and of some fiber-type-specific proteins in muscle appears to depend on the innervation that the muscle fiber receives. For example, specific neuronal activity seems to be necessary for the synthesis of adult fast and slow myosin heavy-chain isozymes,[23] the adult patterns of expression of troponin[6] and tropomyosin,[18] and the synthesis of the fast-twitch-fiber-specific Ca^{2+}-binding protein, parvalbumin.[13] Denervation of the neonatal rat soleus muscle prevents the emergence of type 1 fibers and the synthesis of slow myosin,[19] and denervation of maturing fast-twitch muscles suppresses the increase in parv-

NICHOLAS D. CARTER • Department of Child Health, St. George's Hospital Medical School, University of London, London SW17 ORE, United Kingdom.

albumin usually seen after birth.[13] In addition, chronic nerve stimulation induces fast-to-slow fiber conversions of the major functional systems of the muscle fibers (for reviews, see references 16 and 20).

Two features of CA III expression suggest that the gene may be under the control of slow fiber neurons. One is the striking relationship between fiber type and CA III protein distribution in adult muscle, and the other is the pattern of distribution of CA III during development. For example, in humans, in which all adult muscles are a mixture of type 1 and type 2 fibers, CA III protein is detectable in fetal muscle at 10 weeks of gestation but remains low until about 25 weeks of gestation, when a steep rise is detected, giving 50% of the adult level at birth.[10,14] Similarly, in the rat soleus, the amount of CA III 3 weeks after birth is about 25% of the adult value and rises rapidly to the adult value during the next 10 weeks. In both species, the major increase in CA III occurs at a stage in development when fetal neuronal innervation has regressed and specific nerve impulse patterns are being established. We have investigated these interrelationships further by examining the changes that occur, after nerve resection, to rat CA III protein and also to the associated CA III mRNA content in slow- and fast-twitch muscles.[5]

As far as endocrine influences on muscle are concerned, thyroid disorders are also known to strikingly alter the fiber composition of human muscle.[24] Thyroidectomy apparently increases the number of type 1 fibers in rat soleus and increases the amount of 30 kDa protein in this muscle.[7] In our studies, we specifically investigated the effect of thyroidectomy on CA III concentrations and distribution in AT, extensor digitorum longus (EDL), and soleus muscles of rat (see below).

1. Effects of Denervation on Muscle

In this laboratory we investigated the expression of the muscle-specific CA III gene under the influence of neuronal activity; CA III protein and mRNA contents in fast- and slow-twitch rat muscles have been examined after resection of the sciatic nerve, and marked changes were observed. The most striking alterations occur in the fast-twitch AT and EDL muscles, which show marked elevation of CA III protein after denervation, and the effects are greater than can be attributed to the consequences of muscle atrophy after denervation (Fig. 1A and B). Furthermore, since the CA III protein and the specific CA III mRNA are both increased severalfold 16 days after denervation, it seems likely that the increases are regulated by changes in gene transcription rather than by increased translatability of the mRNA. This regulation appears to involve the switching on of CA III expression in the normally nonexpressing type 2 fast-twitch muscle fibers, and the results agree with those of immunohistochemical studies which have demonstrated significant amounts of CA III protein in type 2 fibers after denervation[25] (Fig. 2).

Although the overall conclusion is straightforward, detailed examination of the data reveals complexities due to differences in the response to denervation shown by the two fast-twitch muscles, AT and EDL. The protein and mRNA data for these two muscles are not in close agreement when compared 16 days after denervation. In particular, the amount of CA III mRNA in EDL is low compared with the protein content, whereas in the AT the mRNA content is relatively high (Fig. 1C). This may reflect differences in the timing of denervation-induced responses in AT and EDL or variations in the CA III protein/mRNA turnover times in the two muscles. In addition, the CA III mRNA apparently decreases relative to other mRNAs in the EDL mRNA pool after denervation, whereas in the AT there is a threefold increase in the proportion of CA III mRNA. This finding implies that the relative rates of synthesis of other member mRNAs in the muscle mRNA pools are changing and that the patterns of change are different in the two muscles. This conclusion is further evidenced by the greater loss of actin mRNA in EDL (92% of control) compared with AT (52% of control).

It is not clear why the AT and EDL muscles should respond differently to denervation. However, there may be some effect of fiber composition. Both muscles comprise predominantly type 2 fibers, but there are some type 1 fibers in the AT and a higher proportion of type 2A (intermediate) than 2B (fast) in AT than in EDL.[11,25] It is noteworthy that a similar variation in the extent of myoglobin induction among denervated muscles of similar fiber composition has also been reported.[1]

If the differentiation of fiber-type-specific metabolism and biochemistry is dependent on a particular pattern of nerve activity, then it might be supposed that in the absence of neuronal stimulation, mature fibers would gradually lose their fiber-specific features and take on the characteristics of an undifferentiated immature muscle fiber. However, the induction of expression of CA III in EDL and AT to values close to those found in the normal adult soleus muscle suggests that the change perhaps reflects a fast-to-slow-fiber transition. This interpretation agrees with various other reports in the literature; for example, myoglobin, a protein usually confined to type 1 fibers, is increased threefold in denervated EDL and AT,[1] to values higher than those found in fetal tissue[22] and similar to those of adult soleus. In addition, the synthesis of parvalbumin, characteristic of type 2 fibers, is decreased after denervation to a value characteristic of type 1 slow-twitch muscle.[13]

There is some evidence that denervation may lead to the reverse, i.e., slow-to-fast, transition in soleus. For example, 30 days after cord section in the rat there is a histochemical evidence of an increased proportion of fast fibers in the soleus, and this normally slow muscle acquires myosin light chains of the fast-muscle type,[19] although it is not clear whether these are the adult fast isoforms. However, the rapid appearance of fast-fiber-specific proteins is not a general feature. The low contents of parvalbumin in fetal and adult soleus are unaffected by denervation.[13] Furthermore, the denervated soleus is apparently slow to lose proteins characteristic of type 1 fibers, since little change is seen in the CA III or myoglobin protein content during the 32 days after denervation.[1,5] Eventual loss of CA III protein from denervated soleus is presumably heralded by the decrease of CA III mRNA to 30%

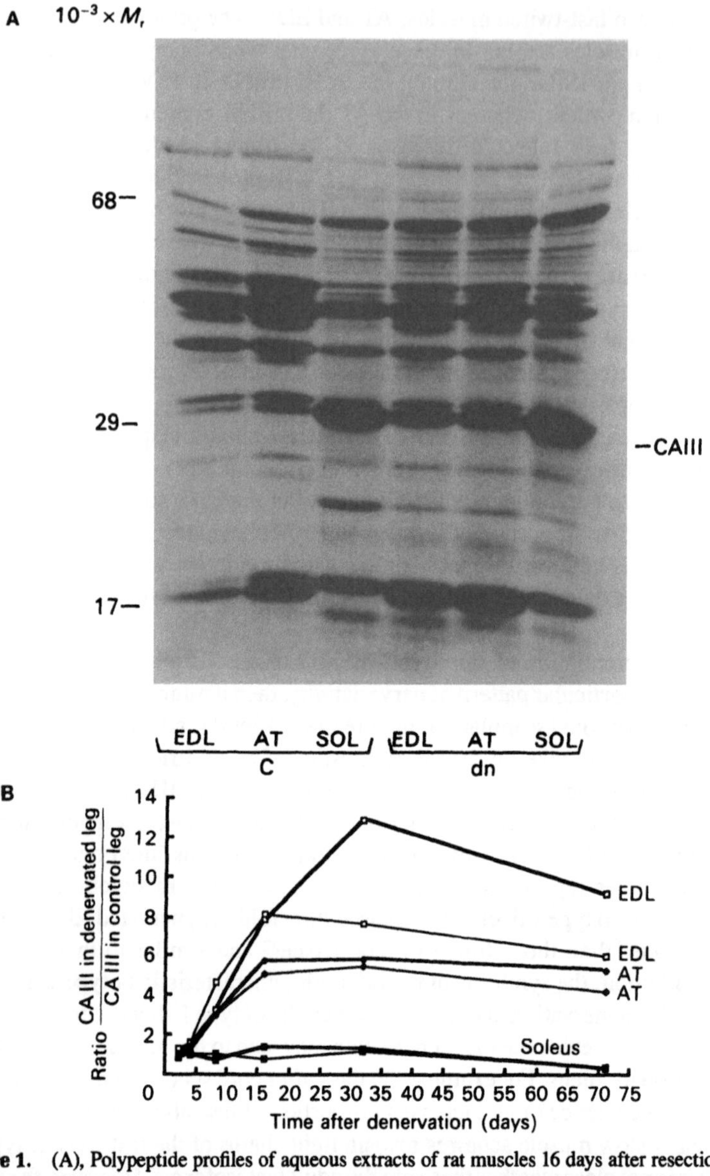

Figure 1. (A), Polypeptide profiles of aqueous extracts of rat muscles 16 days after resection of the sciatic nerve (dn) and in contralateral control muscles (C). SOL, soleus. (Reprinted with permission from *Biochemical Journal.*) (B) Changes in the amounts of CA III protein in rat muscle after sciatic nerve resection assessed by radioimmunoassay. CA III protein is expressed per milligram wet weight of tissue (——) or per microgram of soluble protein (**——**) and plotted as a ratio of the denervated/control muscles. Each point represents the average results for two animals except at 16 days, where data were collected from three animals for EDL and from a single animal for AT and soleus. (Reprinted with permission from *Biochemical Journal.*) (C) Northern blots of CA III mRNA in individual rat leg muscles [control (c) and denervated (d) for 15 days]. Loading/mRNA (2 μg) was the same for each slot. SOL, soleus. (Reprinted with permission from *Biochemical Journal.*)

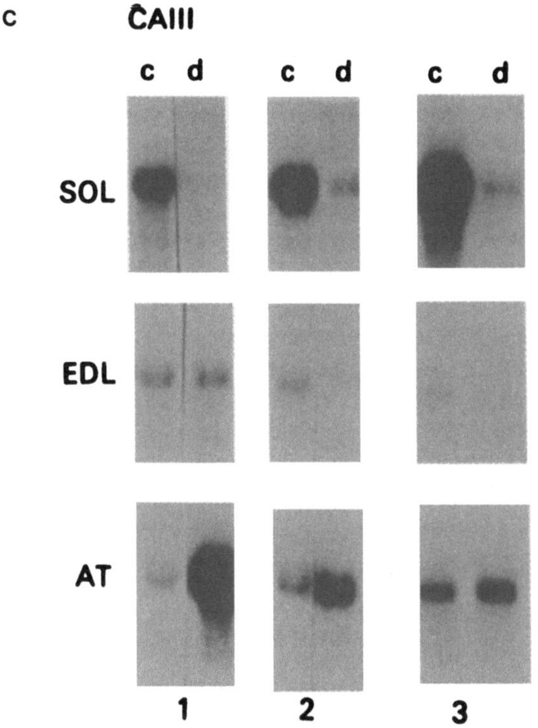

Figure 1. (*Continued*)

of control values. An important corollary to the studies described above has been reported.[3] The mRNAs of four genes expressed in rabbit AT and EDL were measured after subjection to direct chronic electrical stimulation. The mRNA for CA III rose significantly, showing that the direct effect of low-level electrical stimulation is on gene transcription, but how neural impulses are decoded to give gene activity is unknown. The molecular signal for switching on muscle CA III synthesis may be a growth factor secreted by the nerve ending or could relate directly to a neurotransmitter. Alternatively, the situation may be the result of hormonal control and similar to that found in hypophysectomized female rats, in which the absence of growth hormone appears to switch on the synthesis of CA III protein in the liver.

As discussed below, a highly analogous switch-on for CA III is initiated by removal of the influence of the thyroid.

2. Effects of Thyroidectomy on Muscle CA III

Thyroidectomy is known to affect rat muscle, with much research concentrated on slow-twitch muscle, particularly the soleus.[15] It has recently been shown[7]

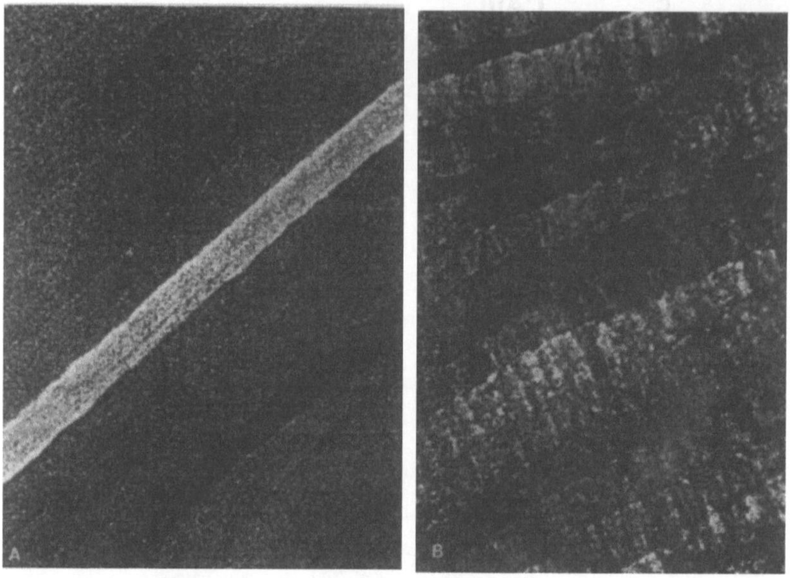

Figure 2. Immunocytochemical localization of CA III in control AT (A) and in AT muscle 15 days postdenervation (B). (Reprinted with permission from *Comparative Biochemistry and Physiology*.)

that in thyroidectomized rats there are more type 1 fibers in soleus than in the controls, with a concomitant increase of 30-kDa protein (now identified as CA III; P. A. Rogers, personal communication). Since the age of the rats at the time of thyroidectomy was not given in that study, it was not clear whether thyroidectomy actually induced the formation of type 1 fibers or whether it maintained the initial level in these fibers. It is important to know how that number of type 1 fibers varies in soleus throughout the lifetime of a rat. Our study did not show which alternative is true for soleus, since the thyroidectomized animals and their age-matched controls both contained only type 1 fibers. Study of AT and EDL was far more informative, with a clear increase in type 1 fibers in the thyroidectomized animals, together with a highly significant increase in CA III concentrations (Figs. 3 and 4). Interestingly, the concentrations of CA III in soleus in both male and female rats rose substantially after thyroidectomy, although the fiber types in control and experimental animals were the same.

As stated above, a marked change in fiber distribution was found in conjunction with increased CA III concentration in AT and EDL after thyroidectomy, there were far more lighter 2A-like fibers in the thyroidectomized muscle, and the majority of these stained for CA III, with a level of staining much greater than that in similar fibers in the control. It seems probable that these fibers are in the process of changing to type 1 and that this accounts for the increasing number of type 1

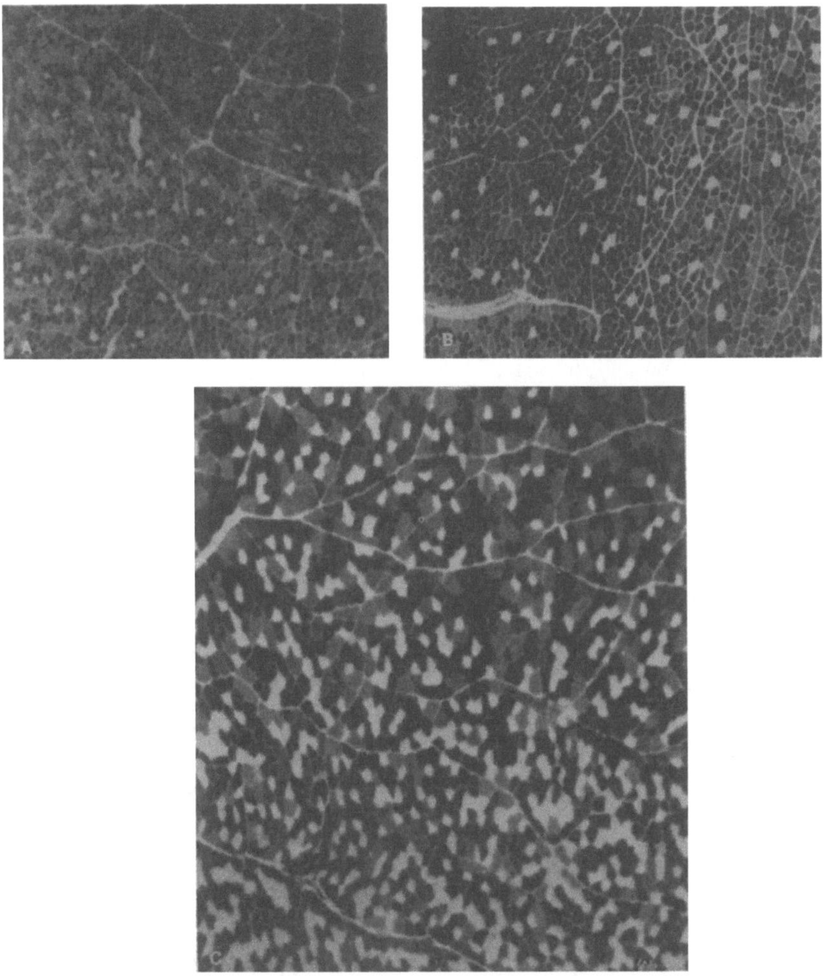

Figure 3. (A and B) Sections of AT from control male rat at beginning (A) and end (B) of experiment. Staining is for ATPase; type 1 fibers are white. (C) Section from thyroidectomized male stained for ATPase. Note vast increase in type 1 fibers with sections compared to A and B. (Reprinted with permission from *Journal of Histochemistry and Cytochemistry*.)

fibers in these muscles. It is not clear whether type 2B-like fibers change as a result of thyroidectomy and only a much longer-term experiment, with serial sampling of the population of thyroidectomized rat muscles and counting of each fiber type, would answer this question.

Male and female rats were used in this study because CA III in rat liver exhibits a sexual dimorphism based on the differing patterns of growth hormone

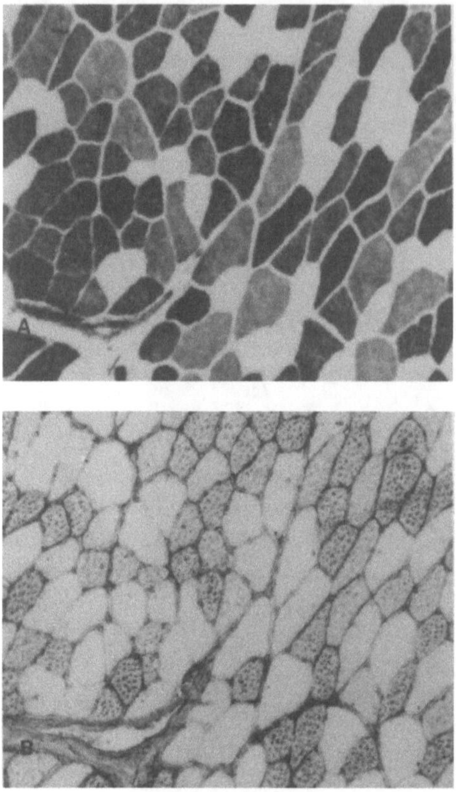

Figure 4. Thyroidectomized AT muscle: enlargements to highlight the intense immunostaining of CA III in type 1 and "light-type 2A" fibers. (A) ATPase staining; (B) CA III immunostaining. (Reprinted with permission from *Journal of Histochemistry and Cytochemistry*.)

secretion between the sexes,[12] and thyroidectomy is known to alter the release of growth hormone from the pituitary to a more female-type pattern. There was in fact no such dimorphism detectable in CA III concentrations in muscle either of controls or of thyroidectomized animals.

Whether the effects observed after thyroidectomy are due to the absence of thyroxine or are caused in part by the lowering of serum calcium because of parathyroid removal is not certain. Previous work suggests the former, as fiber type changes in the soleus after thyroidectomy are reversed by injection of T3.[15]

The evidence from SDS–gels is that myoglobin is not induced in parallel with CA III, but immunocytochemical and radioimmunoassay data are required to substantiate this point (Fig. 5). Thyroid hormone is known to affect the production of RNA for myosin heavy chain in rat soleus and cardiac muscle, and it will be interested to discover whether mRNA levels for CA III increase in the three muscle types after thyroidectomy. The fact that there was an increase in CA III concentrations in the soleus of thyroidectomized animals, with no alteration in fiber type,

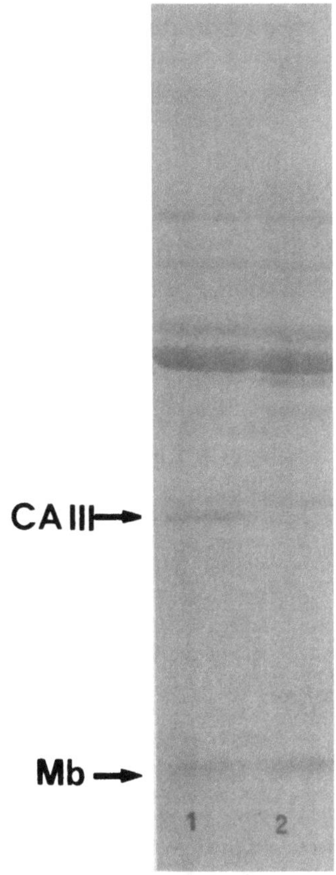

Figure 5. Electrophoretic separation of rat muscle proteins on SDS–gel. Lane 1 is thyroidectomized muscle showing increased CA III but myoglobin level similar to that of control muscle (lane 2). Other soluble proteins are similar in both lanes. (Reprinted with permission from *Journal of Histochemistry and Cytochemistry*.)

shows that histochemical data alone are not sufficient to show alterations in the function of muscle fibers. Since the major soluble protein in rat soleus is known to be CA III,[4] this protein seems likely to be an extremely useful marker for investigations into hormonal and neurological muscle changes.

It seems that both thyroxine and neuronal innovation suppress CA III synthesis in fast-fiber muscles, an effect that seem likely to be transcriptional repressor related. Removal of the influence of this putative repressor by denervation or thyroidectomy produces a tantalizingly similar increase in CA III in each case. To what extent those two phenomena act through similar pathways remains to be elucidated.

One impressive recent study on actin and cytochrome expression in atrophied skeletal muscle and the alleviating effect of the anabolic drug Clembuterol provides preliminary evidence for a control pathway.[2] For example, the usual striking loss of actin mRNA during atrophy was completely prevented by Clembuterol. Although this effect may relate to agonist activity, it has been suggested that the drug may have a more direct effect on the transcriptional or translational machinery and the effect of Clembuterol on CA III expression is under investigation. This and other lines of approach should define likely repressor or inducer molecules.

References

1. Askmark, H., Coulson, M., and Roxin, L.-E., 1984, *Muscle Nerve* 7:656–661.
2. Babij, P., and Booth, F. W., 1988, *Am. J. Physiol.* **254**:C657–C660.
3. Brownson, C. A., Isenberg, H., Brown, W., Salmons, S., Edwards, Y., *Muscle Nerve*, (in press).
4. Carter, N. D., Jeffery, S., and Shiels, A., 1982, *FEBS Lett.* **139**:265.
5. Carter, N. D., Wistrand, P. J., Isenberg, H., Askmark, H., Jeffery, S., Hopkinson, D. A., and Edwards, Y., 1988, *Biochem. J.* **256**:147–152.
6. Dhoot, G. K., and Perry, S. V., 1980, *Exp. Cell. Res.* **127**:75–87.
7. Gagnon, J., Ho-Kim, M. A., Champagne, C., Temblay, R. R., Rogers, P. A., 1985, *FEBS Lett.* **180**:335.
8. Holmes, R. S., 1977, *Comp. Biochem. Physiol.* **B57**:117–120.
9. Jeffery, D., Edwards, Y. H., Jackson, M. J., Jeffery, S., and Carter, N. D., 1982, *Comp. Biochem. Physiol.* **B73**:971–975.
10. Jeffery, S., Edwards, Y. H., and Carter, N. D., 1980, *Biochem. Genet.* **18**:843–849.
11. Jeffery, S., Carter, N. D., and Smith, A., 1986, *J. Histochem. Cytochem.* **34**:513–516.
12. Jeffery, S., Wilson, C., Mode, A., Gustafsson, J. A., and Carter, N., 1986, *J. Endocrinol.* **110**:123–126.
13. Leberer, E., and Pette, D., 1986, *Biochem. J.* **235**:67–73.
14. Lloyd, J., McMillan, S., Hopkinson, D., and Edwards, Y. H., 1986, *Gene* **41**:233–239.
15. Nwyoe, L., Monmaetts, W. F. H. M., Simpson, D. R., Scraydarian, K., and Maruisch, M., 1982, *Am. J. Physiol.* **242**:R401.
16. Pette, D., 1984, *Med. Sci. Sports* **16**:517–528.
17. Register, A. M., Koester, M. K., and Noltmann, E. A., 1978, *J. Biol. Chem.* **253**:4143–4152.
18. Roy, R. K., Sreter, F. A., and Sankar, S., 1979, *Dev. Biol.* **69**:15–30.
19. Rubinstein, N. A., and Kelly, A. M., 1978, *Dev. Biol.* **62**:473–485.
20. Salmons, S., and Henricksson, J., 1981, *Muscle Nerve* **4**:94–105.
21. Väänänen, H. K., Takala, T., and Morris, D. C., 1986, *Histochemistry* **86**:175–179.
22. Weller, P. A., Price, M., Isenberg, H., Edwards, Y. H., and Jefferys, A., 1986, *Mol. Cell. Biol.* **6**:4539–4547.
23. Whalen, R. G., Sell, S. M., Butter-Browns, G., Schwartz, K., Bouveret, P., and Pinset-Harstrom, I., 1981, *Nature* (London) **292**:805–809.
24. Wiles, C. M., Young, A., Jones, D. A., Edwards, R. H. T., 1979, Muscle relaxation rate, fiber-type composition and energy turnover in hyper- and hypo-thyroid patients. *Clin. Sci.* **57**:375.
25. Wistrand, P. J., Carter, N. D., and Askmark, H., 1987, *Comp. Biochem. Physiol.* **A86**:177–184.

CHAPTER 22

Skeletal Muscle Mitochondrial Carbonic Anhydrase

A Case of Species Specificity

BAYARD T. STOREY

Much of the carbonic anhydrase (CA) activity in skeletal muscles is CA III, which makes up a substantial proportion of the muscle cell protein.[4,5,20,21,25,31] Skeletal muscle contain CA activity with properties attributable to the well-characterized isozymes, CA I and CA II.[17,23,27,32,36] The subcellular distribution of these isozymes in muscle has remained a matter of some controversy, which has been exacerbated by clear indications of distribution differences between mammalian species.[6,16,23,30]

1. CA Activity in Sarcoplasmic Reticulum and Mitochondria

An important question relating to muscle CA II activity concerns its compartmentation in organelles within muscle cells. Eighty-five percent of assayable CA II activity in blood-free homogenates from predominantly white skeletal muscle of rabbit was found in the 100,000 × *g* supernatant fraction, and 15% was associated with the sedimented membrane fractions.[35] Since the homogenization was carried out on frozen tissues, it was not possible to tell to what extent the apparently soluble enzyme was sequestered in the intracellular organelles. A gentler mode of homogenization was later used to isolate sarcoplasmic reticulum vesicles from both red and white rabbit skeletal muscle.[2] These preparations contained substantial CA activity which, from its high sensitivity to sulfonamide inhibitors, could be classified as CA II. Mitochondrial CA activity in both types of rabbit skeletal muscle was negligible.

BAYARD T. STOREY • Department of Obstetrics and Gynecology and Department of Physiology, University of Pennsylvania School of Medicine, Philadelphia, Pennsylvania 19104-6085.

In contrast to the lack of mitochondrial CA activity in rabbit skeletal muscle, substantial CA activity has been reported in guinea pig skeletal muscle.[10] In this species, only mitochondria isolated from liver and skeletal muscles had CA activity; mitochondria from heart, brain, and kidney showed none. The probability of species specificity with regard to this isozyme was increased by the observation that rat kidney mitochondria have substantial CA activity.[8] In contrast, the CA activity of rat liver mitochondria was less than half that observed in guinea pig liver mitochondria.[8] Species specificity with regard to skeletal muscle mitochondrial CA activity was demonstrated more recently,[38] confirming both earlier results[28] with Percoll gradient-purified mitochondria from guinea pig and rabbit leg muscle.

2. Guinea Pig Skeletal Muscle Mitochondrial CA

Skeletal muscle mitochondria prepared by the Percoll gradient method[29] were shown by electron microscopy to contain 5% or less contamination by sarcoplasmic reticulum vesicles.[38] Sarcoplasmic reticulum vesicles were prepared by the same method and recovered in the heavy fraction in a band well separated from the mitochondria for comparative determinations of CA activity. The activities of these organelles determined by the changing pH methods[28] at 4°C in the presence and absence of Triton X-100 are shown in Table I for both rabbit and guinea pig. The CA activities of heavy sarcoplasmic reticulum vesicles from both species are comparable. The increase seen on addition 0.5% Triton X-100 indicates that the CA is in the vesicle interior.[22] No CA activity was detected in rabbit skeletal muscle mitochondria above the background of the uncatalyzed CO_2 hydration rate. The value of k_{enz} observed for guinea pig muscle mitochondria is low in the absence of Triton X-100 but is raised by a factor of 6.0 in its presence. A similar result was obtained at 28°C, using the spectrophotometric method of Khalifah.[24] In the absence of Triton X-100, $k_{enz} = 0.024 \pm 0.001$ (50 SD, $n = 8$); in its presence, $k_{enz} = 0.134 \pm 0.009$ (30 DS, $n = 8$), for a 5.7-fold increase. The same increase at the two temperatures indicates that the same enzyme is involved. The mitochondrial value of k_{enz} is at 4°C is fourfold that of the sarcoplasmic reticulum. One may conclude from those results that the mitochondrial activity is localized in the mitochondrial matrix and is not affected by contamination from sarcoplasmic reticulum vesicles.

Compartmentalization of the mitochondrial CA activity in the matrix would place it in a compartment bounded by a membrane with limited permeability to HCO_3 or H^+. As a result, the activity of CA within the compartment would not be observed by determination of H^+ formation from H_2CO_3 formed from the hydration of CO_2[1]. Removal of the permeability barrier would expose the latent CA activity. One way to measure CA activity when it is thus compartmentalized is by the mass spectrometric ^{18}O exchange method.[22] This method has the interesting diagnostic feature that, if CA activity is within a compartment bounded by an

TABLE I
CA Activity, Expressed as k_{enz}, for Guinea Pig and Rabbit Skeletal Muscle Organelles in the Presence and Absence of Triton X-100[a]

Species	k_{enz} (mg/ml)$^{-1}$ sec^{-1} (mean ± SD)	
	Mito[b] (n = 8)	SR[c] (n = 6)
Guinea pig		
0%[d]	0.0037 ± 0.0003	0.0017 ± 0.0002
0.5%	0.022 ± 0.002	0.0052 ± 0.0007
Rabbit		
0%	ND	0.0033 ± 0.0006
0.5%	ND	0.0077 ± 0.0004

[a]Assays were done at 4°C and pH 7.4. The rate constant k_{enz} is for the enzyme-catalyzed CO_2 hydration normalized to organelle protein concentration and is obtained by dividing the catalyzed rate by the initial concentration of added CO_2.[8] See Storey *et al.*[38] for experimental details.
[b]Mitochondrial protein concentration (Mito) was 0.5–1.1 mg/ml. ND, Not detectable above uncatalyzed rate.
[c]Sarcoplasmic reticulum (SR) protein concentration was 1.0–1.5 mg/ml.
[d]Triton X-100 concentration.

impermeable membrane, the disappearance of ^{18}O-labeled CO_2 is biphasic in time, with an initial rapid decrease in labeled CO_2 resembling a step function. This step was observed by using this technique with guinea pig liver mitochondria; the step was greatly attenuated in freeze-thawed samples.[10] Precisely the same step, nearly abolished by Triton X-100, was observed with guinea pig skeletal muscle mitochondria.[38] This finding further supports the conclusion that mitochondrial CA, when it is present in a given mammalian tissue, is a matrix enzyme.

Characterization of the skeletal muscle mitochondrial enzyme shows that its kinetic parameters are those expected of the CA II isozyme. The CA activity is highly sensitive to sulfonamide inhibitor,[5] as expected for CA II.[34] The K_i for ethoxzolamide inhibition was found to be 8.7 nM at pH 7.4[38] by use of the Easson–Stedman plot.[13] The temperature dependence of k_{enz} is high and comparable to that of CA II. Determination of the Arrhenius activation energy, E_A, for k_{enz} gave the value 12.8 kcal/mole,[38] close to the previously determined values of activation enthalpy for CA II ranging between 7.9 and 11.7 kcal/mol.[14,33] The latter values were based on the first order rate constant k_{cat}; k_{enz} is a second-order constant. The temperature dependence of the two types of rate constants will differ somewhat, since k_{enz} contains the temperature-dependent rate constant k_{cat} and rate constant ratio k_{-2}/k_{-1}.[22] The difference should not exceed 2–3 kcal/mol as a result of compensation in the rate constant ratio. The relatively high value of ΔH for CA II is in marked contrast to those obtained for CA I, which range from 3.7 to 4.9 kcal/

mol.[33] No significant difference for the values of k_{enz} at 25 and 37°C was found for CA V in guinea pig liver mitochondria,[8,10] indicating that this isozyme has ΔH comparable to or less than that of CA I. Based on high sensitivity to ethoxzolamide and high activation energy, the guinea pig muscle mitochondrial enzymes resembles the CA II isozyme in kinetic properties.

The Easson–Stedman plot which yielded K_i for ethoxzolamide also gave the mitochondrial content of enzymes as 0.002 nmol/mg protein.[38] This content is lower than but comparable to those of the matrix dehydrogenase, e.g., glutamate and isocitrate dehydrogenase,[26] and implies a high-activity CA isozyme. A rough estimate of the first-order constant k^0_{cat}, corresponding to V_{max} and giving the maximum turnover of the enzyme, may be calculated from k_{enz} at 25°C and pH 7.4 and the mitochondrial content, for comparison with the values obtained for human CA I and CA II at 25°C and pH 7.5.[33] The estimate assumes an apparent K_m of 5 or 10 mM for CO_2 in the hydration reaction, compared with the apparent K_m for guinea pig liver CA V of the 2.5 mM and rat liver CA V of 7.5 mM,[8] and K_m values of 12 and 4 mM reported[33] for human CA II and CA I, respectively. The term apparent K_m is derived from rates calculated as k_{enz} (CO_2) for the hydration reaction, as discussed,[8] and is close to, but not theoretically the same as, the conventional K_m. For an initial CO_2 concentration equal to the apparent K_m, the CO_2 hydration rate should be half the maximal rate, V_{max}. At 25°C and pH 7.4, $k_{enz} = 0.118 \pm 0.014$ ($n = 4$) for the muscle mitochondrial CA activity.[38] This leads to a calculated half-maximal rate at 5 mM CO_2 of 0.59 μmol/mg of protein per sec and thus a value of k^0_{cat} of 0.6×10^6 sec^{-1}. For an apparent K_m of 10 mM, this value would be 1.2×10^6 sec^{-1}. These values are comparable to the value of k^0_{cat} for human red blood cell CA II of 1.1×10^6 sec^{-1} at 25°C and pH 7.5.[33] The activity of the muscle mitochondrial enzyme, based on this estimate, also corresponds to CA II.

3. Possible Biosynthetic Role of Muscle Mitochondrial CA

The role of CA V as a participant in the mitochondrial biosynthesis of citrulline at the level of carbamyl phosphate synthetase has been documented.[7,9,11,12] One plausible candidate for a biosynthetic reaction of this type in skeletal muscle mitochondria would the ATP-dependent carboxylation of pyruvate. This reaction has been shown to occur in rat skeletal muscle mitochondria.[37] The reaction was greatly stimulated by acetyl-L-carnitine and yielded malate from intact and citract from sonicated mitochondria. The rates reported for malate production were 2.5 nmol/mg of protein per min. The much lower rate of malate synthesis seen in intact mitochondria, in which CA would be expected to participate, indicates that this role for CA is not operative. All attempts to show participation of guinea pig skeletal muscle mitochondria in this biosynthetic pathway were negative (B. T. Storey, unpublished data). The question of the function of the CA present in guinea pig skeletal muscle mitochondria and the parallel question of the loss of any such function due to the absence of CA in rabbit skeletal muscle mitochondria remain unanswered.

4. Envoi

Consideration of the species specificity governing the presence or absence of CA activity in mitochondria from liver, kidney, and skeletal muscle in guinea pig, rat, and rabbit raises an interesting problem regarding the regulation of intracellular enzymes. The liver enzyme in guinea pig has been shown to be CA V, a genetically distinct isozyme.[19] While this has not yet been demonstrated for the rat liver enzyme, the observation that rat liver mitochondria also have CA activity suggests that a CA V isozyme is present, although the dependence of activity on pH is considerably less than for the guinea pig liver enzyme.[8] The two anomalous situations are rat versus guinea pig kidney and guinea pig versus rabbit skeletal muscle, in which mitochondria from the first species have CA activity whereas those from the second lack it. The similarity of the guinea pig muscle mitochondria CA to CA II suggests that a mode of intracellular regulation of enzyme targeting may occur at the posttranslational level rather than at the level of the gene.[18] In the example here, CA II in muscle, which is normally targeted to the sarcoplasmic reticulum in most if not all species may be targeted postsynthesis in certain species to other organelles, in this case the mitochondria. This form of regulation would allow for differences between species to occur at the developmental level, with minimal changes in the gene for the enzyme under consideration. This suggestion is accessible to experimentation by more extensive study of certain key enzymes, in particular CA, among a greater number of mammalian species.

References

1. Booth, V. H., 1938, *J. Physiol.* (London) **93**:117–128.
2. Bruns, W., Dermietzel, R., and Gros, G., 1986, *J. Physiol.* (London) **371**:354–364.
3. Carter, M. J., 1972, *Biol. Rev.* **47**:465–513.
4. Carter, N., Jeffery, S., Shiels, A., Edwards, Y., Tipler, T., and Hopkinson, D. A., 1979, *Biochem. Genet.* **17**:837–854.
5. Carter, N., this volume, Chapter 21.
6. Dermietzel, R., Leibstein, A. Siffert, W., Zamboglou, N., and Gros, G., 1985, *J. Histochem. Cytochem.* **33**:93–98.
7. Dodgson, S. J., 1987, *J. Appl. Physiol.* **63**:2134–2141.
8. Dodgson, S. J., and Contino, L. C., 1988, *Arch. Biochem. Biophys.* **260**:334–341.
9. Dodgson, S. J., and Forster, R. E., 1987, *J. Appl. Physiol.* **60**:646–652.
10. Dodgson, S. J., Forster, R. E., II, Storey, B. T., and Mela, L., 1980, *Proc. Natl. Acad. Sci. USA* **77**:5562–5566.
11. Dodgson, S. J., Forster, R. E., and Storey, B. T., 1982, *J. Biol. Chem.* **257**:1705–1711.

12. Dodgson, S. J., Forster, R. E., II, Schwed, D., and Storey, B. T., 1983, *J. Biol. Chem.* **258**:7696–7701.
13. Easson, L. H., and Stedman, E., 1936–7, *Proc. R. Soc. London Ser. B* **121**:142–164.
14. Ghannam, A. F., Tsen, W., and Rowlett, R. S., 1986, *J. Biol. Chem.* **261**:1164–1169.
15. Geers, C., and Gros, G., this volume, Chapter.
16. Gros, G., and Dodgson, S., 1988, *Annu. Rev. Physiol.* **50**:664–694.
17. Gros, G., Siffert, W., and Schmid, A., 1980, in: *Biophysics and Physiology of Carbon Dioxide* (C. Bauer, G. Gros, and H. Bartels, eds.), Springer Verlag, Berlin, pp. 410–416.
18. Hewett-Emmet, D., Hopkins, P. J., Tashian, R. E., and Czelusniak, J., 1984, *Ann. N.Y. Acad. Sci.* **429**:338–358.
19. Hewett-Emmet, D., Cook, R. G., and Dodgson, S. J., 1986, *Isozyme Bull.* **19**:13.
20. Holmes, R. S., 1976, *J. Exp. Zool.* **197**:289–195.
21. Holmes, R. S., 1977, *Eur. J. Biochem.* **78**:511–520.
22. Itada, N., and Forster, R. E., 1977, *J. Biol. Chem.* **252**:3881–3890.
23. Jeffery, S., Carter, N. D., and Smith, A., 1986, *J. Histochem. Cytochem.* **34**:513–516.
24. Khalifah, R. G., 1971, *J. Biol. Chem.* **246**:2561–2573.
25. Koester, M. K., Register, A. M., and Noltmann, E. A., 1977, *Biochem. Biophys. Res. Commun.* **76**:196–204.
26. Kroeger, A., and Klingenberg, M., 1967, in: *Current Topics In Bioenergetics*, Volume 2 (D. R. Sandi, ed.), Academic Press, New York, pp. 151–193.
27. Maren, T. H., 1967, *Physiol. Rev.* **47**:595–781.
28. Maren, T. H., and Couto, E. O., 1979, *Arch. Biochem. Biophys.* **196**:501.
29. Mickelson, J. R., Greaser, M. L., and Marsh, B. B., 1984, *Anal. Biochem.* **109**:255–260.
30. Moyle, S., Jeffery, S., and Carter, N. D., 1984, *J. Histochem. Cytochem.* **32**:1262–1264.
31. Moynihan, J. B., 1977, *Biochem. J.* **168**:567–589.
32. Riley, D. A., Ellis, S., and Bain, J., 1982, *J. Histochem. Cytochem.* **30**:1225–1288.
33. Sanyal, G. and Maren, T., 1981, *J. Biol. Chem.* **256**:608–612.
34. Sanyal, G., Swenson, E. R., Pessah, N. I., and Maren, T. H., 1982, *Mol. Pharmacol.* **22**: 211–220.
35. Siffert, W., and Gros, G., 1982, *Biochem. J.* **205**:559–566.
36. Siffert, W., Teske, W., Gartner, A., and Gros, G., 1983, *J. Biochem. Biophys. Methods* **8**: 331–338.
37. Spydevold, O., Davis, E. J., and Brewer, J., 1976, *Eur. J. Biochem.* **71**:155–165.
38. Storey, B. T., Lin, L. C., Tompkins, B., and Forster, R. E., 1989, *Arch. Biochem. Biophys.* **270**:144–152.

PART VI

Carbonic Anhydrases of the Gastrointestinal Tract

CHAPTER 23

Distribution and Functions of Carbonic Anhydrase in the Gastrointestinal Tract

ERIK R. SWENSON

1. Introduction

Carbonic anhydrase (CA) is found along the entire gastrointestinal tract, from salivary glands to the colon. This review focuses on the distribution of CA and its role in gastric, pancreatic, biliary, and intestinal function. Numerous roles subserved by the enzyme in the GI tract—acid–base transport, neutral salt absorption, cell volume, and intracellular pH regulation—have in large part been explored with the use of specific inhibitors. However, as Maren[104] has argued, strict guidelines must be followed in their use. Emphasis will be given to those studies in which these precepts have been followed, most notably in the choice of inhibitor concentration.

2. Distribution

Biochemical, histochemical, and immunocytochemical studies in mammals reveal a general pattern of distribution (Table I). There is a considerable heterogeneity between organs in enzyme activity, isozyme content, and cellular localization. CA activity (expressed as whole-cell activity) may be either in apparent high excess of physiological need[102] or low and concentrated in membrane fractions. It is possible that the more relevant enzymatic activity is that in the microenvironment surrounding the ion-translocating membranes, since there are large differences in

ERIK R. SWENSON • Department of Medicine, University of Washington and Veterans Administration Medical Center, Seattle, Washington 98108.

TABLE I
Activity and Isoenzyme Characterization of CA in the Gastrointestinal Tract

	Activity (enzyme units/g)		
Location	Rat[b]	Human[c]	Isoenzymes[a]
Stomach	430	280	
Parietal cell	7200		II
Surface cell	720		II
Duodenum	77		II
Jejunum	26	80	I, II
Ileum	19	23	I, II
Colon	345	160	I, II, IV
Appendix	(+)		I, II
Pancreas	10		
Ductal cells	(+)		II, IV
Islet cells	(−)		
Liver (hepatocytes)	20		I, II, III, IV, V
Biliary tract			
Bile ducts	(+)		II
Gall bladder	(+)		II
Red cells	2400		I, II

[a]Taken from Kumpulainen,[83] Lönnerholm *et al*,[98] and Carter *et al.*[21a]
[b]Numerical data taken from Maren[102]; presence (+) or absence (−) taken from Kumpulainen.[83]
[c]Taken from Lönnerholm *et al.*[98]

total cellular enzyme content between organs and cell types involved in equal acid–base or salt transport. The seeming paradox of modest changes in total enzymatic activity in response to physiological mediators in those organs with very high CA content (such as stomach) may hinge on this concept.

3. Stomach and Duodenum

Gastric mucosa was the first epithelial tissue found to have CA, which is present in mammalian gastric surface epithelial, parietal, and chief cells of the stomach and in surface epithelial cells and Brunner's glands of the duodenum.[98,149] Except for a rare duodenal enterocyte staining for CA I,[96] CA II is the principal isozyme found in stomach and duodenum. The distribution of gastric CA in a series of nonmammalian vertebrates is similar to that in mammals; CA is present in surface epithelial cells and in the acid-secretory oxyntic–peptic cells.[119] The specialized bovine rumen contains CA, which is similar to CA I in activity, histochemical staining, and sensitivity to inhibitors.[20,88] Changes in gastric CA activity have been attributed to a wide variety of agents, such as histamine, gastrin, bile salts, thiocyanate, ethanol, and aspirin, which are known to alter gastric acid secretion or cause ulceration. No data, however, exist to suggest that any act directly on the enzyme *in vitro* at relevant concentrations.[113,139,156]

3.1. Gastric Acid Secretion

The parietal or oxyntic cell is responsible for the secretion of gastric acid, a biological fluid of unsurpassed acidity (~150 mM HCl). Davenport[31] first proposed that CA was necessary to supply sufficient H^+ from the catalyzed hydration of CO_2 but later rejected this idea.[32] Davies concluded otherwise, calculating that the uncatalyzed rate would be insufficient and showing in the frog that CA inhibition with sulfanilamide reduced acid output.[33] The advent of acetazolamide allowed a convincing demonstration[66] that catalysis is required for high rates of acid secretion. Salient aspects of CA inhibition are best documented in the quantitative dose–response study with acetazolamide in the cat,[36] showing a smooth dose–response curve from 10 mg/kg (no effect) to 50 mg/kg (maximal effect: 50–80% suppression, depending on the pre-drug treatment stimulated rate of acid output). In these experiments and others, measured residual acid secretion can be accounted for by the uncatalyzed rate when reasonable estimates for parietal cell volume (~10% of stomach volume) are taken (reviewed in reference 102). Data in dogs and humans are comparable. In humans, 10 mg/kg has no effect, 50 mg/kg intravenously gives maximal suppression (60–90%) of stimulated secretion,[67,106] and equivalent oral doses give a similar dose response.[89,112]

Local and systemic acid–base changes occur with and alter the response to CA inhibitors. Metabolic acidosis in dogs[16] stimulates acid secretion, whether induced by HCl infusion or by prior drug administration (which by its renal effect causes HCO_3^- loss in urine). $NaHCO_3$ infusion (metabolic alkalosis) had the opposite effect. Acetazolamide had no effect on stimulated acid secretion in metabolic acidosis, and its correction by $NaHCO_3$ restored the inhibitory effect of acetazolamide. Doses of acetazolamide that inhibit gastric CA lead to tissue CO_2 retention and respiratory acidosis.[102] The surprising finding[16] that 5 mg/kg modestly inhibits gastric acid secretion in the dog may be explained by a lack of respiratory acidosis, which at the higher doses (20–30 mg/kg) may cause some partial refractoriness of the parietal cell to the drug that is overcome with greater doses (>30 mg/kg). Respiratory acidosis, like metabolic acidosis, may augment acid secretion, but this has not been studied *in vivo*. The isolated frog gastric mucosa responds to an imposed respiratory acidosis with increased acid secretion.[71] These findings regarding acid–base status may help to explain much of the variability in earlier data and also the relatively minor role that CA inhibitors have assumed in treatment of ulcer disease (see below).

Recent advances in our understanding of gastric acid secretion, while providing great insight into parietal cell membrane biology, have not greatly altered the original role proposed for CA by Davenport and others.[137] Figure 1 places CA

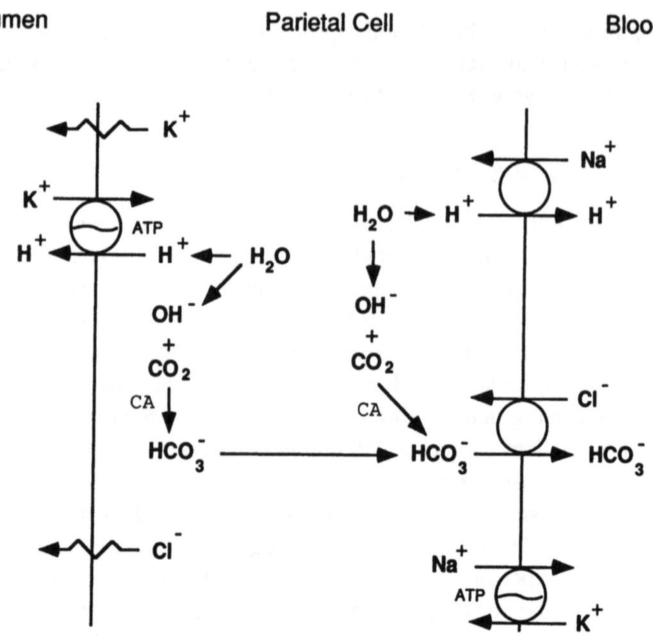

Figure 1. Membrane ion transport events in gastric acid secretion.

functionally with apical membrane-associated H^+–K^+ ATPase, the well-characterized gastric proton pump. The marked rise in gastric HCl output stimulated by food or secretagogues is preceded by the insertion of cytoplasmic vesicles containing H^+–K^+ ATPase,[136] Cl^- and K^+ conductance channels, and CA.[149] There is a 50% increase in specific CA II mRNA within minutes of secretogogue application to isolated canine parietal cells.[17] This regulation of CA II gene expression with stimulation explains earlier findings of increased gastric CA activity in animals treated with secretogogues.[113,138] Roughly 35% of total CA activity in parietal cells is firmly bound to purified membranes rich in H^+–K^+ ATPase.[92,161] This close association may be critically important, since the rate of ATP phosphorylation by H^+–K^+ ATPase is inhibited by increasing pH.[91,147] Basolateral Na^+–H^+ exchange appears to mediate an early intracellular alkalinization that may be necessary to activate pH-sensitive basolateral Cl^-–HCO_3^- exchange in advance of high rates of HCl secretion.[137] The importance of both of these basolateral exchangers comes from experiments showing >90% inhibition of acid secretion when both are blocked.[111]

As acid output by stimulated parietal cell rises, only a minimal increase in intracellular pH occurs, attesting to a remarkable efficiency in OH^- dissipation.[120] The H^+–K^+ ATPase pump generates an equivalent amount of OH^- which can rapidly combine with CO_2 to form HCO_3^- for disposal by basolateral Cl^-–HCO_3^-

exchange. These two processes help to forestall a limiting alkalosis in the microenvironment of the apical pump. Inhibition of gastric CA causes greater intracellular alkalosis,[63] as observed in other acid-secreting epithelia such as the renal proximal tubule.[75] The effects of acidosis and alkalosis on acid secretion and response to CA inhibition are consistent with a pump whose activity is pH dependent. The enzyme does not appear to supply H^+ from the hydration of CO_2 since (1) the hydrolysis of H_2O with ATP consumption provides sufficient protons for this purpose,[91] (2) H^+ production is two to three times greater than the rate of CO_2 production or O_2 consumption,[33] and (3) removal of all exogenous CO_2 under certain conditions does not reduce H^+ secretion in the frog stomach.[140] CA is associated with basolateral membranes of all stomach cells,[149] where it may be involved with intracellular pH regulation by supplying H^+ and HCO_3^- for Na^+–H^+ and Cl^-–HCO_3^- exchange.[142]

3.2. Gastroduodenal Bicarbonate Secretion

It has long been recognized that gastric and duodenal mucosa are capable of resisting acid–peptic digestion, but only within the last decade have the important characteristics of this defense been elucidated.[47] The resistance to acid is largely a result of parallel mucus and bicarbonate secretion by surface epithelial cells of the stomach and duodenum. Abundant evidence supports a functional role for CA in HCO_3^- secretion by both epithelia. CA appears principally in the cytoplasm of gastric and duodenal surface cells.[95,149] Activity, inhibition, and immunological measurements show it to be CA II.[84,96] In contrast to the parietal cell, there is little CA activity in membrane fractions[148] which contain other important enzymes involved in HCO_3^- secretion. Since intraluminal acid is one factor that stimulates HCO_3^- secretion, chronic acid suppression with omeprazole (an H^+–K^+ ATPase inhibitor) and ranitidine (an H_2 blocker) were studied as the cause of down regulation of epithelial CA content. Neither gastric nor duodenum CA activity or distribution was altered in rats with drug treatment.[96]

Secretion of HCO_3^- by surface cells (Fig. 2) into the overlying mucus gel maintains a neutral pH at the luminal surface despite a low bulk intraluminal pH.[46] An active energy-dependent secretory process moves HCO_3^- against an electrochemical gradient. The calculated HCO_3^- concentration of human nonparietal secretions is 60–90 mM.[41] The mechanism by which gastric and duodenal HCO_3^- is actively secreted is obscure. There appear to be both apical electroneutral Cl^-–HCO_3^- exchange and electrogenic HCO_3^- transport.[44,155] A passive paracellular pathway also exists for HCO_3^- movement that is sensitive to variations in transmucosal hydrostatic pressure and transepithelial HCO_3^- gradient.[44] This paracellular route may be important in conditions in which the integrity and function of the surface epithelial cells are impaired.

Inhibitors of CA reduce HCO_3^- production by surface mucosa of the stomach and duodenum. This was first shown in the amphibian gastric mucosa[43] and then in

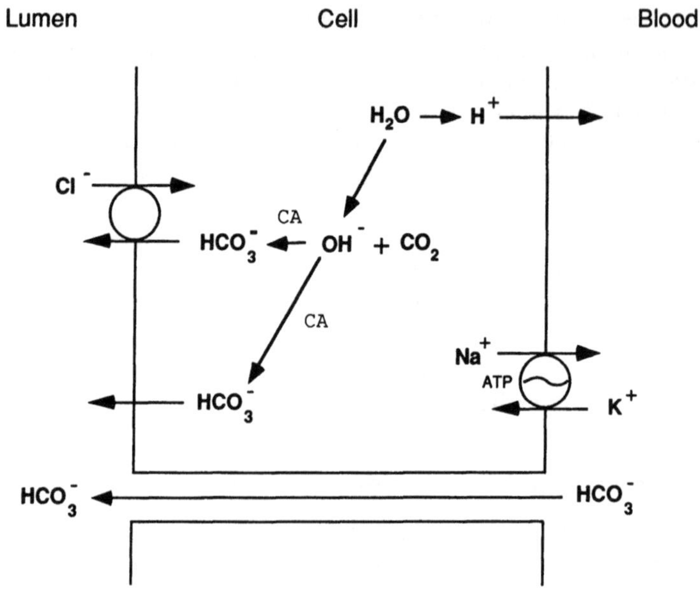

Figure 2. Membrane ion transport events in gastric and duodenal bicarbonate secretion.

the duodenum.[45] Basal and stimulated HCO_3^- secretion are reduced by acetazolamide when these rates are high in the amphibian. Removal of all exogenous CO_2–HCO_3^- *in vitro* mimics CA inhibition. In mammalian stomach and duodenum, inhibition is generally seen only under conditions of stimulated HCO_3^- output.[46,70,131] These studies together show that stimulated HCO_3^- secretion is reduced 40–70% by acetazolamide. There are unfortunately no good dose–response data with acetazolamide or any other CA inhibitor. The vast majority of studies used only very high single doses or concentrations ($>10^{-3}$ M *in vitro*) of acetazolamide. Although these levels represent the top end of the dose response and are consistent with CA inhibition, lower doses or concentrations would make the case more convincingly.[104] This has been shown with 0.5 mM acetazolamide in the stimulated guinea pig gastric surface cells, in which there is a marked reduction in alkaline secretion and the transepithelial potential difference (PD).

The cytoprotective function of CA-mediated HCO_3^- secretion is evident from numerous studies in which acetazolamide (1) directly induces gastric and duodenal ulcerations,[29,33,73,81,163] (2) enhances indomethacin-related ulceration,[76,134] and (3) markedly aggravates ulceration induced by hemorrhagic shock.[74] Kumagai *et al.*[81] showed in the rat that correction of systemic metabolic acidosis caused by acetazolamide did not attenuate the severity of ulceration, thus implicating CA inhibition in gastric mucosa. Paradoxically, it has been reported that acetazolamide is cytoprotective against aspirin and ethanol-induced ulceration.[77,134] One study concluded that this effect was a function of its sulfhydryl content, since ethoxzolamide a more powerful inhibitor was less effective.[86]

3.3. CA Inhibitors in Ulcer Disease

The power of CA inhibitors to suppress acid secretion led early to their use in peptic ulcer disease; however, their therapeutic promise has not been realized. An early uncontrolled trial[50] showed modest effects with acetazolamide (15 mg/kg per day) on gastric acid output in ulcer patients, but symptoms were improved. Their dose was likely too low, and the strong placebo effect noted in all major trials of therapy in ulcer disease probably accounted for any benefit. The only studies using sufficient dose ranges to achieve near-maximal acid suppression are those from Romania utilizing acetazolamide and ethoxzolamide.[126,127] Endoscopic documentation of ulcer healing and measurements of gastric acid suppression showed that CA inhibitor therapy was superior to placebo or antacid–anticholinergic therapy. However, they have not tested whether CA inhibitors are better than more modern therapy employing H_2 blockers and omeprazole, which have many fewer side effects. Complete suppression of stimulated acid secretion is seldom achieved with CA inhibitors. Furthermore, systemic metabolic and respiratory acidoses induced by renal CA inhibition oppose the action of drug at the stomach. Lastly, inhibition of CA involved in gastric and duodenal cytoprotective HCO_3^- secretion may more than offset the benefits of reduced acid secretion. In fact, patients with duodenal ulcer may have impaired duodenal HCO_3^- secretion,[65] placing them at greater risk to this deleterious consequence of CA inhibitor therapy.

4. Pancreas

The stimulated exocrine pancreas secretes large volumes of HCO_3^--rich fluid that neutralize gastric acid and activate certain pancreatic enzymes. CA II is present in the HCO_3^--secreting ductular epithelium[82] and is found in both cytoplasm and membranes.[15] Isozyme IV has also been found in membranes.[21a] Lesser amounts of cytoplasmic CA II are present in the non-HCO_3^--secreting centroacinar cells, with a complete absence in the islets of Langerhans.[15]

Studies using CA inhibitors in humans and other animals demonstrate a functional role for CA in pancreatic HCO_3^- secretion. Suppression of secretin-stimulated HCO_3^- output *in vitro* and *in vivo* varies from 30 to 60%.[9,22,34,79,125,128,129,153] Basal and CA-inhibited rates of stimulated HCO_3^- output are accounted for by uncatalyzed CO_2 reaction rates, and the potential catalytic activity of pancreatic enzyme is 1000–2000 times that required for stimulated HCO_3^- secretion (reviewed in reference 103). As in other systems involving CA-mediated acid–base transport, variations in systemic pH alter pancreatic HCO_3^- output. Metabolic acidosis decreases and metabolic alkalosis increases fluid and HCO_3^- secretion *in vivo* and *in vitro*.[1,128,129] Respiratory acidosis does not affect secretion, in contrast to respiratory alkalosis, which markedly reduces secretion.[129] Acetazolamide further lowers fluid secretion and HCO_3^- output under all acid–base conditions.[129]

Considerable controversy remains about the mechanisms of pancreatic HCO_3^- transport. Novak[115] has extensively reviewed this literature and proposes a model outlined in Fig. 3. The critical elements include (1) apical Cl^-–HCO_3^- exchange and chloride conductance and (2) basolateral Na^+–H^+ exchange, Na^+–K^+ ATPase, and K^+ conductance. Translocation of intracellular protons by Na^+–H^+ exchange yields OH^-, which is rapidly titrated by CO_2 to form HCO_3^- for transport across the apical membrane in exchange for Cl^-. Inhibition of CA causes an increase in OH^- and a reduced production rate of HCO_3^-, both of which slow the rate of apical and basolateral membrane ion exchange and lead to a diminished HCO_3^- output. If other anions are substituted for HCO_3^-, such as acetate,[79,153] fluid secretion is minimally affected and is not inhibited by acetazolamide. This finding suggests that the fundamental process is H^+–OH^- separation at the basolateral membrane and not HCO_3^- formation or uptake from blood. In the presence of CO_2, OH^- is preferentially titrated by CO_2, but in its absence, other titratable and diffusible small organic anions if available may readily substitute.[85,141]

An energy-consuming proton pump (proposed from studies with DCCD, a nonspecific inhibitor of H^+ATPases located at the basolateral membrane may directly extrude H^+ out of the cell into blood,[55] since secretin stimulation causes disappearance of cytoplasmic vesicles similar to those seen with induction of

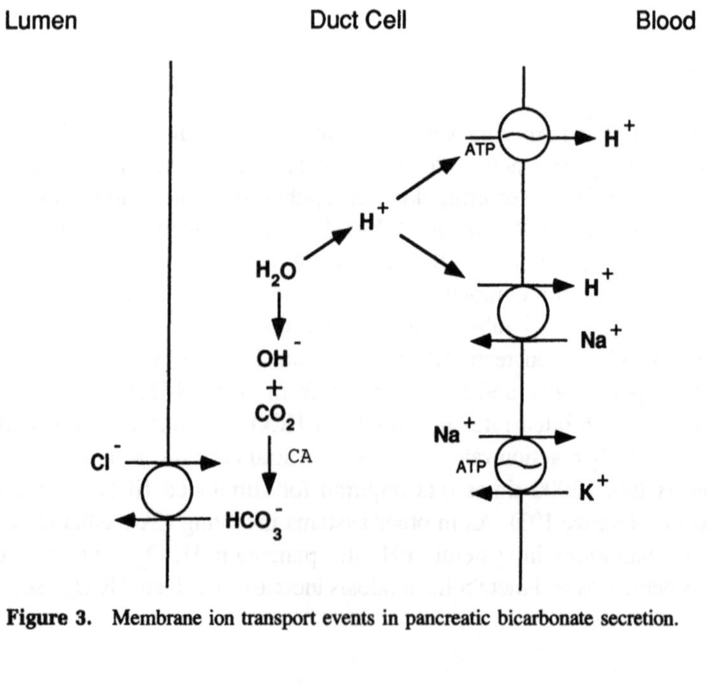

Figure 3. Membrane ion transport events in pancreatic bicarbonate secretion.

respiratory acidosis.[14] These vesicles may contain similar H^+ ATPase pumps, thought to be responsible for acid secretion in the distal nephron and amphibian bladder. CA would then function (as in stomach) to both hasten production of HCO_3^- and prevent a critical pH increase in the vicinity of the proton pump. In contrast, it has been proposed that ionic HCO_3^- is taken up across the basolateral membrane rather than there being direct H^+ secretion back into blood.[1,79] This hypothesis is based partly on the observations that 95% of secreted bicarbonate is derived from blood[23] and that in isolated perfused gland, HCO_3^- secretion is directly related to arterial HCO_3^- concentration and independent of pH.[1] These findings do not rule out direct H^+ secretion, since they can be explained by the known inhibitory effect of metabolic acidosis[129] and the opposing effects of differing directional changes in pCO_2 and OH^- when pH is held constant on the rate of HCO_3^- formation. Intracellular pH measurements in stimulated duct cells might resolve this question, since CA inhibition would be predicted to further raise intracellular pH if H^+ secretion were occurring and not if direct HCO_3^- uptake were the primary mechanism.

The clinical consequence of suppressed pancreatic HCO_3^- secretion in humans by use of CA inhibitors is unknown. Human and animal dose–response data predict less than maximal suppression at usual oral doses.[34] The residual stimulated rate may be sufficient in those with normal pancreatic function. In the acute treatment of pancreatitis, Anderson and Copass[2] used acetazolamide to successfully reduce pancreatic output by 25–50% and hasten recovery. This therapy, however, has not gained widespread use, possibly because of the drug's diuretic effect and its deleterious effects in liver disease, a common coexisting condition in many patients with pancreatitis.

5. Intestines

In addition to salt, water, and nutrient absorption, the intestines transport more than 1000 mEq of HCO_3^- per day. This major acid–base transport capacity is intimately linked to NaCl adsorption, short-chain fatty acid uptake, and systemic acid–base homeostasis.[27] Marked differences in CA content, total ion transport rates, direction of HCO_3^- flux, and response to systemic acid–base disorders[26] exist between jejunum, ileum, and colon.

5.1. Jejunum

The jejunum contains little CA (Table I), chiefly in basolateral membranes of both villus and crypt cells. Depending on the species studied, there is some staining in the cytoplasm and apical brush border membranes.[95] Both CA I and CA II have been detected.

Considerable HCO_3^- reabsorption occurs across the jejunum, since summed HCO_3^- output from gastric, duodenal, biliary, and pancreatic secretions exceeds gastric acid production. Intuitively, one would anticipate a role for CA in jejunal HCO_3^- reabsorption. In the only mammalian studies using acetazolamide, no effect on jejunal acidification *in vitro* was found,[11] in contrast to *in vivo* results for rats[121] and humans,[158] in which NaCl and HCO_3^- absorption decreased by 40–50%. Unfortunately, both *in vivo* studies used high concentrations (>1 mM), which may cause nonspecific effects[102] such as inhibition of stilbene-sensitive Cl^-–HCO_3^- exchange and possibly $NaHCO_3$ cotransport. A convincing case has been made for CA-mediated HCO_3^- reabsorption in the amphibian jejunum; the rate of serosal alkalinization (or mucosal acidification) is halved[167] by 100 μM acetazolamide, and cell depolarization in response to withdrawal of serosal HCO_3^- is reduced.[165]

A number of mechanisms capable of absorbing luminal HCO_3^- have been postulated, but their existence and relative importance remain uncertain (Fig. 4). Apical Na^+–H^+ exchange, which in many tissues is subserved by CA and is found in rat jejunal apical membrane vesicles,[24] is invoked to account for (1) a low pH in the microclimate of the rat jejunal villus,[144] (2) a Na^+ dependence of HCO_3^- absorption,[158] and (3) a high luminal pCO_2 of jejunal fluid.[124] However, these findings can be theoretically attributed to direct HCO_3^- absorption (possibly by apical $NaHCO_3$ cotransport) and generation of high pCO_2 by tissue metabolism and buffering of absorbed HCO_3^- by blood.[39] Further data against an apical Na^+–H^+ exchange are that jejunal Na and HCO_3^- reabsorption are not affected by changes in systemic or luminal pH or pCO_2,[26] results generally found in other

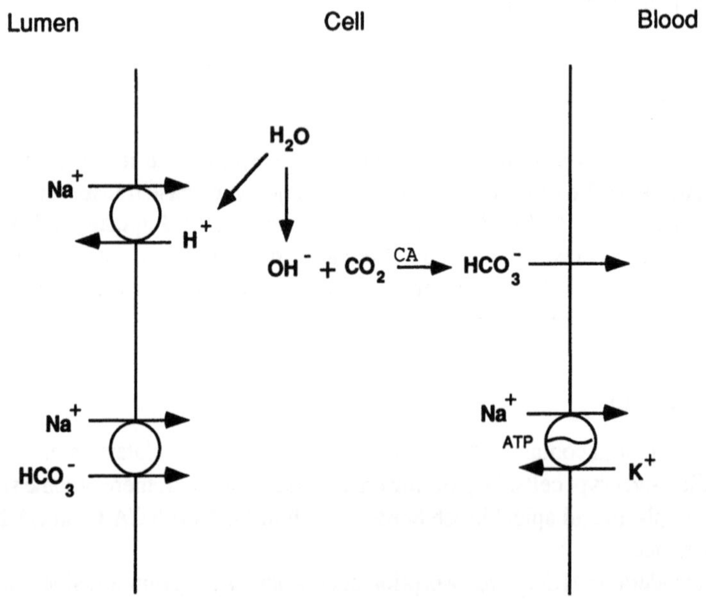

Figure 4. Membrane ion transport events in jejunal acidification and bicarbonate absorption.

tissues with unequivocal Na^+–H^+ exchange. Lastly, results with amiloride are controversial since high concentrations were used, which may not solely reflect inhibition of Na^+–H^+ exchange.[144] Since not all HCO_3^- reabsorption is Na^+ dependent, other mechanisms may operate. Amphibian jejunum contains an omeprazole-sensitive, K^+-dependent, H^+ ATPase pump[166] similar to that in mammalian parietal cells. This has not been sought for in mammalian jejunum, but if it were demonstrated, a role for CA would be readily apparent.

In summary, CA appears to mediate HCO_3^- absorption in the amphibian jejunum, subserving apical H^+ extrusion via the K^+–H^+ ATPase pump. However, until good dose–response data are obtained with CA inhibitors in the mammal, a stronger argument can be made against a role for CA in jejunal absorption than in support of it.

5.2. Ileum

The ileum is a relatively CA-poor organ with a considerable capacity to alkalinize intestinal chyme.[27] This apparent HCO_3^- secretion is tied to NaCl absorption and maintenance of a favorable luminal pH for the normal resident microflora of the lower gut. As in jejunum, a definitive role for CA remains somewhat controversial.

In a variety of mammals (human, monkey, and guinea pig), less than 5% of ileal enterocytes stain for CA.[98] In contrast, there is more homogeneous but minimal staining in rat ileum.[99] Both CA I and CA II have been detected. Histochemical staining in the absence of immunofluorescence to anti-CA I and II antibodies suggests the presence of CA IV,[98] as does low but demonstratable CA activity in apical brush membranes of rabbit ileum.[78]

Studies with CA inhibitors, principally acetazolamide, yield a confusing picture. In the rabbit, 0.1–8 mM acetazolamide had no effect on *in vitro* HCO_3^- ileal secretion,[145] suggesting that CA is unnecessary for basal HCO_3^- secretion linked to luminal alkalinization. In contrast, HCO_3^- secretion by the amphibian ileum, which has high CA activity, was almost halved by 0.1 mM acetazolamide.[64]

In brush border membrane vesicles of rabbit ileum, 1 mM acetazolamide did not block Cl^-–HCO_3^- exchange but did reduce Na^+-stimulated uptake of Cl^- in the presence of CO_2.[78] At concentrations of greater than 5 mM, acetazolamide clearly diminishes NaCl absorption.[49,114] In early *in vivo* studies, use of 1 mM or greater concentrations caused reduced NaCl absorption and HCO_3^- secretion in rats,[121] dogs,[72] and humans.[157] The effect of 0.1 mM acetazolamide in the rat was reported under normal conditions and during respiratory acidosis and alkalosis: HCO_3^- secretion was not affected, but Na^+ and Cl^- absorption were reduced 30–40% in all acid–base states.[28] Somewhat paradoxically, the absolute magnitude of increase in NaCl reabsorption observed with hypercapnia was not affected.

Figure 5 represents a model for ileal ion transport that incorporates current knowledge regarding membrane ion transport processes in the mammal.[27] The apical brush border contains both Na^+–H^+ and Cl^-–HCO_3^- exchangers, with CA

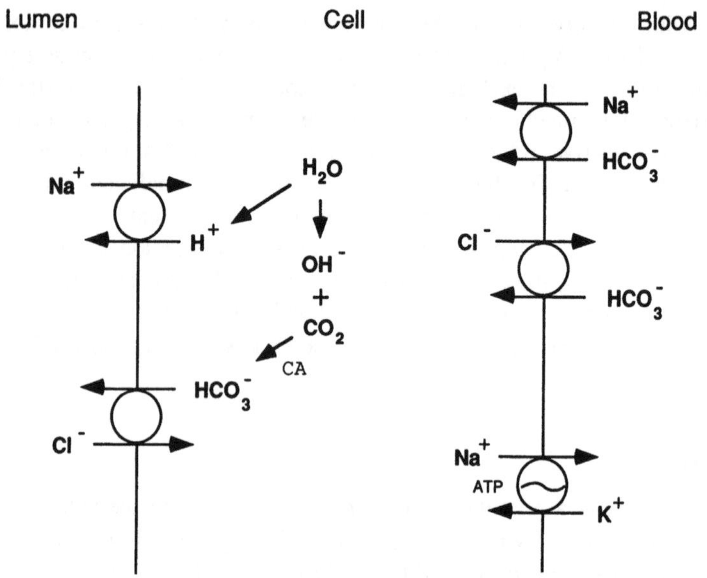

Figure 5. Membrane ion transport events in ileal NaCl absorption and bicarbonate secretion.

present in cytoplasm and at the apical border. Na^+–K^+ ATPase and either Cl^-–HCO_3^- exchange or a $NaHCO_3$ cotransporter are located on the basolateral membrane. The finding that 0.1–1.0 mM acetazolamide reduces NaCl absorption both *in vivo* and *in vitro* suggests that intracellular CA catalyzes the rapid generation of HCO_3^- from CO_2 and OH^-. This has the dual effect of generating HCO_3^- rapidly for apical Cl^-–HCO_3^- exchange and consuming OH^- produced by the operation of the Na^+–H^+ exchanger. In this fashion, CA can function strictly to augment neutral transepithelial NaCl transport, as has been proposed for the mammalian renal cortical thick ascending limb of Henle.[48] Two findings, however, are not easily explained in this model. One is the failure of acetazolamide to reduce measured HCO_3^- secretion. It may be that this fraction of HCO_3^- flux, not functionally linked to NaCl absorption but rather linked to luminal alkalinization, proceeds at a slow enough rate to not require any catalysis or is mediated by a non-CA-dependent process such as $NaHCO_3$ cotransport. Second is the finding[28] that the absolute increase in NaCl absorption observed with hypercapnia is not inhibited by acetazolamide. These authors concluded that CA does not participate in the augmentation of NaCl absorption induced by hypercapnia, since acetazolamide does reduce a similar increase in the colon, which has very high CA activity. This seeming paradox in the ileum might be explained by an increase in the uncatalyzed rate ($OH^- + CO_2 \rightarrow HCO_3^-$), since the rise in pCO_2 occurs without a significant drop in intracellular pH.[162] The opposite finding in colon may simply reflect much higher stimulated absorptive rates that are not completely supported by the uncatalyzed CO_2 reactions.

In summary, CA likely functions in amphibian ileal ion transport and alkalinization. In the mammal, data are consistent with a role in NaCl reabsorption but not as yet in luminal alkalinization.

5.3. Colon

CA has been found in the colons of all mammals, and a general consensus exists as to its functional role in various salt and acid–base movements. The cecum and proximal colon have more activity than distal colon, both CA I and CA II are present, and the enzyme tends to be concentrated in the surface epithelium and upper portions of glands.[96] CA II appears to be more concentrated in apical (luminal) membranes, in contrast to CA I, which is cytoplasmic and does not appear to be concentrated in membranes. CA III may be present in small amounts, and, recently, CA IV has been detected in human colon.[21a] Any discussion of the role of CA in the colon must begin with an appreciation for the existence of considerable species and segmental differences in ion and acid–base transport.[8] Figure 6 includes all putative CA roles in the colon, although these different functions probably do not exist in all colonic cells.

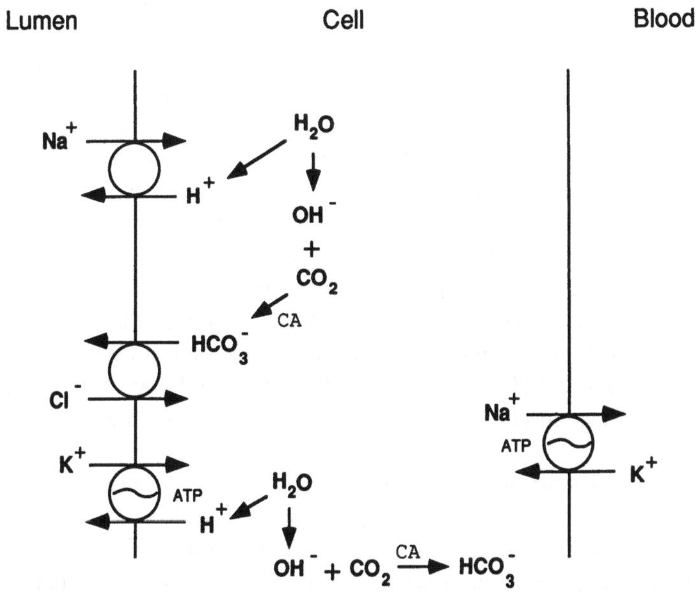

Figure 6. Membrane ion transport events in colonic NaCl absorption, bicarbonate secretion, and acid secretion.

The best-defined role for colonic CA is that of electroneutral NaCl reabsorption, as already described for the ileum. In the colon, studies with acetazolamide are in strong agreement and are less troubled by use of inordinately high concentrations. *In vitro*, 0.1 mM acetazolamide reduces colonic NaCl reabsorption by 40–60%.[6,7,54] Similar findings have been found *in vivo* at 0.1–1.0 mM.[28,123,121,100] Furthermore, acetazolamide markedly reduced the increase in NaCl absorption induced by hypercapnia.[28] Taken together, these studies in the rat are consistent with a role for CA and CO_2 promoting a rapid rate of NaCl absorption via the synchronous operation of apical Na^+–H^+ and Cl^-–HCO_3^- exchange.[25] In several mammalian species, including human but not rat, a large fraction of distal colonic NaCl reabsorption is mediated by electrogenic Na^+ transport.[8] In rabbit, a model of colonic electrogenic Na^+ reabsorption, 8 mM acetazolamide reduced Cl^- reabsorption.[62] This finding is consistent with a role for CA generating HCO_3^- for Cl^-–HCO_3^- exchange, but a nonspecific effect of acetazolamide on anion exchange or $NaHCO_3$ cotransport may be postulated. Of interest is the recent finding that colonic CA activity in the rat is sensitive over a four-fold range to aldosterone,[152] which is known to increase electrogenic Na^+ reabsorption in the colon.[8] In the ileum, which lacks electrogenic Na^+ reabsorption, there was no change in CA activity with aldosterone or adrenalectomy.[150]

The colon also secretes HCO_3^- and alkalinizes its luminal contents. This alkalinizing HCO_3^- secretion (like that fraction of HCO_3^- flux linked to NaCl reabsorption) is dependent on apical Cl^-–HCO_3^- exchange.[41a] Under these conditions, net alkalinization would require unequal and greater Cl^-–HCO_3^- exchange than Na^+–H^+ exchange. Acetazolamide reduces HCO_3^- secretion *in vitro* by 50% at 0.1 mM and by 90% at 1 mM in rat colon.[40] In contrast, *in vivo* data in the rat showed little or no effect on HCO_3^- secretion by 0.1 mM acetazolamide[28,123] or 1.0 mM[100] when drug was perfused through the colon. The failure of intraluminal acetazolamide in these experiments may relate to its poor lipid solubility and failure to penetrate intestinal mucus. When acetazolamide was administered both orally and perfused through the colon in rats, HCO_3^- secretion was reduced 60%.[121]

Middle and distal portions of guinea pig colon possess an ability to acidify *in vitro*, at rates equivalent to those measured for HCO_3 secretion[151]; 0.1 mM methazolamide reduces H^+ secretion (and serosal alkalinization) by almost 60%. The effects of K^+ and Na^+ removal,[121] vanadate, and ouabain were all consistent with a K^+-dependent ATPase at the apical (luminal) membrane. A similar proton–potassium exchange has been detected in colonic apical membrane vesicles[60] and *in vivo*.[121a]

The exact physiological significance of colonic proton secretion is uncertain, but it may facilitate passive nonionic diffusive fatty acid uptake, by subserving apical Na^+–H^+ exchange[142a] or a proton ATPase pump.[60] CA-mediated short-chain fatty acid absorption in rumens of bovine species and large intestines of all mammals was first proposed to depend on CA I.[18] The finding of less colonic CA I activity in the sterile fetal gastrointestinal tract[99] lent further support to this idea. However, no difference was found in distribution or amounts of CA I and II in colons of normal and germfree rats.[97] There seems to be no compelling reason to invoke CA I in this process rather than CA II, or IV, as their apical location in the surface epithelium puts them in a more strategic position for regulation of the extracellular microclimate.

An entirely speculative role for CA in intestinal NaCl reabsorption is that delineated[154] for the osmoregulatory salt-secreting elasmobranch rectal gland, a possible phylogenetic predecessor of the mammalian appendix which also contains CA.[84] Very high rates of stimulated NaCl secretion are blocked by CA inhibitors as a consequence of severe intracellular respiratory acidosis. The enzyme serves to facilitate the dissipation of metabolic CO_2 by augmenting the contribution of intracellular HCO_3^- to carry CO_2 from its site of production in mitochondria to the capillary. It is conceivable that at high rates of intestinal NaCl reabsorption (and thus metabolic activity), CA may function in this same capacity.

The consequences of enzyme inhibition in patients treated with CA inhibitors appear minimal. Diarrhea and stool HCO_3^- loss might be anticipated; however, aside from mild nausea, gastrointestinal side effects are rare. It may be that other non-CA-dependent mechanisms of salt and acid–base transport (in combination with the residual uncatalyzed contribution) compensate fully. Of possible equal importance is the well-documented increase in NaCl absorption in both the ileum and colon caused by systemic metabolic and respiratory acidosis.[26,38] Since both acidoses occur as a result of renal and red cell CA inhibition, these systemic effects may completely counter the local effect of these drugs.

6. Liver and Secretion of Bile

CA in the intermediary metabolism of urea and glucose (CA V in the mitochondria) and the unknown functions of CA I and CA III are discussed elsewhere in this volume. Compared with other secretory organs, mammalian liver contains relatively low levels of total CA activity (Table I). The majority of activity is cytoplasmic,[53,105] with less than 20% found in nuclear, mitochondrial, and membrane fractions. Histochemical and immunocytochemical studies reveal that CA I accounts for a small degree of staining; the majority is CA II except in the case of male rodent liver, where CA III is predominant.[21,68] There is little CA at the cell membrane, which contrasts with distinct membrane staining in pig hepatocytes.[15] Plasma membranes from rat hepatocytes have shown high CA activity that was insensitive to anions and easily inhibited by sulfonanides.[53] The isolated enzyme has a molecular weight of 47,000[159] and thus appears to be CA IV.[21a] CA II shows a striking localization to perivenous hepatocytes.[12,21,93] The epithelium of bile ducts also stains intensely for CA II.[15,21,146]

Hepatic bile is moderately alkaline, and HCO_3^- concentrations may reach as high as 55 mM in the human.[37] Its importance in bile formation is suggested by data that (1) perfusion of rat livers with HCO_3^--free solution reduces bile flow by 50%,[61,160] (2) administration of secretin and other choleretics to dogs increases biliary HCO_3^- output,[5] and (3) infusion of certain nonphysiological bile acids such as ursodeoxycholate (UDC) leads to a marked hypercholeresis rich in HCO_3^-.[35] In general, HCO_3^- secretion appears linked to that fraction of bile flow termed bile acid-independent flow (BAIF) except for the special case of UDC-stimulated hypercholeresis.

Effects of CA inhibition in studies of biliary flow and composition depend on the species studied and whether hepatic (canalicular) or ductular secretion is measured. In general, early studies collected mixed bile, and they showed that acetazolamide did not affect unstimulated bile secretion but reduced biliary $[HCO_3^-]$ and pH and raised $[Cl^-]$ in bile acid-infused dogs. Bile flow was usually increased[42,105,118,164] so that HCO_3^- output fell <10%. The fall in biliary $[HCO_3^-]$ and rise in bile flow elicited by CA inhibitors with bile acid infusion is unique, since inhibition in other HCO_3^--secreting tissues leads either to both a fall in fluid flow and $[HCO_3^-]$ or to a fall in fluid flow without change in $[HCO_3^-]$.[102] One explanation is certainly the complex composition and origin of bile, of which only a fraction is dependent on HCO_3^- secretion. In marked contrast, acetazolamide is without effect in rats[10,105] since CA III is present. Secretin-stimulated ductular biliary $[HCO_3^-]$ and bile flow are clearly reduced by CA inhibition.[12,118] In the pig, it is possible to selectively stimulate ductular HCO_3^- secretion (with secretin infusion in bile acid-deprived animals) and canalicular HCO_3^- secretion (with taurocholate infusion in secretin-deprived animals.[12] Using a high dose of acetazolamide (150 mg/kg) the authors reduced ductular HCO_3^- secretion by 67% under normal acid–base conditions but showed no change in canalicular HCO_3^- secretion. CA inhibition also reduces UDC-stimulated hypercholeresis by 20–25% in the pig[12] and rat.[30,51] The site of UDC-stimulated HCO_3^- output is not clearly established, but it may occur both at the canalicular and ductular levels.

Ductular (secretin-stimulated) secretion is sensitive to changes in arterial pCO_2 and HCO_3^-.[11a,57,102,118] Both rising pCO_2 and HCO_3^- increase secretion. In the dog, acetazolamide has minimal effects in metabolic acidosis and respiratory alkalosis but does reduce HCO_3^- secretion in metabolic alkalosis.[102,118] Resting and taurocholate-stimulated canalicular biliary HCO_3^- secretion in the rat are affected proportionally by changes in plasma HCO_3^-.[52] HCO_3^- secretion in UDC-stimulated canalicular hypercholeresis is sensitive to acid–base perturbations, similar to the findings for ductular secretion.[30,52]

The mechanisms by which HCO_3^- is secreted by hepatocytes are better understood than in the ductular epithelium. Figure 7 shows the membrane ion-translocating processes thought responsible for HCO_3^- secretion into canalicular bile. The critical elements include basolateral (sinusoidal) Na^+–K^+ ATPase, Na^+–H^+ exchange,[110] Na–bile acid cotransport,[37] and $NaHCO_3$ cotransport[132] The apical (canalicular) membrane contains Cl^-–HCO_3^- exchange[107] and an organic anion carrier possibly capable of bile acid–inorganic anion exchange.[108] Although originally suspected, it does not appear that any other proton-translocating activity (such as a H^+ ATPase) at the basolateral membrane is necessary to account for

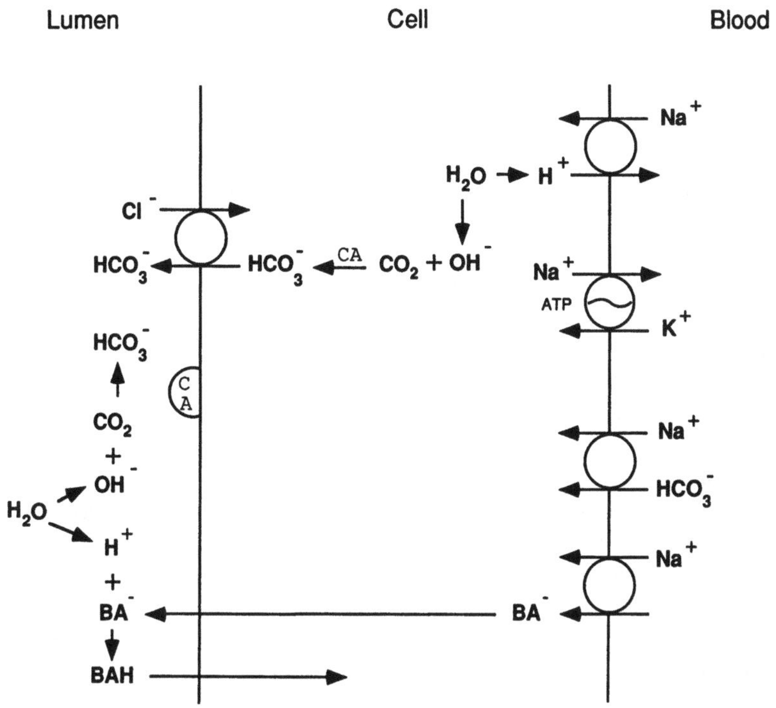

Figure 7. Membrane ion transport events in hepatic canalicular bicarbonate secretion.

measured H^+ fluxes in the hepatocyte.[4] HCO_3^- secretion is primarily the result of H^+–OH^- separation because DMO can substitute for HCO_3^- in the perfused rat liver to maintain normal BAIF.[61] Since the rate of HCO_3^- secretion is not decreased by CA inhibitors, it would appear that Na^+–H^+ exchange and Cl^-–HCO_3^- exchange are not so fast as to require catalysis of HCO_3^- formation from CO_2 and OH^-.[12,102]

Ductular HCO_3^- secretion is inhibited by acetazolamide,[12] with no response to amiloride[11a,58] and reduction by DCCD, a proton pump inhibitor,[56] suggesting features in common with pancreatic HCO_3^- secretion (see Fig. 3). As in the pancreas, secretin causes disappearance of intracytoplasmic vesicles containing proton pumps,[13] which are thought to fuse with and increase the basolateral surface area. In this model, intracellular or basolaterally placed CA would subserve high rates of proton exit and intracellular HCO_3^- formation, as in the stomach and possibly the pancreas. Uncatalyzed reaction rates (assuming that ductular cells account for about 2.5% of total liver water) can account for only about 30% of normal secretin-stimulated HCO_3^- output,[12] in agreement with the finding that acetazolamide reduces secretion by 70%.

Biliary HCO_3^- secretion during UDC-stimulated hypercholeresis is three to six times higher than basal HCO_3^- secretion[3,30] and is reduced by CA inhibitors (see above). This result can be explained in one of two ways. First, intracellular CA may be necessary to generate HCO_3^- for canalicular exit and to prevent intracellular pH from rising too greatly in the face of stimulated Na^+–H^+ exchange that is known to occur with UDC administration.[109] Primary H^+–OH^- separation in this process is suggested by the finding that DMO or acetate can substitute for HCO_3^- in the isolated perfused liver and still maintain high rates of stimulated bile flow.[52] A second role for CA role would be to catalyze rapid intraluminal protonation of UDC, which is secreted in its anionic form (pK~5). Since nonionized UCD is very lipophilic, its reabsorption (and thus recycling) results in intraluminal generation of HCO_3^-. This concept[168] requires the presence of canalicular membrane-bound CA[159] with access to luminal fluid. CA bound to the luminal membrane of the ductular epithelium could also contribute to further HCO_3^- generation as canalicular bile is delivered into the biliary ducts.

The predominant perivenous to periportal gradient of CA activity observed histochemically (see above) cannot be easily related to bile formation. Despite the fact that perivenous hepatocytes may have a greater role in BAIF,[59] the data suggest little if any effect of CA inhibition on normal canalicular HCO_3^- secretion. Thus, the heterogeneous distribution of CA in the liver must be attributed to some other function for CA, such as urea or glucose metabolism. Perhaps the perivenous

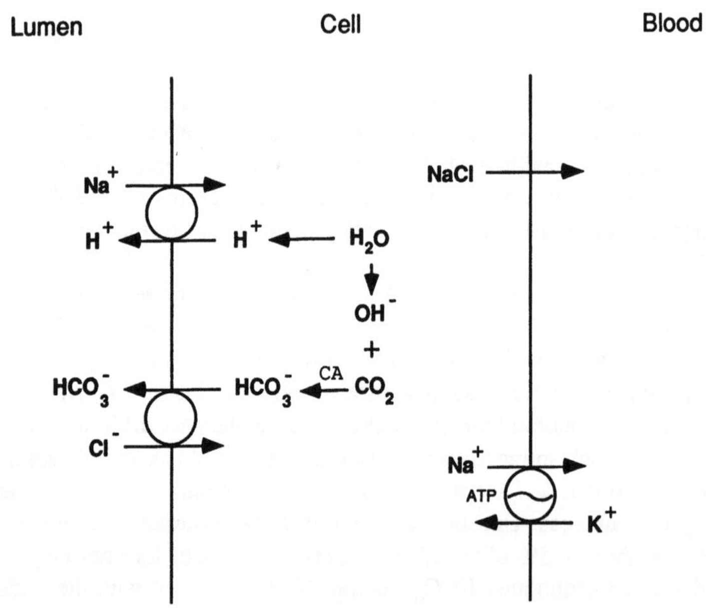

Figure 8. Membrane ion transport events in gallbladder NaCl absorption and acidification.

concentration of CA is necessary for normal elimination of metabolic CO_2 in these cells, whose capillary blood is more hypoxic[69] and probably more hypercapnic than incoming periportal blood. This is supported by the finding that hepatic CA activity is directly accessible to plasma.[90a,141a] Such activity is consistent with basolateral membrane CA IV whose function would be to facilitate CO_2 movement from hepatocytes to the blood.

The gallbladder contains CA in significant concentration, the majority of which is associated with cell membranes.[122] Enzyme inhibition with proper concentrations of acetazolamide or other CA inhibitors decreases NaCl reabsorption,[102,122] reduces the short-circuit current,[116] impairs cell volume regulation[122] and blocks acid secretion.[122a] Gallbladder concentrates bile by active NaCl reabsorption and partial acidification.[133] Both processes are crucial to storing bile and preventing gallstone formation.[129a,140a] A large fraction of NaCl reabsorption is mediated by apical membrane-coupled Na^+–H^+ and Cl^-–HCO_3^- exchange [162a] as described for the intestines, in which CA functions to dispose of OH^- generated by Na^+–H^+ exchange (Fig. 8). A smaller fraction of Na uptake is electrogenic, possible mediated by apical $NaHCO_3$ cotransport, and could account for some gallbladder bile acidification.[116] Its reduction by 0.1 mM acetazolamide implicates CA in this process, but exactly how escapes simple explanation.

References

1. Ammar, E. M., Hutson, D., and Scratcherd, T., 1987, *J. Physiol.* (London) **388**:395–504.
2. Anderson, M. C., and Copass, M. K., 1966, *Am. J. Digest. Dis.* **11**:367–376.
3. Anwer, M. S., Hondalus, M. K., and Atkinson, J. M., 1989, *Am. J. Physiol.* **257**:G371–G379.
4. Anwer, M. S., and Nolan, K., 1988, *Hepatology* **8**:728–734.
5. Barnhart, J. L., and Combes, B., 1978, *J. Pharmacol. Exp. Ther.* **206**:190–197.
6. Binder, H. J., Foster, E. S., Budinger, M. E., and Hayslett, J. P., 1987, *Gastroenterology* **93**: 449–455.
7. Binder, H. J., and Rawlins, C. L., 1973, *Am. J. Physiol.* **225**:1232–1239.
8. Binder, H. J., and Sandle, G. I., 1987, in: *Physiology of the Gastrointestinal Tract* (L. R. Johnson, ed.), Raven Press, New York, pp. 1389–1394.
9. Birnbaum, D., and Hollander, F., 1953, *Am. J. Physiol.* **174**:191–195.
10. Bizard, G., Vanlerenberghe, J., Guerrin, F., and Godchauz, R., 1958, *J. Physiol.* (Paris) **50**: 155–156.
11. Blair, J. A., Lucas, M. L., and Matty, A. J., 1975, *J. Physiol.* **245**:333–350.

11a. Blot-Chabaud, M., Dumont, M., Corbic, M., and Erlinger, S., 1990, *Am. J. Physiol.* **258**:6863–6872.

12. Buanes, T., Grotmol, T., Veel, T., Landsverk, T., Ridderstrale, Y., and Raeder, M. G., 1988, *Acta Physiol. Scand.* **133**:535–544.
13. Buanes, T., Grotmol, T., Landsverk, T., and Raeder, M. G., 1988, *Gastroenterology* **95**: 417–424.
14. Buanes, T., Grotmol, T., Landsverk, T., and Raeder, M. G., 1987, *Acta Physiol. Scand.* **131**: 55–62.
15. Buanes, T., Grotmol, T., Landsverk, T., Ridderstrale, Y., and Raeder, M. G., 1986, *Acta Physiol. Scand.* **128**:437–444.

16. Byers, F. M., Jordan, P. H., and Maren, T. H., 1962, *Am. J. Physiol.* **202**:429–436.
17. Campbell, V. W., Valle, J. D., Hawn, M., Park, J., and Yamada, T. 1989, *Am. J. Physiol.* **256**:G631–G636.
18. Carter, M. J., 1972, *Biol. Rev.* **47**:465–513.
19. Carter, M. J., and Parsons, D. S., 1972, *J. Physiol.* **220**:465–478.
20. Carter, M. J., and Parsons, D. S., 1971, *J. Physiol.* (London) **215**:71–94.
21. Carter, N., Wistrand, P. J., and Lonnerholm, G., 1989, *Acta Physiol. Scand.* **135**:163–167.
21a. Carter, N. D., Fryer, A., Grant, A. G., Hume, R., Strang, R. G., and Wistrand, P. J., 1990, *Biochim. Biophys Acta* **1026**:113–116.
22. Case, R. M., Harper, A. A., and Scratcherd, T., 1969, *J. Physiol.* (London) **201**:335–348.
23. Case, R. M., Scratcherd, T., and Wynne, R. D. A., 1970, *J. Physiol.* (London) **210**:1–15.
24. Cassano, G., Stieger, B., and Murer, H., 1984, *Pfluegers Arch.* **400**:309–317.
25. Charney, A. N., and Egnor, R. W., 1989, *Am. J. Physiol.* **256**:C584–C590.
26. Charney, A. N., and Feldman, G. M., 1984, *Am. J. Physiol.* **247**:G1–G12.
27. Charney, A. N., and Goldfarb, D. S., 1990, in: *Secretory Diarrheas* (E. Lebenthal and M. E. Duffy, eds.), Raven Press, New York, pp. 95–108.
28. Charney, A. N., Wagner, J. D., Birnbaum, G. J., and Johnstone, J. N., 1986, *Am. J. Physiol.* **251**:G682–G687.
29. Cho, C. W., Chen, S. M., Chen, S. W., Chow, C. K., Lai, K. H., and Pfeiffer, C. J. P., 1984, *Digestion* **29**:5–13.
30. Corbic, M., Munoz, C., Dumont, M., DeCouet, G., and Erlinger, S., 1985, *Hepatology* **4**: 494–599.
31. Davenport, H. W., 1939, *J. Physiol.* (London) **97**:32–43.
32. Davenport, H. W., 1946, *Physiol. Rev.* **26**:560–573.
33. Davies, R. E., and Edelman, J., 1951, *Biochem. J.* **50**:190–194.
34. Dreiling, D. A., Henry, M. D., Janowitz, H. D., and Halpern, M., 1955, *Gastroenterology* **29**:262–279.
35. Dumont, M., Erlinger, S., and Uchman, S., 1980, *Gastroenterology* **79**:82–89.
36. Emas, S., 1962, *Gastroenterology* **43**:557–563.
37. Erlinger, S., 1987, in: *Physiology of the Gastrointestinal Tract*, 2nd ed. (L. Johnson, ed.), Raven Press, New York, pp. 1557–1580.
38. Feldman, G. M., 1989, *Am. J. Physiol.* **256**:G1036–G1040.
39. Feldman, G. M., Arnold, M. A., and Charney, A. N., 1984, *Am. J. Physiol.* **246**:G687–G694.
40. Feldman, G. M., Bernan, S. F., and Stephenson, R. L., 1988, *Am. J. Physiol.* **254**:C383–C390.
41. Feldman, G. M., 1983, *J. Clin. Invest.* **72**:295.
41a. Feldman, G. M., Koethe, J. D., and Stephenson, R. L., 1990, *Am. J. Physiol.* **258**:G825–G832.
42. Fink, S., 1956, *N. Engl. J. Med.* **254**:258–262.
43. Flemstrom, G., 1977, *Am. J. Physiol.* **233**:E1–12.
44. Flemstrom, G., 1987, in: *Physiology of the Gastrointestinal Tract* (L. R. Johnson, ed.), Raven Press, New York, pp. 1011–1029.
45. Flemstrom, G., 1980, *Am. J. Physiol.* **239**:G198–G203.
46. Flemstrom, G., and Kivalaakso, E., 1983, *Gastroenterology* **84**:787–794.
47. Flemstrom, G., and Turnberg, L., 1984, *Clin. Gastroenterol.* **13**:327–354.
48. Friedman, P. A., and Andreoli, T. E., 1982, *J. Gen. Physiol.* **80**:683–711.
49. Frizzell, R. A., Nellans, H. N., Rose, R. C., Markscheid-Kaspi, L., and Schultz, S. G., 1973, *Am. J. Physiol.* **224**:328–337.
50. Gailitis, R. J., and Schreiber, W., 1960, *Am. J. Digest. Dis.* **5**:473–478.
51. Garcia-Marin, J. J., Dumont, M., Corbic, M., DeCouet, G., and Erlinger, S., 1985, *Am. J. Physiol.* **248**:G20–G27.
52. Garcia-Marin, J. J., Corbic, M., Dumont, M., and DeCouet, G., 1985, *Am. J. Physiol.* **249**:G335–G341.

53. Garcia-Marin, J. J., Perez-Barriocanal, F., Garcia, A., Serrano, M. A., Regueiro, P., and Esteller, A., 1988, *Biochem. Biophys. Acta* **945**:17–22.
54. Goldfarb, D. S., Egnor, R. W., and Charney, A. N., 1988, *J. Clin. Invest.* **81**:1903–1910.
55. Grotmol, T., Buanes, T., and Raeder, M. G., 1986, *Acta Physiol. Scand.* **128**:547–554.
56. Grotmol, T., Buanes, T., Raeder, M. G., 1987, *Scand. J. Gastroenterol.* **22**:207–213.
57. Grotmol, T., Buanes, T., Raeder, M. G., 1987, *Acta Physiol. Scand.* **131**:183–193.
58. Grotmol, T., Buanes, T., Raeder, M. G., 1987, *Acta Physiol. Scand.* **130**:447–455.
59. Gumucio, J. J., Balabaud, C., Miller, D. L., DeMason, L. J., Appelman, H. D., Stoecker, T. J., and Franzblau, D. R., 1978, *J. Lab. Clin. Med.* **91**:350–362.
60. Gustin, M., Goodman, D. B. P., 1981, *J. Biol. Chem.* **256**:10651–10656.
61. Hardison, W. G. M., and Wood, C. A., 1978, *Am. J. Physiol.* **235**:E158–E164.
62. Hatch, M., Freel, R. W., Goldner, A. M., Earnest, D. L., 1984, *Gut* **25**:232–237.
63. Hersey, S. J., and High, W. L., 1971, *Biochem. Biophys, Acta* **233**:604–609.
64. Imon, M. A., and White, J. F., 1981, *J. Physiol.* (London) **314**:429–443.
65. Isenberg, J. I., Selling, J. A., Hogan, D. L., and Koss, M. A., 1987, *N. Engl. J. Med.* **316**: 374–379.
66. Janowitz, H. D., Colcher, H., and Hollander, F., 1952, *Am. J. Physiol.* **171**:325–330.
67. Janowitz, H. D., Dreiling, D. A., Roblin, H. L., and Hollander, F., 1957, *Gastroenterology* **133**:378–384.
68. Jeffery, S., Carter, N. D., and Wilson, C., 1984, *Biochem. J.* **221**:927–929.
69. Jungermann, K., 1986, *Enzyme* **34**:161–180.
70. Kauffman, G. L., and Steinbach, J. H., 1981, *Surgery* **89**:324–328.
71. Kidder, G. W., and Montgomery, C. W., 1974, *Am. J. Physiol.* **227**:300–304.
72. Kinney, V. R., and Code, C. F., 1964, *Am. J. Physiol.* **207**:98–1004.
73. Kivilaakso, E., Barzilai, A., Schiessel, R., Fromm, D., and Silen, W., 1981, *Gastroenterology* **40**:77–83.
74. Kivilaakso, E., and Silen, W., 1981, *Scand. J. Gastroenterol.* **67**:219–221.
75. Kleinman, J. G., Brown, W. W., Ware, R. A., and Schwartz, J. H., 1980, *Am. J. Physiol.* **239**:F440–F444.
76. Kollberg, B., Aly, A., Rubio, C., and Johansson, C., 1981, *Scand. J. Gastroenterol.* **16**: 385–487.
77. Konturek, S. J., Brzozowski, T., Piastuchi, J., and Radechi, T., 1983, *Digestion* **28**:125–132.
78. Knickelbein, R., Aronson, P. S., and Schron, C. M., 1985, *Am. J. Physiol.* **249**:G236–G245.
79. Kuijpers, G. A. J., Vannooy, I. G. P., Depont, J. J., and Bonting, S. L., 1984, *Biochim. Biophys. Acta* **778**:324–331.
80. Korhonen, L. K., Korhonen, E., and Hyyppa, M., 1966, *Histochemie* **6**:168–172.
81. Kumagai, J., Kaneko, E., and Honda, N., 1986, *Gastroenterol. Jpn.* **21**:204–207.
82. Kumpulainen, T., and Jalovaara, P., 1981, *Gastroenterology* **80**:796–799.
83. Kumpulainen, T., 1984, *Ann. N.Y. Acad. Sci.* **429**:359–368.
84. Kumpulainen, T., 1981, *Histochemistry* **72**:425–31.
85. Kuroshima, T., Himeno, S., Kurokawa, M., Tsuji, K., and Tarui, S., 1986, *Am. J. Physiol.* **250**:G398–G404.
86. Kusterer, K., and Szabo, S., 1987, *Eur. J. Pharmacol.* **141**:7–13.
87. Lakshmanan, M. C., Kahn, M. W., and Jones, R. T., 1979, *Enzyme* **24**:107–112.
88. Laurent, G., Venot-Graud, N., Marriq, C., Limozin, N., Garcon, D., and Lilippi, D., 1977, *Biochim. Biophys. Acta* **481**:222–226.
89. Lindner, A. E., Cohen, N., Berkowitz, J., Janowitz, H. D., 1964, *Gastroenterology* **46**:273–275.
90. Lindner, A. E., Cohen, N., Dreiling, D. A., and Janowitz, H. D., 1962, *J. Appl. Physiol.* **17**: 514–518.
90a. Lipsen, B., and Effros, R. M., 1988. *J. Appl. Physiol.* **65**:2736–2743.
91. Ljungstrom, M., Vega, F. V., and Mardh, S., 1984, *Biochim. Biophys. Acta* **769**:220–230.

92. Ljungstrom, M., Norberg, L., Olaisson, H., Wernstedt, C., Vega, F. V., Arvidson, G., and Mardh, S., 1984, *Biochim. Biophys. Acta* **769**:209–219.
93. Lonnerholm, G., 1980, *Cytochem.* **28**:427–433.
94. Lonnerholm, G., 1977, *Acta Physiol. Scand.* **99**:53–61.
95. Lonnerholm, G., 1983, *Acta Physiol. Scand.* **117**:273–279.
96. Lonnerholm, G., Knutson, L., Wistrand, P. J., and Flemstrom, G., 1989, *Acta Physiol. Scand.* **136**:253–262.
97. Lonnerholm, G., Midtvedt, T., Schenholm, M., Wistrand, P. J., 1988, *Acta Physiol. Scand.* **131**:159–166.
98. Lonnerholm, G., Selking, O., and Wistrand, P. J., 1985, *Gastroenterology* **88**:1151–1161.
99. Lonnerholm, G., and Wistrand, P. J., 1983, *Biol. Neonate* **44**:166–176.
100. Lubcke, R., Haag, K., Berger, E., Knauf, H., and Gerok, W., 1986, *Am. J. Physiol.* **25**:G132–G139.
101. Lutjen-Drecol, E., Eichhorn, M., and Barany, E. H., 1985, *Acta Physiol. Scand.* **124**:295–307.
102. Maren, T. H., 1967, *Physiol. Rev.* **47**:595–781.
103. Maren, T. H., 1988, *Annu. Rev. Physiol.* **50**:695–717.
104. Maren, T. H., 1977, *Am. J. Physiol.* **232**:F291–297.
105. Maren, T. H., Ellison, A. C., and Fellner, S. K., 1966, *Mol. Pharmacol.* **2**:144–157.
106. McGowan, J. A., Stanley, M. M., and Powell, J. J., 1952, *Bull. N. Engl. Med. Ctr.* **15**:117.
107. Meier, P., Knickelbein, R., Moseley, R. H., Dobbins, J. W., and Boyer, J. L., 1985, *J. Clin. Invest.* **75**:1256–1263.
108. Meier, P. H., Valantinas, J., 1987, *Am. J. Physiol.* **253**:G461–G468.
109. Moseley, R. H., Ballatori, N., Smith, D. J., and Boyer, J. L., 1987, *J. Clin. Invest.* **80**:684–690.
110. Moseley, R. H., Meier, P. J. H., Aronson, P. S., and Boyer, J. L., 1986, *Am. J. Physiol.* **250**: G35–G43.
111. Muallem, S., Blissard, D., Cragoe, E. J., and Sachs, G., 1988, *J. Biol. Chem.* **263**:14703–14711.
112. Nadell, J., 1953, *J. Clin. Invest.* **32**:622–629.
113. Naurmi, S., and Kanno, M., 1973, *Biochem. Biophys. Acta* **311**:80–89.
114. Nellans, H. N., Frizzell, R. A., and Schultz, S. G., 1975, *Am. J. Physiol.* **228**:1808–1814.
115. Novak, I., 1988, in: *pH Homeostasis: Mechanisms and Control* (D. Haussinger, ed.), Academic Press, London, pp. 447–470.
116. O'Grady, S. M., and Wolters, P. J., 1989, *Am. J. Physiol.* **257**:C45–C51.
117. Omaille, E. R. L., and Richards, T. G., 1977, *J. Physiol.* (London) **265**:855–866.
118. Pak, B. H., Hong, S. S., Pak, H. K., and Hong, S. K., 1966, *Am. J. Physiol.* **210**:624–628.
119. Palatroni, P., Gabrielle, M. G., and Grappasonni, I., 1988, *Acta Histochem.* **84**:1–14.
120. Paradiso, A. M., Townsley, M. C., Wenzl, E., Machen, T. E., 1989, *Am. J. Physiol.* **257**:C554–C561.
121. Parsons, D. S., 1956, *Q. J. Exp. Physiol.* **41**:410–420.
121a. Perrone, R. D., and McBride, D. E., 1990, *Pfluegers Arch.* **416**:632–638.
122. Persson, B. E., and Larson, M., 1986, *Acta Physiol. Scand.* **128**:501–507.
122a. Petersen, K. U., Wehner, F., and Winterhager, J. M., 1990, *Pfluegers Arch.* **416**:312–321.
123. Phillips, S. F., and Schmalz, P. F., 1970, *Proc. Soc. Exp. Biol. Med.* **135**:116–122.
124. Podestra, R. B., and Mettrick, D. F., 1977, *Am. J. Physiol.* **1**:E62–E68.
125. Pratt, E. B., and Aikawa, J. K., 1962, *Am. J. Physiol.* **202**:1083–1086.
126. Puscas, I., and Buzas, G., 1986, *Int. J. Clin. Pharmacol. Ther. Toxicol.* **24**:97–99.
127. Puscas, I., Paun, R., Ursea, N., Dragomir, N., and Roscin, M., 1976, *Arch. Fr. Mal. Appar. Dig.* **65**:577–583.
128. Raeder, M., and Mathisen, O., 1982, *Acta Physiol. Scand.* **114**:97–102.
129. Rawls, J. A., Wistrand, P. J., and Maren, T. H., 1963, *Am. J. Physiol.* **205**:651–657.
129a. Rege, R. V., and Moore, E. W., 1987, *Gastroenterology* **92**:281–289.
130. Rehm, W. S., Canosa, C. A., Schlesinger, H. S., Chandler, W. K., and Dennis, W. H., 1961, *Am. J. Physiol.* **200**:1074–1082.

131. Reichstein, B. J., and Cohen, M. M., 1984, *J. Lab. Clin. Med.* **104**:797–804.
132. Renner, E. L., Lake, J. R., Scharschmidt, B. F., Zimmerli, B., and Meier, P. J., 1989, *J. Clin. Invest.* **83**:1225–1235.
133. Reuss, L., and Stoddard, J., 1987, *Annu. Rev. Physiol.* **49**:35–49.
134. Robert, A., Lancaster, C., Davis, J. P., Kolbasa, K. P., and Nezamis, J. E., 1985, *Eur. Pharmacol.* **118**:193–201.
135. Rutten, M., Rattner, D., and Silen, W., 1985, *Am. J. Physiol.* **249**:C503–C513.
136. Sachs, G., Chang, H. H., Raon, E., Schackmann, R., Lewin, M., and Saccomani, G., 1976, *J. Biol. Chem.* **251**:7690–7698.
137. Sachs, G., Muallem, S., and Hersey, S. J., 1988, *Comp. Biochem. Physiol.* **90A**:727–731.
138. Salganik, R. I., Argutinskaya, S. V., and Bersimbaev, R. I., 1972, *Experientia* **28**:1190–1191.
139. Salomoni, M., Zuccato, E., Granelli, P., Montorsi, W., Doldi, S. B., Germiniani, R., and Mussini, E., 1989, *Scand. J. Gastroenterol.* **24**:28–32.
140. Sanders, S. S., Hayne, V. B., Rehm, W. S., 1973, *Am. J. Physiol.* **225**:1311–1321.
140a. Schiffman, M. L., Sugarman, H. J., and Moore, E. W., 1990, *Gastroenterology* **99**:1452–1459.
141. Schulz, I., Strover, F., and Ullrich, K. J., 1971, *Pfluegers Arch.* **323**:121–130.
141a. Schwab, A. J., Goresky, C. A., and Rose, C. P., 1989, *Circulation Res.* **65**:1646–1656.
142. Seidler, V., Carter, K., Ito, S., Silen, W., *Am. J. Physiol.* **256**:G466–475.
142a. Sellin, J. H., and DeSoigne, R., 1990, *Gastroenterology* **99**:676–683.
143. Shigehiko, N., and Kanno, M., 1973, *Biochim. Biophys. Acta* **311**:80–89.
144. Shimada, T., 1987, *J. Physiol.* (London) **392**:113–127.
145. Smith, P. L., Cascairo, M. A., and Sullivan, S. K., 1985, *Am. J. Physiol.* **249**:G358–G368.
146. Spicer, S. S., Sens, M. A., and Tashian, R. E., 1982, *J. Histochem. Cytochem.* **30**:864–873.
147. Stewart, B., Wallmark, G., and Sachs, G., 1981, *J. Biol. Chem.* **256**:2682–2690.
148. Stiel, D., Murray, D. J., and Petters, T. J., 1984, *Am. J. Physiol.* **247**:G133–G139.
149. Sugai, N., and Ito, S., 1980, *J. Histochem. Cytochem.* **28**:511–525.
150. Suzuki, S., and Ozaki, N., 1984, *Ann. N.Y. Acad. Sci.* **429**:306–308.
151. Suzuki, Y., and Kaneko, K., 1987, *Am. J. Physiol.* **16**:G155–G164.
152. Suzuki, S., Rent, L. J., and Chen, H., 1989, *J. Steroid Biochem.* **33**:89–99.
153. Swanson, C. H., and Solomon, A. K., 1975, *J. Gene Physiol.* **65**:22–45.
154. Swenson, E. R., and Maren, T. H., 1984, *Am. J. Physiol.* **247**:F86–F92.
155. Takeuchi, K., Merhav, A., and Silen, W., 1989, *Am. J. Physiol.* **243**:G377–G388.
156. Takeuchi, K., Ohtsuki, M., and Okabe, S., 1986, *Digest. Dis. Sci.* **31**:306–411.
157. Turnberg, L. A., Bieberdorf, F. A., Morawski, S. G., and Fordtran, J. S., 1970, *J. Clin. Invest.* **49**:557–567.
158. Turnberg, L. A., Fordtran, J. S., Carter, N. W., and Rector, F. C., 1970, *J. Clin. Invest.* **49**: 548–556.
159. Valantinas, J., Meier, P. J., 1987, *Hepatology* **7**:1038.
160. VanDyke, R. W., Stephens, J. E., and Scharschmidt, B. F., 1982, *J. Clin. Invest.* **70**:505–517.
161. Vega, F. V., Olaisson, H., and Mardh, S., 1985, *Acta Physiol. Scand.* **124**:573–579.
162. Wagner, J. D., Kurtin, P., and Charney, A. N., 1986, *Am. J. Physiol.* **250**:G588–G593.
162a. Weinman, S. A., and Reuss, L., 1982, *J. Gen. Physiol.* **80**:299–321.
163. Werther, J. L., Hollander, F., and Altamirano, M., 1965, *Am. J. Physiol.* **209**:127–133.
164. Wheeler, H. O., and Ramos, O. L., 1960, *J. Clin. Invest.* **39**:161–170.
165. White, J. F., 1989, *Am. J. Physiol.* **257**:C252–C260.
166. White, J., 1985, *Am. J. Physiol.* **248**:G256–G259.
167. White, J. F., and Imon, M. A., 1981, *Am. J. Physiol.* **241**:G389–G396.
168. Yoon, Y. B., Hagey, L. R., Hofmann, A. F., Gurantz, D., Michelotti, E. L., and Steinbaach, J. H., 1986, *Gastroenterology* **90**:837–852.

CHAPTER 24

Hormonal Regulation and Localization of Rat Liver Carbonic Anhydrase

STEPHEN JEFFERY

1. Hormonal Regulation

The first reports that rat liver contained endogenous carbonic anhydrase (CA)[4,11] indicated that the CO_2 hydrase activity of liver CA was poorly inhibited by acetazolamide compared with the strong inhibition of rat red cell CAs (CA I and CA II). A partial purification of CA from the liver of male rats showed that the specific CO_2 hydrase activity of this acetazolamide-resistant form was approximately 1% of CA II purified from rat red cells.[5] The livers of male rats contained far more of the sulfonamide-refractory form of CA than did the livers of female rats. However, female rat livers were richer than male livers in a "high-activity" sulfonamide-sensitive CA. Testosterone apparently induced the male type CA.[4]

At the time these discoveries were made, it was not clear whether the effects of testosterone were due to activation of an existing isozyme population or the instigation of new synthesis. The isozyme CA III was then unknown; its discovery in 1976[7] led to the realization that the major CA isozyme in male rat liver was CA III.[1] Raising a specific antiserum against rat CA III enabled a radioimmunoassay for this protein to be established.[16] From this development there flowed a great deal of information on rat liver CA III. There is 10–20 times more CA III in the livers of male rats than in the livers of females.[15] Prepubertal concentrations of CA III are low, and in geriatric males concentrations are down to those found in females.[15] Castration of male rats causes a reduction of CA III concentrations of 75%, a process that is partially reversed by a subcutaneous testosterone implant.[15] Tes-

STEPHEN JEFFERY • Department of Child Health, St. George's Hospital Medical School, University of London, London SW17 ORE, United Kingdom.

tosterone implants in ovariectomized females cause an increase in liver CA III concentrations of about 300%, but male levels are not reached. It seemed clear that androgens played a major role in the regulation of hepatic CA III in rat, and it was hypothesized that a direct action on the hepatic testosterone-binding protein might be involved. However, the results of hypophysectomy experiments on male and female rats suggested that this idea was not correct.

Concurrently with these experiments, it was discovered that rat liver CA II also showed a sexual dimorphism.[8] It became clear that the "high activity" isozyme of Garg[4] was CA II. The situation was the reverse of that found for CA III, with greater concentrations in female than in male livers. The differences were far smaller than for CA III, with around three times more CA II in the livers of females compared with males. Hypophysectomy also had an effect on the concentration of this isozyme. The primary finding from the hypophysectomy experiments was that the operation produced a male pattern of hepatic CA isozymes. Thus, male CA II and CA III concentrations were not affected by the operation, whereas female CA II concentrations were reduced and CA III concentrations were raised, both to male levels. Since serum testosterone is reduced to zero by this operation, it was clear that the androgen could not be exerting its effect directly on the liver to produce a "masculine" pattern.

A possible regulatory influence on CA synthesis in rat liver was suggested by the work of Gustafsson and his group[12] on the effect of growth hormone (GH) on liver steroid-metabolizing enzymes. Gustafsson and co-workers showed that patterns of GH release from the pituitary had a major influence on the synthesis of these enzymes. A female pattern of release induced a female pattern of enzymes in the liver; a male type of release from the pituitary gave a masculine isozyme profile in the liver.[12] Although the amounts of GH secreted over a 24-hr period were the same in both male and female rats, male pituitaries gave regular pulses at intervals of approximately 3 hr, whereas females showed irregular pulses (Fig. 1). A further

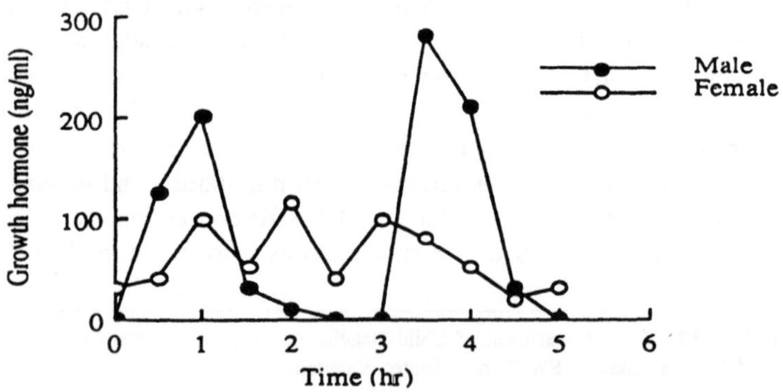

Figure 1. Typical GH profiles in the plasma of male and female rats.

important difference between the sexes was that male levels of GH declined to zero in the plasma between peaks, whereas in females there was a continued secretion of GH into the plasma between peaks. As with CA, hypophysectomy of rats produced a male pattern of steroid-metabolizing enzymes in the liver, while constant infusion of GH into male rats using subcutaneous minipumps produced a female distribution of these enzymes.[12] A similar experiment was conducted to look at the effect on CA II and CA III in male rats.[9] The result of infusion of GH into male rats was a decrease in the concentration of hepatic CA III, though levels did not fall to those in female controls and there was an increase in CA II concentrations (Table I). From these data, it appeared that GH might well have an important regulatory effect on rat liver CA synthesis. Either production of regular peaks of GH in males or the reduction to basal levels between peaks might be a factor in the establishment of the differences between males and females in hepatic CA isozyme synthesis. The hypophysectomy experiments suggested the latter possibility, but removal of the pituitary is such a drastic operation in the examination of hormonal pathways that these results need to be treated with caution.

In an attempt to produce a male-type pattern of GH release in female rats without recourse to hypophysectomy, experiments were carried out using an infusion of growth hormone-releasing factor (GRF) and/or somatostatin into the jugular vein of male and female rats.[10] CA III concentrations were examined in the clarified supernatants of liver homogenates, using radioimmunoassay. Female rats with indwelling cannulae were given doses of somatostatin at 20 μg/hr, with interruptions in the treatment for 10, 30, or 90 min at intervals of 3 hr. GRF was administered by injection 2, 6, or 24 times a day at a total daily dosage of 8 μg. The GRF-treated females showed no change in CA III concentrations compared with the control animals. For the somatostatin experiment, CA III concentrations were increased 8-fold with a 90-min break in infusion but only 1.5-fold with a 10-min gap. Concentrations of CA III reached 40% of those in control males. Although both GRF and somatostatin treatment produced regular pulses of GH in the female animal (Fig. 2), in the former group there was still endogenous production of GH

TABLE I
Effect of GH on Hepatic CA II and CA III in Male Rats[a]

Group	CA III	CA II
Normal females	2.81 ± 0.76	160.4 ± 33.7
Normal males	39.04 ± 12.75**	99.88 ± 16.6*
GH-treated males	10.16 ± 3.4**	138.0 ± 28.51

[a]Results are expressed as a means ± SD nanograms of CA II and micrograms of CA III per milligram of soluble protein. There were six animals in each group. *, $P < 0.01$; **, $P < 0.001$ compared with normal females; $P < 0.02$; $P < 0.01$ compared with normal males (Scheffe's test after one-way analysis of variance).

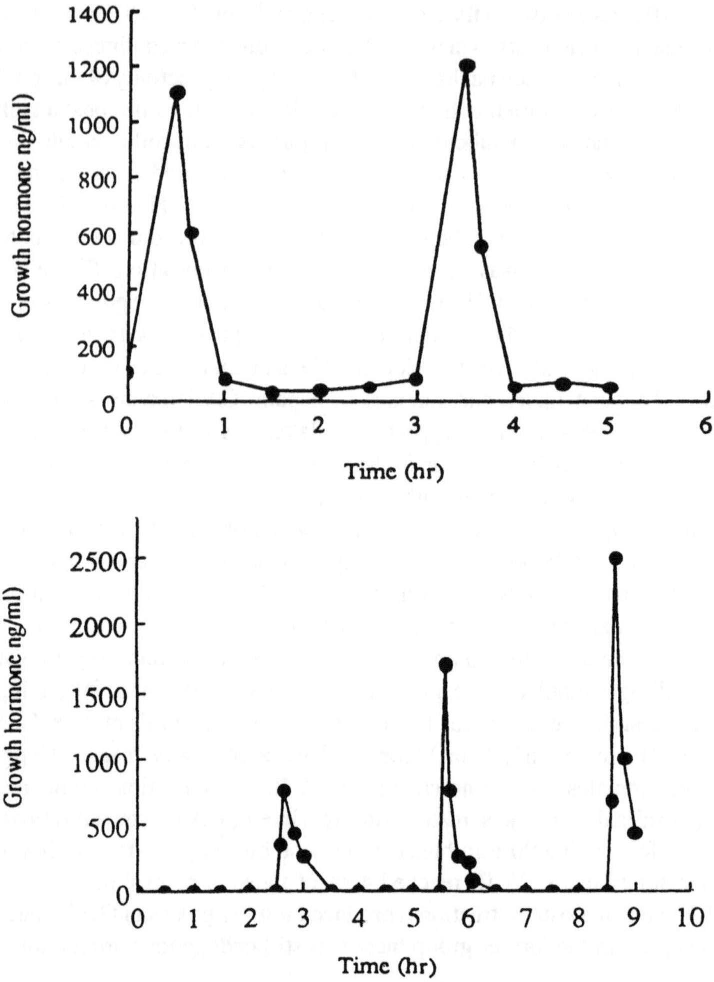

Figure 2. (A) Effect of GRF every 3 hr on GH profiles in female rats. (B) Effect of somatostatin infusion with a 30-min break every 3 hr on GH profiles in female rats.

between peaks, whereas in the somatostatin-treated animals there was no GH produced between peaks. These data suggested that it is indeed the absence of GH between peaks in the males which is responsible, at least in part, for the high levels of CA III found in male rat liver.

Infusion experiments using minipumps on mice have shown that concentrations in excess of 3 μg of bovine GH per hr are necessary to produce a feminizing effect. If 1.5 μg of GH per hr is used, then there is no effect.[13] Dwarf rats have now been examined to discover whether there is a sexual dimorphism for CA III.[10]

These rats have very little pituitary GH and show no apparent difference in secretion between males and females.[3] However, the hepatic CA III concentrations are identical to those in normal males and females. It is possible that there is a difference in GH release between male and female dwarf rats, but if so it is below the threshold of the radioimmunoassay used. There would therefore be a large

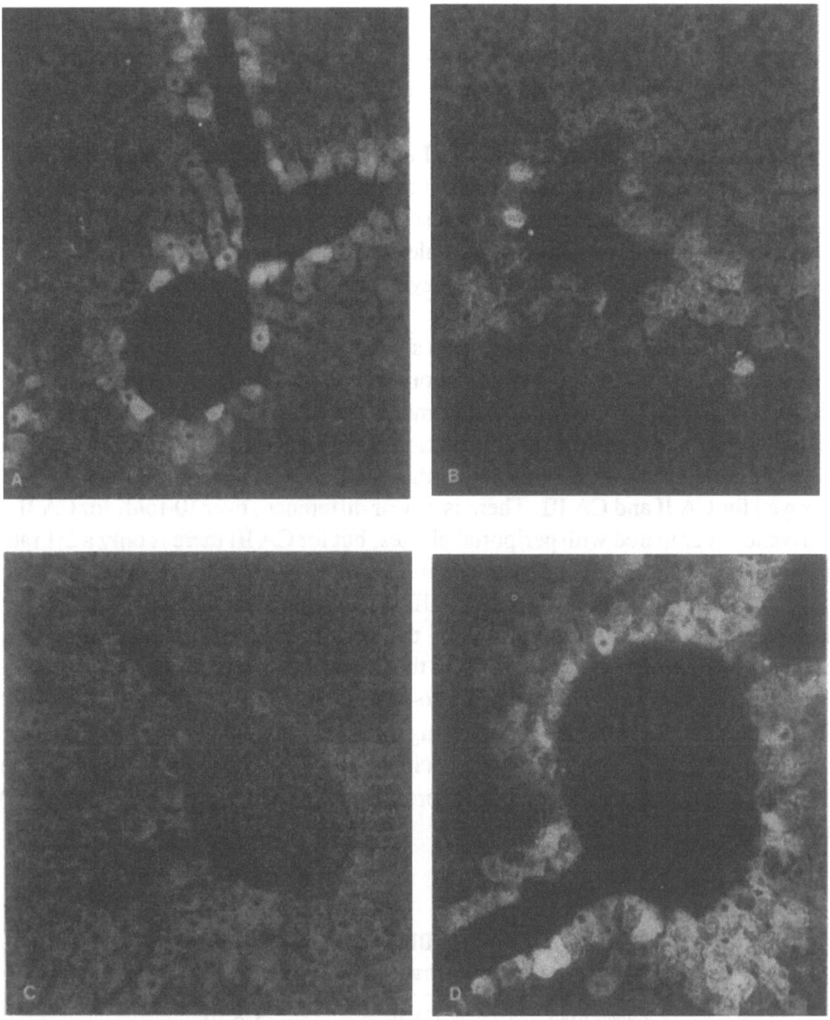

Figure 3. Rat livers stained with anti-CA II or anti-CA III, visualized by using fluorescent second antibody. (A) Female livers stained for CA II; (B) male liver stained for CA II; (C) female liver stained for CA III; (D) male liver stained for CA III. (Taken from N. D. Carter, P. J. Wistrand, G. Lönnerholm, 1989, *Acta Physiol. Scand.* **135**:163–167.)

discrepancy between the *in vivo* differences in GH concentration required to produce CA III sexual dimorphism and those necessary with minipump infusion. Alternatively, the results from dwarf rats might mean another reappraisal of the regulatory control of CA synthesis in rat liver.

2. Location

Hepatocytes were once considered a homogenous cell type, but it is now clear that there is a differentiation between those in the perivenous and periportal regions of the liver.[6] This difference is reflected in the enzymes that are involved in ammonia metabolism.

The first indications that rat CA II and CA III show a compartmentalization within liver as well as a sexual dimorphism came from this laboratory,[2] using specific antibodies and immunofluorescence on sections of perfused liver. The results showed the expected male–female differences for CA II and CA III but also demonstrated localization of both isozymes in the perivenous area of the liver (Fig. 3).

More recently, digitonin perfusion studies and experiments using a rat cDNA clone for CA III for *in situ* hybridization have given results differing from those discussed above. The perfusion experiments were carried out by Dr. Kai Lindros in Finland and involved the specific elution of perivenous or periportal areas of the liver.[14] Table II shows the results of these experiments when the eluates were assayed for CA II and CA III. There is a clear difference, over 10-fold, for CA II in perivenous compared with periportal eluates, but for CA III there is only a 2:1 ratio for the same samples. Figure 4 shows the result of *in situ* hybridization of a liver section from a male rat, using a rat CA III cDNA probe labeled with ^{32}P. Although there are concentrations of grains in the perivenous region, there are also a considerable number of silver grains in the periportal area indicating a gradient of CA III expression across the whole porto-venous axis. This contrasts strongly with the results from the same laboratory using a cDNA probe for glutamine synthetase, in which case the grains are found specifically in the perivenous region (Fig. 4). Work is now in progress to establish more exactly the distribution of CA II mRNA in rat liver.

TABLE II
CA Concentrations in Liver

Eluate from digitonin perfusion	CA II (ng/mg of protein)	CA III (μg/mg of protein)
Perivenous	588 ± 326 (range, 224–1108)	81.8 ± 14.4
Periportal	36 ± 2.5	42.9 ± 5.6

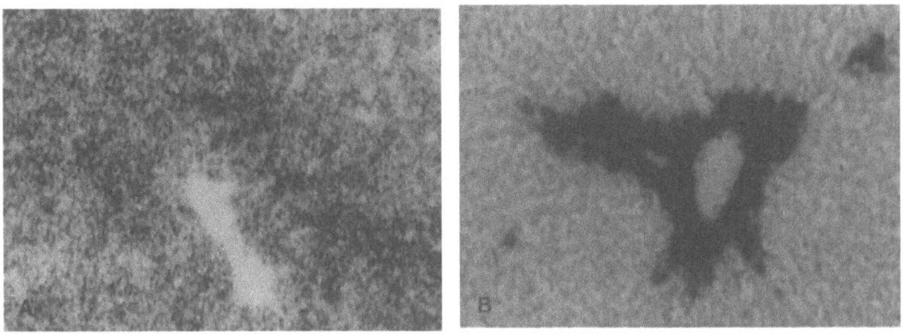

Figure 4. Rat liver treated with ^{35}S-labeled CA III cDNA (A) and ^{35}S-labeled glutamine synthetase cDNA (B).[17]

References

1. Carter, N. D., Hewett-Emmett, D., Jeffery, S., and Tashian, R. E., 1981, *FEBS Lett.* **128**:114–118.
2. Carter, N., Jeffery, S., Legg, R., Wistrand, P., and Lönnerholm, G., 1987, *Biochem. Soc. Trans.* **15**:667–668.
3. Charlton, H. M., Clark, R. G., Robinson, I. C. A. F., Porter-Goff, A. E., Cox, B. S., Bugnon, C., and Block, B. A., 1988, *J. Endocrinol.* **119**:51–58.
4. Garg, L. C., 1974, *Pharmacol. Exp. Ther.* **189**:557–562.
5. Garg, L. C,. 1975, *J. Pharmacol. Exp. Ther.* **192**:297–302.
6. Haüssinger, D., Gerok, W., and Sies, H., 1984, *Trends Biochem. Sci.* **9**:300.
7. Holmes, R. S., 1976, *J. Exp. Zool.* **197**:289–295.
8. Jeffery, S., Carter, N. D., and Wilson, C. A., 1984, *Biochem. J.* **221**:927–929.
9. Jeffery, S., Wilson, C. A., Mode, A., Gustafsson, J. A., and Carter, N. D., 1986, *J. Endocrinol.* **110**:123–126.
10. Jeffery, S., Carter, N. D., Clark, R. G., and Robinson, I. C. A. F., 1990, *Biochem J.* **266**:69–74.
11. King, R. W., Garg, L. C., Huckson, J., and Maren, T. H., 1974, *Mol. Pharmacol.* **10**:335–343.
12. Mode, A., Gustafsson, J. A., Jansson, J. O., Eden, S., and Isaksson, O., 1982, *Endocrinology* **111**:1692–1697.
13. Norstedt, G., and Palmiter, R., 1984, *Cell* **36**:805–812.
14. Pösö, A. R., Penttilä, L. E., Soulima, E. M., and Lindros, K. O., 1986, *Biochem. J.* **239**:263.
15. Shiels, A. Jeffery, S., Phillips, I. R., Shephard, E. A., Wilson, C. A., and Carter, N. D., 1983, *Biochem. Biophys. Acta* **760**:335–342.
16. Shiels, A., Jeffery, S., Wilson, C. A., and Carter, N. D., 1984, *Biochem. J.* **218**:281–284.
17. Kelly, C., Carter, N., DeBoer, P., Jeffery, S., Moorman, A., and Smith, A., 1991, submitted.

CHAPTER 25

Liver Mitochondrial Carbonic Anhydrase (CA V), Gluconeogenesis, and Ureagenesis in the Hepatocyte

SUSANNA J. DODGSON

1. Evidence for Carbonic Anhydrase Activity in Mitochondria from Guinea Pig and Rat Liver

In the 1970s, there was great interest in the functioning of isolated mitochondria, which led in 1978 to the awarding of the Nobel Prize for Physiology and Medicine to Dr. Peter Mitchell for his chemiosmotic theory. A few weeks later, the reviewer arrived in Philadelphia as the postdoctoral fellow of Dr. Robert E. Forster II. Dr. Forster has had a successful career devoted to studying anomalies in the theories of gas diffusion, pH disequilibrium, and carbonic anhydrase (CA)[18,22]; he believed that there was something wrong in Dr. Mitchell's theory and that a mitochondrial CA would dissipate the proton gradient required. A collaboration was started with two experts in mitochondrial bioenergetics: Dr. Leena Mela, who knew how to isolate mitochondria from several organs, and Dr. Bayard Storey, who knew how to isolate skeletal muscle mitochondria.[42] By ^{18}O mass spectrometric CA analysis,[18,30] the reviewer found abundant CA activity in guinea pig liver and skeletal muscle but none in brain, kidney, or heart.[15] The work on skeletal muscle that has been done since then is reviewed elsewhere,[42] as is the work with the CA V-containing rat kidney mitochondria.[6] Exhaustive literature searches indicated that liver fractionation by previous workers had given some evidence of CA activity in mitochondria, but several believed that this was the result of contamination from the cytosol of hepatocytes or erythrocytes (reviewed in reference 21). Work in which sulfonamide inhibition reduced HCO^{3-}-linked Ca^{2+} transport across the mitochondrial membrane did, however, convince the investigators of the existence

---

SUSANNA J. DODGSON • Department of Physiology, University of Pennsylvania School of Medicine, Philadelphia, Pennsylvania 19104-6085.

of a mitochondrial CA.[17,24] With one notable exception,[45] there has been until recently little interest outside of this laboratory in studying this lovely isozyme directly but considerable interest in studying the effects of rendering it nonfunctional by sulfonamide CA inhibition. Mitochondrial CA inhibition has been concluded to be responsible for decreased urea and glucose synthesis by alligators and chameleons *in vivo*,[5] decreased urea synthesis by isolated perfused rat livers,[25,34] decreased urea and glucose synthesis by isolated rat hepatocytes,[3,35,39] and by isolated guinea pig hepatocytes,[7,10,11,14] and decreased citrulline synthesis by intact isolated liver mitochondria from guinea pigs[7,12] and rats[46] (S. J. Dodgson and A. J. Meijer, unpublished results). Reviews of ongoing work in the field have appeared in the past decade.[14,21,22]

The proof of the existence of CA V in mitochondria has not altered the acceptance of Dr. Mitchell's chemiosmotic theory. However, evidence has accumulated in the past decade which indicates this theory is not as generally applicable as first believed.[40]

2. Mitochondrial Carbonic Anhydrase Is a Unique Isozyme, CA V

The first clue that guinea pig liver mitochondria contain a unique CA was seen by the extraordinary pH sensitivity of the CA activity of disrupted mitochondria.[15] In guinea pig, there is an eightfold increase in disrupted liver mitochondrial CA activity from pH 7.0 to 8.0, compared with a twofold increase in disrupted erythrocytic CA activity.[9] With disrupted rat liver mitochondria, the increase in activity over this pH range is only three- to fourfold.[8]

The second clue that mitochondrial CA was unique was finally seen by the reviewer after a fruitless year of trying to run a nondenaturing polyacrylamide gel with purified enzyme. Beautiful bands were seen for rat and guinea pig blood isozymes; no matter which acrylamide concentration CA V migrated as a comet. After assuming for some time that incompetence was the problem, the reviewer finally accepted that CA V was probably far too lipophilic for this procedure.[7]

Guinea pig CA V has been partially sequenced[29] (Table I). Guinea pig CA V was purified by affinity chromatography and shown to be one isozyme by overloading an SDS-PAGE gel.[7,14] Similarly, rat CA V has been purified (this laboratory), and polyclonal antibodies have been raised to it in rabbits (by Dr. Nicholas D. Carter). This antibody reacts only with soluble extracts from mitochondria.[3a]

3. CA V Kinetics in Rat and Guinea Pig

In this laboratory, both guinea pig liver CA V and rat CA V have been studied.[7,8,13,15,16,21,29] Total CA activity of freeze-thawed mitochondria is meas-

TABLE I
Amino Acid Sequence of Guinea Pig Liver CA V N-Terminal Polypeptide[a]

	Sequence														
Isozyme	1	2	3	4	5	6	7	8	9	10	11	12	13	14	15
CA V	Val-	Pro-	Arg-	Gly-	Thr-	Arg-	Gln-	Ser-	Pro-	Ile-	Asn-	Ile-	Gln-	Arg-	Asp-
	22+			25					30					35	36
Most CA I	Ile-	Ala-	Asn-	GLY-	Asn-	Asn-	GLN-	SER-	PRO-	Val-	Asp-	ILE-	Lys-	Thr-	Ser-
Most CA II	Ile-	Ala-	Lys	GLY-	Glu-	ARG-	GLN-	SER-	PRO-	Val-	Asp-	ILE-	Asp-	Thr-	Lys-
Most CA III	Asn-	Ala-	Lys-	GLY-	Glu-	Asn-	GLN-	SER-	PRO-	ILE-	Glu-	Leu-	His-	Thr-	Lys-

	Sequence														
Isozyme	16	17	18	19	20	21	22	23	24	25	26	27	28	29	30
Most CA I	Asp-	?-	Ile-	Try-	Asp-	?-	?-	Leu-	Pro-	?-	?-	Lys-	Val-	?-	Tyr
	37			40					45						51
Most Ca I	Glu-	Thr-	Lys-	His-	ASP-	Thr-	Ser-	LEU-	Lys-	Pro-	#-	Ser-	VAL-	Ser-	TYR-
Most CA II	Ala-	Ala-	#-	His-	ASP-	Pro-	Ala-	LEU-	Lys-	Pro-	Leu-	Ala-	VAL-	Ser-	TYR-
Most CA III	ASP-	Ile-	#-	His-	ASP-	Pro-	Ser-	LEU-	Lys-	Pro-	Trp-	Thr-	Ala-	Ser-	TYR-

[a]Taken from reference 29.

ured at 37°C, pH 7.4, with the ^{18}O mass spectrometric assay; measurement at physiological temperature without a big pH change is not possible by any other technique.[18] It is not strictly correct to obtain Michaelis–Menten and kinetic constants by the mass spectrometric assay[18]; for this reason, these and other kinetic constants are asterisked in the text that follows.

Rat and guinea pig liver CA V *velocity decreases with increasing pCO_2.[8] *$K_m(CO_2)$ is 7.5 mM for guinea pig liver CA V and 2.2 mM for rat liver CA V. Turnover number (*TON) calculated for rat kidney CA V is 24,000 sec^{-1}[6]; since it has been concluded that rat liver CA V is identical, then this should also be the *TON for liver CA V.

4. CA V as Intramitochondrial Indicator of pH in the Guinea Pig

The unusual pH sensitivity of guinea pig CA V was used to determine intramitochondrial changes in pH in metabolizing mitochondria.[12,13,21] CA activity of intact mitochondria was calculated, and the pH was estimated from a pH-versus-CA activity curve calculated for disrupted mitochondria. The rat liver CA V is not so sensitive to pH, and so it is not possible to use this technique in this species. The other requirement is an isotope ratio mass spectrometer to continuously monitor concentrations of $C^{16}O_2$ (mass 44), $C^{18}O^{16}O$ (mass 46), and $^{16}O_2$ (mass 32).[18]

5. Synthetic Pathways for which CA V Provides HCO_3^-

Inclusion of the CA inhibitor acetazolamide resulted in decreased citrulline synthesis by mitochondria incubating in an isotonic HCO_3^-–PIPES buffer under an atmosphere of 5% CO_2–95% O_2 at 25°C.[12] It was concluded that CA V had a metabolic function in providing the substrate HCO_3^- for the first urea cycle enzyme carbamyl phosphate synthetase I (CPS I) (Fig. 1). This was the first time that a CA isozyme has been implicated in intermediary metabolism. The statement that acetazolamide was as effective in decreasing citrulline synthesis as in decreasing CA activity of disrupted mitochondria was found after publication to be incorrect[12]; however, there is no question that higher concentrations of acetazolamide decrease citrulline synthesis. Decrease in citrulline synthesis by acetazolamide at 37°C reaches a maximum at approximately 120 μM (Dodgson and Meijer, unpublished results), whereas disrupted mitochondria have all of their CA activity inhibited by less than 1 μM. A better CA inhibitor for demonstrating the relationship between CA inhibition and citrulline inhibition is ethoxzolamide. Figure 1B is a plot of citrulline synthesis by intact rat liver mitochondria as a function of ethoxzolamide concentration, Fig. 1C is a plot of inhibition of CA

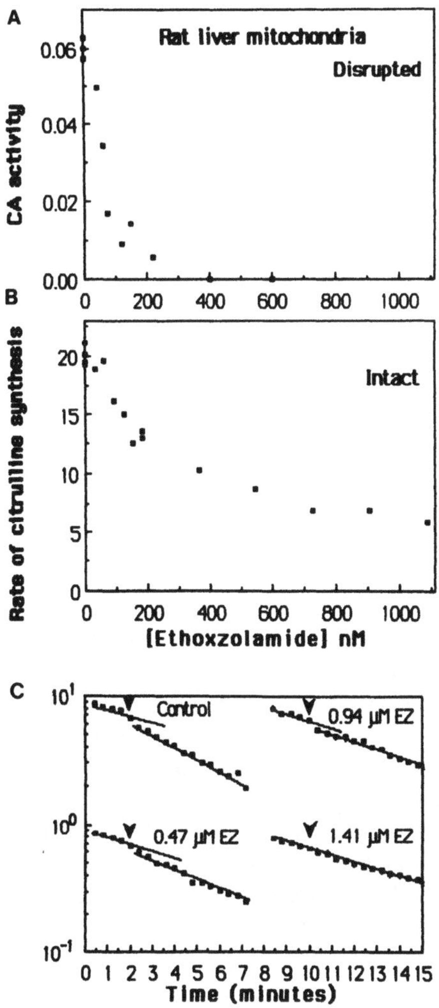

Figure 1. (A) CA activity of disrupted rat liver mitochondria when increasing concentrations of ethoxzolamide are included in the reaction chamber of the mass spectrometer. (B) Plot of citrulline synthesis as a function of increasing ethoxzolamide concentration in rat liver mitochondria. (C) CA activity of intact, citrulline-synthesizing rat liver mitochondria when increasing concentrations of ethoxzolamide are included in the reaction chamber of the mass spectrometer. Intact isolated mitochondria (1 mg/ml) were incubated in 50 mM MOPS-KCl buffer with 10 mM pyruvate, 10 mM ornithine, 10 mM NH_4Cl, and 25 mM $KHCO_3$ at 37°C, pH 7.4, under an atmosphere of 5% CO_2–95% O_2 for 12 min; aliquots were removed and added to $HClO_4^-$ at 4-min intervals.

activity of intact rat liver mitochondria as a function of ethoxzolamide concentration (Dodgson and Meijer, unpublished results), and Fig. 1A is a plot of CA activity of disrupted rat liver mitochondria, assayed by the mass spectrometer exchange technique[13,30] at 37°C.

The effect of CA inhibitors in decreasing citrulline and urea synthesis has been demonstrated also in other laboratories (reviewed in reference 21). An effect on the decrease of urea synthesis by human liver slices has also been reported.[26]

It has been demonstrated in this laboratory[11] that glucose synthesis by the intact hepatocyte is dependent on CA V when glucose is synthesized obligatorily through a bicarbonate-dependent pathway (by pyruvate carboxylation) but not through a bicarbonate-independent pathway (when glutamine is the oxidizable substrate). Endogenous hepatic glucose was removed by withholding food from guinea pigs for 48 hr before sacrifice. Substrates for urea synthesis as well as glucose synthesis were included with the incubating hepatocytes; the sulfonamide CA inhibitor ethoxzolamide decreased urea synthesis with increasing concentration whether glucose was inhibited (when pyruvate was carboxylated) or whether glucose was not affected (when glutamine was metabolized). It was thus concluded that HCO_3^- can also ratelimit gluconeogenesis. An earlier report had also shown that glucose synthesis by intact male and female rat hepatocytes was sensitive to the sulfonamide CA inhibitor acetazolamide.[35]

6. Origin of CO_2 Hydrated by CA V for Intramitochondrial Carboxylation

It was the question of the origin of the CO_2 required for urea synthesis and glucose synthesis that interested several researchers in the mitochondrial CA in the past decade. It seems unlikely that the CO_2 fixed by CPS I all arises from intramitochondrial decarboxylations when the pCO_2 in the liver cytosol is so high. It has been demonstrated that a raised external pCO_2, compensated with a raised external bicarbonate concentration, overrides acetazolamide inhibition of urea synthesis in perfused male rat livers.[25] There has been recent evidence for metabolite channeling in the urea cycle in the hepatocyte cytosol and in the mitochondrion.[4,34,41,47] It thus appears likely that the CA isozymes function by "handing" a CO_2 generated by one decarboxylating enzyme to another enzyme as HCO_3^-. This hypothesis may explain why CA V is required for urea or glucose synthesis.

7. Localization of CA V in the Liver

Over the past few years, it has become possible to separate hepatocytes from both the periportal and perivenous regions of the liver by the digitonin pulse technique.[32,37] If the only function of liver CA V were to provide HCO_3^- for CPS I, then it would be predicted that CA V would be present only in the area of the liver capable of synthesizing urea and glucose, the periportal hepatocytes.[19] CA concentration and activity as well as isozyme type are heterogeneous throughout the liver, with more CA activity concentrated in the perivenous hepatocytes.[31]

CA V is also present in the perivenous hepatocytes.[3a] The functions of CA V in this region are unknown.

8. Diabetes Increases CA V Concentration

A collaboration with Dr. Malcolm Watford resulted in the finding that the livers of diabetic male rats (6 days of administration of streptozotocin, 70 mg/kg) have approximately twice as much CA V activity and approximately half as much CA III activity as do livers of control rats[6] (Table II). This study is the first in which an increase of liver CA V activity has been found; an earlier study from this laboratory showed that rat kidney CA V (but not liver CA V) is increased during mild acidosis.[8]

The observation that CA V activity is doubled in diabetic rats must result in a rethinking of the concept that the activity of CA V is greatly in excess. If the activity is not close to that needed, then there should be no alteration in the concentration of this enzyme. However, if CA V is synthesized extramitochondrially and subsequently located inside the mitochondrion at great energetic cost to the hepatocyte, then perhaps only as much CA V as the hepatocyte needs is present at any time.

9. Changes in CA V Activity in Disrupted Rat Liver Mitochondria with Age and Sex

Measurements of CA V activity of rats from 18 days postconception up to 90 days after birth have shown that the activity increases after birth. There is no difference in the CA V activity of age-matched male and female rats at any stage in their development (S. J. Dodgson and K. Cherian, unpublished results). These data contrast with the great changes in CA II and CA III activity of rat liver in males and females; these changes are also age dependent.[31]

TABLE II
CA Activity of Preparations of Perfused Livers Isolated from Normal and Diabetic Rats[a]

	k_{enz}	
CA isozyme	Control[b]	Diabetic[a]
V	0.08 ± 0.01	0.20 ± 0.04
V + III + II + I	0.79 ± 0.08	0.40 ± 0.10
III	0.48 ± 0.02	0.15 ± 0.02

[a]Isolated mitochondria (CA V) and hepatic homogenates in the presence (III) and absence (V + III + II + I) of 7 μM acetazolamide were assayed for CA activity at 37°C, pH 7.4, with 25 mM $NaHCO_3$ (1% labeled with ^{18}O).
[b]Results of three experiments, using two rats in each.
[c]Results for CA V are from four experiments, using four rats in each; other results are from three experiments, using three rats in each.

10. Molecular Biology of CA V

Less than 10% of mitochondrial proteins are encoded for by the mitochondrial genome; the remainder are specified by nuclear genes. CA V is not synthesized inside the mitochondrion; perusal of the oligonucleotide sequence estimated for the determined polypeptide sequence for guinea pig CA V led Drs. David Hewett-Emmett and Richard Tashian to this conclusion.[28,44] The two mitochondrial urea cycle enzymes CPS I and ornithine transcarbamylase are synthesized on free ribosomes in the cytosol as precursors containing N-terminal extensions which are released into the cytosol[25] (Fig. 2). The N-terminal pre-piece contains information to target the protein to the mitochondrion.[36] It is likely that CA V has a similar origin.

A. Ureagenesis

Mitochondrion:

CARBAMYL PHOSPHATE SYNTHETASE I

HCO_3 + NH_4^+ + 2ATP ----------------------> carbamyl phosphate + 2ADP

ORNITHINE TRANSCARBAMYLASE

carbamyl phosphate + ornithine-----------------------> citrulline

Cytosol:

citrulline + ATP-------->----->------> urea + ornithine

B. Gluconeogenesis

Mitochondrion:

PYRUVATE CARBOXYLASE

HCO_3^- + pyruvate + ATP --------------------> oxaloacetate

Cytosol (rat and guinea pig) and mitochondrion (guinea pig):

PHOSPHO ENOL PYRUVATE CARBOXY KINASE

oxaloacetate + GDP --------------------------> GTP + phospho enol pyruvate

Cytosol:

Glycolytic Enzymes

phospho enol pyruvate --->--->---->--->----->--->------> glucose 6 phosphate

Endoplasmic Reticulum:

GLUCOSE 6 PHOSPHATASE

glucose 6 phosphate + H^+OH^- ---------------------------> glucose + P_i

Figure 2. The enzymes, their substrates, and products in hepatic ureagenesis and gluconeogenesis.

Transport of precursors across the two mitochondrial membranes occurs in a single step at contact sites between the outer and inner membranes. The insertion into and translocation across the outer and inner membranes are dependent on the electrochemical potential across the inner membrane.[43]

11. Mitochondria and CA IV

Mitochondrial CA, CA V, is by definition that CA isozyme uniquely localized in the mitochondrial matrix. The membrane-bound CA IV, which has been studied mainly in the plasma membranes of kidney, lung, and brain, (discussed in references 6 and 48) may also be present on the mitochondrial membranes. There has not as yet been any report of CA IV on mitochondrial membranes *per se*; however, the reviewer believes that there is a case to be made for its presence in the rat by analyzing data from other laboratories.

The first evidence for a function for mitochondrial CA IV came in reports that acetazolamide decreased the HCO_3^--facilitated Ca^{2+} uptake by rat liver mitochondria.[17,24] These reports were published before mitochondrial CA was quantitated and localized in the matrix.[15] A membrane-bound CA would be more likely to be involved in ion exchange than would a matrix-localized CA. It is now suggested that these authors were, perhaps, describing a function of CA IV, the membrane-bound CA isozyme.

When quantitation of rat liver mitochondrial CA was reported, it was concluded from fractional digitonin disruption of the mitochondria that a large percentage of the CA activity was located in the inner membrane space.[45] These data may also be interpreted as evidence for CA IV activity in the inner membrane.

ACKNOWLEDGMENTS. The work from this laboratory has been entirely supported by grants from the National Institutes of Health; currently these are HL-PO19737 to R. E. Forster II and DK-38041 to S. J. Dodgson.

References

1. Arinze, M. A., Garber, A. J., and Hanson, R. W., 1973, *J. Biol. Chem.* **248**:2266–2274.
2. Balboni, E., and Lehninger, A. L., 1986, *J. Biol. Chem.* **261**:3563–3570.
3. Boon, L., and Meijer, A. J., 1988, *Eur. J. Biochem.* **171**:465–469.
3a. Carter, N. D., Dodgson, S. J., and Quant, P., 1991, *Biochim. Biophys. Acta*, in press.
4. Cheung, C. W., Cohen, N. S., and Raijman, L., 1989, *J. Biol. Chem.* **264**:4038–4044.
5. Coulson, R. A., and Herbert, J. D., 1984, *Ann. N.Y. Acad. Sci.* **429**:505–515.
6. Dodgson, S. J., this volume, Chapter 1.
7. Dodgson, S. J., 1987, *J. Appl. Physiol.* **63**:2134–2141.
8. Dodgson, S. J., and Contino, L. C., 1988, *Arch. Biochem. Biophys.* **260**:334–341.
9. Dodgson, S. J., and Forster, R. E., II, 1983, *J. Appl. Physiol.* **55**:1292–1298.

10. Dodgson, S. J., and Forster, R. E., II, 1986, *J. Appl. Physiol.* **60**:646–652.
11. Dodgson, S. J., and Forster, R. E., II, 1986, *Arch. Biochem. Biophys.* **251**:198–204.
12. Dodgson, S. J., Forster, R. E., II, Schwed, D. A., and Storey, B. T., 1983, *J. Biol. Chem.* **258**:7696–7701.
13. Dodgson, S. J., Forster, R. E., II, and Storey, B. T., 1982, *J. Biol. Chem.* **257**:1705–1711.
14. Dodgson, S. J., Forster, R. E., II, and Storey, B. T., 1984, *Ann. N.Y. Acad. Sci.* **429**:516–524.
15. Dodgson, S. J., Forster, R. E., II, Storey, B. T., and Mela, L., 1980, *Proc. Natl. Acad. Sci. USA* **77**:5562–5566.
16. Dodgson, S. J., and Watford, M., 1990, *Arch. Biochem. Biophys.*
17. Elder, J. A., and Leninger, A. L., 1986, *J. Biol. Chem.* **261**:3563–3570.
18. Forster, R. E., II, this volume, Chapter 6.
19. Gebhardt, R., and Mecke, D., 1983, *EMBO J*, **2**:567–570.
20. Geers, C., and G. Gros, this volume, Chapter 19.
21. Gros, G., and Dodgson, S. J., 1988, *Annu. Rev. Physiol.* **50**:669–694.
22. Gros, G., Forster, R. E., II, and Dodgson, S. J., 1988, in: *pH Homeostasis. Mechanisms and Control* (D. Häussinger, ed.) Academic Press, New York, pp. 203–231.
23. Hallermayer, G., Zimmerman, R., and Neupert, W., 1977, *Eur. J. Biochem.* **81**:523–532.
24. Harris, E. J., 1978, *Biochem. J.* **176**:983–991.
25. Häussinger, D., and Gerok, W., 1985, *Eur. J. Biochem.* **152**:381–386.
26. Häussinger, D., Kaiser, S., Stehle, T., and Gerok, W., 1986, *Biochem. Pharmacol.* **35**:3317–3322.
27. Herbert, J. D., Coulson, R. A., and Hernandez, T., 1975, *Biochem. Biophys. Res. Commun.* **65**:1054–1060.
28. Hewett-Emmett, D., and Tashian, R. E., this volume, Chapter 2.
29. Hewett-Emmett, D., Cook, R. G., and Dodgson, S. J., 1986, *Isozyme Bull.* **19**:13.
30. Itada, N., and Forster, R. E., II, 1977, *J. Biol. Chem.* **252**:3881–3890.
31. Jeffery, S., this volume, Chapter 24.
32. Lindros, K. O., and Penttilä, K. E., 1985, *Biochem. J.* **228**:757–760.
33. Lipsen, B., and R. M. Effros, 1988, *J. Appl. Physiol.* **65**:2736–2743.
34. Marsolais, C., Huot, S., David, F., Garneau, M., and Brunengraber, H., 1987, *J. Biol. Chem.* **262**:2604–2607.
35. Metcalfe, H. K., Monson, J. P., Drew, P. J., Iles, R. A., Carter, N. D., and Cohen, R. D., 1985, *Biochem. Soc. Trans.* **13**:255.
36. Neupert, W., and Schatz, G., 1981, *Trends Biochem. Sci.* **6**:1–4.
37. Quistorff, B., Grunnet, N., and Cornell, N. W., 1985, *Biochem. J.* **226**:289–291.
38. Quistorff, B., and Grunnet, N., 1987, *Biochem. J.* **243**:87–95.
39. Rognstad, R., 1983, *Arch. Biochem. Biophys.* **222**:442–448.
40. Slater, E., 1987, *Eur. J. Biochem.* **166**:489–504.
41. Srivastava, D. K., and Bernhard, S. A., 1986, *Curr. Top. Cell. Regul.* **28**:1–68.
42. Storey, B. T., this volume, Chapter 22.
43. Stuart, R. A., D. W. Nicholson, M. A. Harmey, and W. Neupert, 1988, *Biochem. Soc. Trans.* **16**:701–702.
44. Tashian, R. E., 1989, *Biosci. Rep.* **10**:186–192.
45. Vincent, S. H., and Silverman, D. N., 1982, *J. Biol. Chem.* **275**:6850–6855.
46. Wanders, R. J. A., van Roermund, C. W. T., and Meijer, A. J., 1984, *Eur. J. Biochem.* **142**: 247–254.
47. Watford, M., 1989, *Trends Biol. Sci.* **14**:313–314.
48. Wistrand, P. J., and Knuuttila, K.-G., 1989, *Kidney Int.* **35**:851–859.

PART VII

The Carbonic Anhydrases of Other Organs

CHAPTER 26

Function and Regulation of the Carbonic Anhydrases in the Vertebrate Neural Retina

PAUL J. LINSER and JOEL L. COHEN

1. Localization of CA II in Vertebrate Neural Retina

1.1. Phylogenetic Analysis

Carbonic anhydrase (CA) (EC 4.2.1.1.) is highly concentrated in the vertebrate retina; e.g., in birds CA is 3% of the total tissue protein in the retina.[10,21] The retinas of most major groups of vertebrates have high CA activity[20] with amphibians having the lowest, while birds and reptiles have 25 times as much CA.[20] Most mammals are intermediate. No relationship was found between high CA activity and degree and type of vascularization.

Representatives from each class of extant vertebrates have been examined by histochemical and immunohistochemical techniques.[12] In all species, the archetypical retinal glial cell, the Müller fiber cell, has high concentrations of CA II.[12] No other CA isozyme has been detected in neural retina. In the chicken retina, Müller cells are approximately 50% of the tissue mass. CA II is 3% of the total tissue protein in the chicken retina and therefore is approximately 6% of the total protein in the Müller cell.

In certain terrestrial vertebrates, CA II has been detected in specific subcategories of photoreceptor cells,[23] amacrine neurons,[15] and perhaps horizontal neurons.[12] Immunohistochemical techniques indicate that horizontal neurons are a major compartment for CA in a variety of fish species.[12,14] In the most primitive

PAUL J. LINSER • Whitney Laboratory and the Department of Anatomy and Cell Biology, University of Florida, St. Augustine, Florida 32086-9604; and Mount Desert Island Biological Laboratory, Salsbury Cove, Maine 04672. *JOEL L. COHEN* • Biomedical Sciences Program and Department of Anatomy, School of Medicine, Wright State University, Dayton, Ohio 45435.

type of vertebrate, represented by the cyclostome *Lampetra*, horizontal neurons are the primary CA compartment, and Müller cells have much lower immunoreactivity.[12] From immunohistochemical results, it was proposed that there was an evolutionary shift in the compartmentalization of CA in the retina from the horizontal neurons to the Müller glial cells.[12] Immunohistochemical results do not always parallel results obtained with the enzyme histochemical method; e.g., enzyme histochemistry has failed to detect horizontal cell staining in fish retina.[6]

1.2. Developmental Analysis

In the retina system, the developmental pattern of CA II expression is complex. Figure 1 shows the concentration of CA II in the chick embryo retina as a function of developmental age. Although the chick is the best studied system in this regard, similar results have also been reported for mouse retina development.[15] CA II expression begins very early in eye organogenesis. Eye-specific expression of CA II is detectable just following the initial interaction between neural ectoderm and the overlying skin ectoderm. In the chick, CA II expression begins at 20 somite pairs and occurs in both the neural ectoderm (preretina) and the overlying skin ectoderm (prelens) (Fig. 2). Similar early CA II expression in chick retinas analyzed by using *in situ* hybridization techniques has been reported.[18] CA II expression is one of the earliest biochemical markers of embryonic induction of eye organogenesis yet described. The timing of CA II accumulation immediately following skin ectoderm–neuroectoderm interaction suggests that CA II expression may actually be induced by the interactions of these two tissue layers.

Following the turn on of CA II expression in the eye rudiment, the tissues undergo dramatic changes as the epithelial sheets infold on themselves, gradually

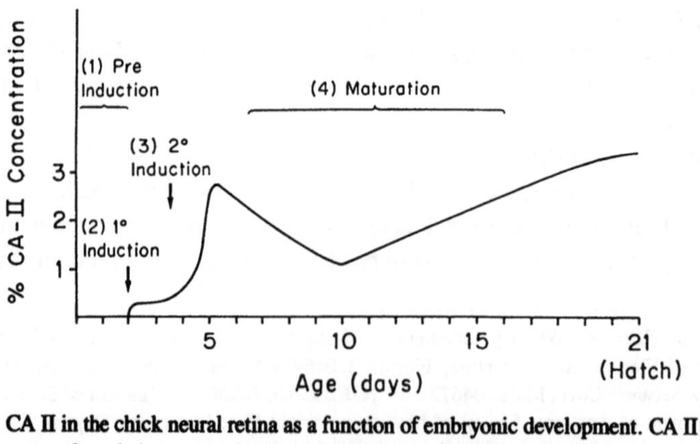

Figure 1. CA II in the chick neural retina as a function of embryonic development. CA II levels are shown as percent of total tissue protein (days 3 through hatching[10]) or in arbitrary units based on immunohistochemistry as in Fig. 2 (days 0 through 3). Four stages in CA II expression in the retina are shown.

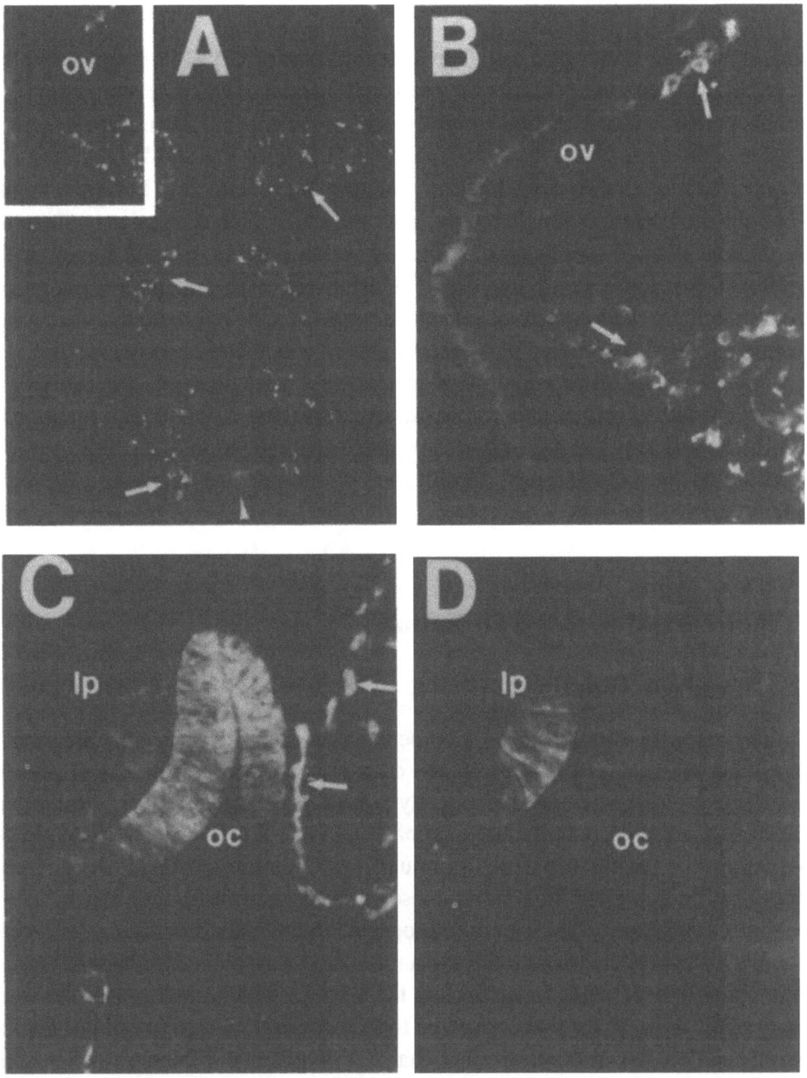

Figure 2. Immunohistochemical localization of CA II (A–C) and the lens marker protein delta-crystallin (D) in the early embryonic chick eye rudiment. (A) Low-power magnification view of the optic vesicle region (OV) of an 18 somite pair stage embryo. CA II staining is not evident in the OV but rather is present in scattered cells (arrows) that border several epithelial structures, including the forming eye. A region of light staining in the floor of the neural tube (arrowhead) is also marked. (B) Higher-magnification view of the region in panel A outlined in white. Note absence of staining in the OV but prominently stained cells in the mesenchymal tissues between the neural ectoderm and skin ectoderm in the region of the forming eye. (C and D) Single section of a 20 somite pair stage embryo at high magnification (as in panel B) immunostained for the simultaneous localization of CA II (C) and delta-crystallin (D). Note bright staining for CA II in the newly folded presumptive neural retina lying between the epithelial layer of the eye rudiment and the head mesenchyme. In panel D, the position of the lens placode is defined by immunostaining for delta-crystallin. This figure demonstrates that the earliest accumulation of CA II in retina begins between the 18 and 20 somite pair stages of development.

producing the layered optic orb. The concentration of CA II in the presumptive retina begins to rise very sharply between days 3 and 5 of chick development (Fig. 1). This period is characterized by the closure of the choroid fissure and hence the creation of an organ with a restricted chamber. Following this sudden rise in CA II, the concentration in the retina declines gradually for several days but again begins to rise and then plateaus after hatching. Four stages in CA II expression in the retina can thus be defined (see legend to Fig. 1) which parallel various stages in the functional development of the tissue. During these several phases in retinal development the cellular compartmentalization of CA II gradually shifts. The enzyme initially is expressed by all retinoblast cells but later becomes restricted to the glial cells as the differentiation of specific cell types proceeds. A corollary of this process is that retinal neurons cease the expression of CA II as a function of their differentiation. The shifting in the compartmentalization of CA II may imply a shifting in the physiological significance of the enzyme in tissue and organ function.

2. Function(s) of CA II in Retina

2.1. Electrical Activity

The majority of reported work on ocular CA II has focused on its function in influencing intraocular pressure by the ciliary apparatus and the CA II present there.[16] Several laboratories have recently examined the influence of CA inhibitors in neurophysiological activity in the retina.[1,7,22] Since CA II is always found in the glial cells of the vertebrate retina, the influence of CA inhibitors on Müller glial cell activity has been studied. In general, these inhibitors have been found to influence the glial component of the electroretinogram, the so-called b wave.[1,7,22] Some discrepancies between laboratories and animal model systems used have led to CA inhibitors both decreasing or increasing the b wave, which underscores the complexity of the ionic fluxes that contribute to the net electrical activity of retina glia. Nevertheless, it has become evident that CA II activity influences retina glial electrical activity.

In addition to examining the effects of CA inhibition on glial physiology, preliminary studies have begun to examine specific retina neurons that possess CA II. As noted earlier, the horizontal neurons of fish retina exhibit immunoreactivity for CA II to varying degrees, with the most primitive fish showing the most intense horizontal cell staining.[12] Horizontal cells are considered a primitive type of vertebrate neuron.[5,14] Among their peculiar characteristics is the fact that horizontal cells communicate laterally with other horizontal cells by electrical (gap) junctions rather than by conventional chemical synapses.[17] The electrical coupling between adjacent horizontal cells is modulated so that the gap junctions open and close in response to physiological signals. The result of this modulation is that the receptive field properties of the retina are modified. Since pH is known to influence electrical coupling between various types of cells, CA II activity may somehow influence electrical coupling between horizontal cells.

The neural retina of the elasmobranch skate (*Raja oscellata*) is among the simplest and most primitive vertebrate retinas. CA II is found primarily in horizontal cells in the skate, as in the lamprey, with lower concentrations in the Müller glial cells.[3] Also, the skate retina possesses only rod photoreceptors, with no cone cells.[5] In general, there are different kinds of horizontal cells that interact with the different kinds of photoreceptors. Thus, a retina which possesses rod photoreceptors and several types of cone cell will also possess a number of different types of horizontal cells. In the skate retina, there may be only one functionally distinct type of horizontal cell.

To assess the possible role of CA II in the electrical communication between horizontal cells, recordings were made from individual cells with intracellular electrodes. The perfused eye cup preparation was used as described.[7] Initially, a spot of light identified the center of the field covered by a specific horizontal cell, and then an annulus of light was applied to indicate signal spread. When 500 μM acetazolamide was applied to the perfusate, the response to the annular stimulus increased relative to the spot stimulus. This result suggests that as a result of inhibition of CA, electrical coupling between horizontal cells was increased. In additional experiments, acetazolamide either hyperpolarized or depolarized the resting potentials of horizontal cells, depending on whether the retina was light or dark adapted, respectively.[3] The differential response of the horizontal cells to acetazolamide depending on the state of light adaptation shows that CA activity is linked to the dynamic properties of the retina. Furthermore, the acetazolamide-mediated increase in apparent electrical coupling between horizontal cells may indicate that CA functions in the modulation of receptive field properties in the neural retina.

2.2. Developmental Role in Eye Morphogenesis

During retinal development, CA II is expressed in a complex and changing pattern. Do the different periods of high CA II expression reflect stages at which the enzyme is important for different processes? This is a question that must be answered by examining the several stages of CA II expression in the embryonic eye individually. The extreme peak in CA II concentration reached at day 5 of development in the chick system is an immediately attractive point to start. What is the importance of CA at this stage of eye development? Again, it is helpful to look at aspects of eye development which also show critical stages at this time in development. One of the major changes in the eye rudiment at this time is that the choroid fissure closes and the eye for the first time becomes a chambered organ with a closed compartment. As a result, the eye also begins morphogenetic spherical expansion. The rapid expansion of the eye at this time in development is driven by intraocular pressure.[4] Furthermore, chronic administration of high levels of acetazolamide to pregnant mice during a critical stage of development can lead to small or missing eyes in the offspring at birth.[19] Since CA activity is involved in the regulation of intraocular pressure in the mature eye,[16] the sudden accumulation of very high concentrations of CA in the embryonic retina as the eye begins rapid, pressure-dependent growth, indicates a role for CA in eye morphogenesis.

Administration of methazolamide to chick embryo eyes during this period of development leads to a significant reduction in eye growth. Inhibitor was applied by injection into the space surrounding the downside eye of an embryo in egg-free culture. As the embryo lies on its side a natural pocket forms around the downside eye as it is the largest protuberance of the embryo. Fluids injected into the extraembryonic spaces around the head naturally flow down into this pocket. Other experiments in which inhibitor solution was continually dripped onto the surface of the upside eye showed similar inhibition of eye growth.[13] Thus, CA activity in the epithelial portion of the eye influences intraocular pressure in development as well as in the mature animal. In the case of the embryo, CA actually plays a role in organ morphogenesis. Whether this role of CA is important in all animals or perhaps only secondary to other mechanisms as proposed by Coulombre[4] remains to be seen. This point is particularly interesting in view of the recently described strain of mutant mouse which fails to express CA II in any of its tissues.[8] These animals appear to have normal eyes, although no in-depth analyses have been performed. It may be that differences in eye size and shape during development and at maturity in different animal models reflect or correlate with different dependence on CA activity during eye development.

3. Regulation of CA II Expression in Retina Development

CA II is expressed as one of the earliest indicators of biochemical differentiation in the optic rudiment. At this early stage, CA II is expressed by all retinoblast cells. Later, as cell differentiation commences, CA II expression is increased in the glial cells but ceases in most neurons as they become distinguishable by position and morphology. The several phases of CA II expression in the retina may be indicative of different regulatory events and mechanisms.

The maturational phase of CA II expression occurs independently of organismal factors. Embryonic retinal cells can be isolated at an early stage of development and placed in rotation-mediated suspension culture. In this environment, retinal cells re-form tissue-typic architecture and undergo cell differentiation.[11] In such cultures, the progression of CA II expression and cellular specification proceeds normally. Thus, it appears that once CA II expression has begun in the optic tissues, successive changes in concentration and cellular distribution of the enzyme can occur without influences external to the eye.

Primary induction of CA II expression at the earliest stage of eye organogenesis is probably critical for subsequent stages and may be subject to controlled perturbations. If eye development is completely blocked by surgically separating neural and skin ectoderm prior to the formation of stable adhesions between these cell layers, no apparent eye development occurs and CA II expression is prevented (unpublished observations). Hence, the interactions between these two ectodermal layers which are known to mediate eye induction may also indúce CA II expression. One such possibly inductive interaction involves epithelial–epithelial interactions through an interposed extracellular matrix.[2] Recently, a cell membrane glycoprotein complex has been identified which mediates the adhesion of the cell surface with certain of the macromolecules in the extracellular matrix. The cell membrane molecule has been named integrin,[2] and probes such as antibodies and synthetic peptides are available which can be used to disrupt cell–matrix interactions. Monoclonal antibodies to integrin were microinjected into the space between the neural ectoderm and the skin ectoderm at the two to five somite pair stage of chick embryo development. This precedes the time at which these two ectodermal layers become mutually adhesive through an extracellular matrix. After several subsequent days of development, the embryos were evaluated for morphological development of the injected and contralateral eye. Control embryos were injected with equal concentrations of additional monoclonal antibodies to integrin which do not interfere with the binding of this molecule to matrix material.[9]

Figure 3 shows a morphological analysis and computer-generated three-dimensional reconstruction of eyes from an experimental embryo. The left eye in the diagram (lower left side in Fig 3a) was injected with the CSAT anti-integrin antibody,[2,9] which blocks integrin–matrix binding. The right eye was not injected. The two eyes were markedly different. The injected eye failed to undergo much of the complex tissue remodeling that results in eye morphogenesis. Figs. 3b and 3c show serial reconstructions of control and injected eyes, respectively. The results of this experiment were typical of numerous such tests. Furthermore, such perturbation of eye morphogenesis was achieved only with probes that block integrin–matrix binding, whereas other antibodies showed no effects. Even though these experiments clearly disrupted the developmental forces that mediate eye morphogenesis, biochemical differentiation was not blocked. CA II expression was triggered in both the skin ectoderm and neural ectoderm even though the physical relationship between these two cell layers was drastically altered.

CA II expression is one of the earliest known markers of embryonic induction in the developing eye, but the actual inductive signal that triggers CA II expression remains unknown. Insight into this question may lead to an understanding of how the complicated pattern of CA II expression is regulated throughout retina development. The approaches to this question discussed briefly above are being continued and combined with molecular analyses of putative regulatory sequences for the CA II gene. Additional analyses of the physiological roles of CA II in retinal function at all stages of development in combination with the investigation of CA II expression

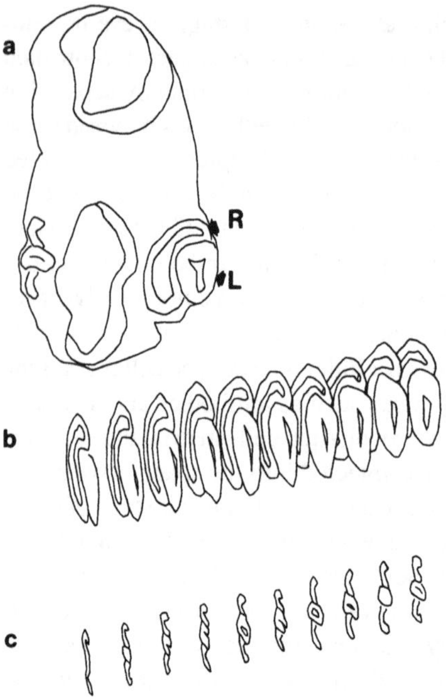

Figure 3. Computer-generated drawing showing the effects of the anti-integrin anibody $CSAT_2$ on morphogenesis of the eye. The embryo shown in cross section in panel a received CSAT antibody between the skin and neural ectoderm at approximately five somite pairs of development in one side (lower left as seen in panel a and serially reconstructed in panel c. The other side of the embryo received no injection. A normal eye developed on the uninjected side, with retina (R) and lens (L) evident (and serially reconstructed in panel b). Although the tissue foldings and interactions that normally occur in eye development were blocked, CA II expression occurred in the perturbed eye as in the control eye. This experiment is typical of many such analyses and seems to indicate that the inductive phenomena which control CA II expression are separable from those that control aspects of eye morphogenesis.

regulation will hopefully yield an integrated picture that can explain the high concentrations of this enzyme in the visual apparatus.

ACKNOWLEDGMENTS. This research was supported by grants from the March of Dimes Birth Defects Foundation (No. 1-1030), the National Science Foundation (No. BNS-84-17797), and the Lucille P. Markey Trust and the Mount Desert Island Biological Laboratory.

References

1. Broeders, G. C., Parmer, R., and Dawson, W. W., 1988, *Ophthalmologica* **196**(2):103–110.
2. Buck, C. A., 1987, *Annu. Rev. Cell. Biol.* **3**:179–205.
3. Cohen, J. L., and Linser, P. J., 1987, *Invest. Ophalmol. Visual Sci. Suppl.* **28**:404.
4. Coulombre, A. J., 1956, *J. Exp. Zool.* **133**:211–225.
5. Dowling, J. E., and Ripps, H., 1970, *J. Gen. Physiol.* **56**:491–520.
6. Flugel, C., Eichorn, M., Lutjen-Drecoll, E., Wiederholt, M., and Zadunaisky, J. A., 1988, *MDIBL Bull.* **27**:120–123.
7. Hensley, S. H., Linser, P. J., and Cohen, J. L., 1987, *Soc. Neurosci. Abst.* **13**(2):1049.
8. Lewis, S. E., Erickson, R. P., Barnett, L. B., Venta, P. J., and Tashian, R. E., 1988, *Proc. Natl. Acad. Sci. USA* **85**:1962–1966.

9. Linser, P. J., Edwards, S. C., Plunkett, J. A., and Strickland, E. R., 1988, *Invest. Ophthalmol. Visual Sci. Suppl.* **29**:377.
10. Linser, P., and Moscona, A. A., 1981, *Proc. Natl. Acad. Sci. USA* **78**:7190–7194.
11. Linser, P. J., and Moscona, A. A., 1984, in: *Gene Expression and Cell-Cell Interactions in the Developing Nervous System* (J. M. Lauder and P. G. Nelson, eds.), Plenum Publishing Corp., New York, pp. 185–202.
12. Linser, P. J., and Moscona, A. A., 1984, *Ann. N.Y. Acad. Sci.* **429**:431–446.
13. Linser, P. J., and Plunkett, J. A., 1989, *Invest. Ophthalmol. Visual Sci.* **30**:783–785.
14. Linser, P. J., Smith, K., and Angelides, K., 1985, *J. Comp. Neurol.* **237**:264–272.
15. Linser, P. J., Sorrentino, M., and Moscona, A. A., 1984, *Dev. Brain Res.* **13**:65–73.
16. Maren, T. H., 1976, *Invest. Opthalmol.* **15**:356–364.
17. Piccolino, M., Neyton, J., Witkovsky, P., and Gerschenfeld, H. M., 1982, *Proc. Natl. Acad. Sci. USA* **79**:3671–3675.
18. Rogers, J. H., and Hunt, S. P., 1987, *Neuroscience* **23**:343–361.
19. Scott, W. J., Lane, D., Randell, J. L., and Schreiner, C. M., 1984, *Ann. N.Y. Acad. Sci.* **429**: 447–455.
20. Trachtenberg, M. C., and Packey, D. J., 1984, *Curr. Eye Res.* **3**:599–604.
21. Vardimon, L., Fox, L. E., and Moscona, A. A., 1986, *Proc. Natl. Acad. Sci. USA* **83**:9060–9064.
22. Wen, R., and Oakley, B., II, 1987, *Sco. Neurosci. Abstr.* **13**(2):1049.
23. Wistrand, P. J., Schenholm, M., and Lönnerholm, G., 1986, *Invest. Ophthalmol. Visual Sci.* **27**:419–428.

CHAPTER 27

Carbonic Anhydrase and Sensory Function in the Central Nervous System

JUDITH A. NEUBAUER

The carbonic anhydrase (CA) (EC 4.2.1.1.) isozymes catalyze the reversible hydration of CO_2. CA activity is present in the central nervous system (CNS).[1,10] In brain tissue, the high-activity isozyme, presence of CA II, and the membrane-bound isozyme, CA IV, has been demonstrated.[4a,8,33]

The cellular distribution of CA isozymes within the CNS is principally restricted to glial cells, with the highest concentrations found in the oligodendrocytes, which have biosynthetic and ion homeostatic functions.[32] The question of whether CA activity is in astrocytes is controversial. Since astrocytes provide an important function in fluid and ion transport in the CNS, the presence of CA is likely. However, the majority of studies have been unable to detect the presence of CA in astrocytes, finding instead that the glial distribution of this enzyme is exclusively restricted to the oligodendrocytes.[7,9,17–19,22] However, there have been a number of reports of studies using immunocytochemical techniques in both brain and spinal cord sections[4,26,31] and primary cell cultures[5,13,15,16,37] which suggest that CAs are present in astrocytes but in much lower concentrations than in oligodendrocytes.

Neurons have also generally been considered to lack CA. However, there seems to be a growing list of exceptions to this widely held notion, including subpopulations of neurons in the retina, peripheral sensory ganglia, and the CNS. Evidence for neuronal CA within the neural retina comes from several sources, as reviewed elsewhere in this volume.[21]

In the peripheral nervous system, the neuronal localization of CA appears to

JUDITH A. NEUBAUER • Department of Medicine, Pulmonary Division, University of Medicine and Dentistry, Robert Wood Johnson Medical School, New Brunswick, New Jersey 08903.

be a general property of primary sensory neurons, whereas motor neurons lack CA II. CA II has now been localized in sensory neurons of the peripheral ganglia of both avian and mammalian species, including dorsal root ganglia, a small percentage (22%) of the trigeminal ganglia neurons which compose the cell bodies of the sensory infraorbital branch of the trigeminal nerve, 2% of the nodose ganglion neurons which distribute entirely to the recurrent laryngeal nerve, sciatic sensory nerves, and a subpopulation of neurons in the celiac ganglion presumably thought to be involved in the peripheral reflex arcs.[2,14,30,43,44] CA II in sensory neurons may be further restricted to medium to large sensory neurons[30,43]; however, some caution should be used in overgeneralizing this morphological distinction, since other cellular or extracellular criteria may influence the phenotypic expression of CA.[14,27] CA activity has not been detected in peripheral motor ganglia and nerves, as evidenced by its absence in ventral root motor nerves, sciatic motor nerves, facial motor nerve, the ciliary and otic parasympathetic motor ganglia, and the superior cervical sympathetic ganglia.[25,30,44]

The functional significance of CA activity in subpopulations of neurons is unknown. Some of the specialized functions proposed, without experimental support, to explain a need for neuronal CA include excessive neuronal production of CO_2, perhaps from a decarboxylation reaction required in synthesis of some neurotransmitter, the need for maintenance of high intracellular Cl^- levels, or better regulation of intracellular pH due either to a high glycolytic activity or an increased Ca^{2+} influx.[43]

One very specialized function for neuronal CA might be ascribed to an involvement in a neuronal ability to transduce CO_2–H^+ into a neuronal signal, i.e., CO_2 chemoreception. In this case, assuming that the signal is H^+–HCO_3^- and not molecular CO_2, the presence of intracellular CA would accelerate the hydration of CO_2 and the change in intracellular pH. Since central CO_2 increases the neural output of the central respiratory centers, it seems likely that CA would be important in the respiratory response to CO_2.

In the 1950s, Leusen[20] localized the ability of CO_2 to stimulate respiration in the absence of the peripheral chemoreceptors to the brain. Later, central chemosensitivity was more specifically localized to distinct bilateral areas of the ventrolateral medulla; a rostral area described earlier,[24] a caudal area,[35] and an intermediate area that functions as an integrating area.[34] With the exception of some evidence of a thicker marginal glial layer[42] and marked vascularity,[6] nothing anatomically distinctive about this region has ever been identified which might distinguish its chemoreceptive function. Electrophysiological recordings from neurons in these regions suggest that only 25% are chemosensitive and that these neurons are dispersed among other nonchemosensitive neurons.[28] Thus, there has been no definitive identification of the chemosensitive cells.

In addition, the site of chemosensitivity, i.e., the issue of whether a change in pH is sensed intracellularly or extracellularly, is still unresolved. If equivalent changes in the extracellular pH in the chemosensitive areas of the ventrolateral medulla are produced by either CO_2 inhalation or infusion of fixed acid, the respiratory response to the CO_2−induced pH change is significantly (two- to threefold) greater than the respiratory response to fixed acid-induced change in pH.[36,40] Since CO_2 readily diffuses across cell membranes whereas fixed acids do not, this suggests that a change in intracellular pH is likely crucial to the transduction process. The presence of an intracellular CA in the chemosensitive cells would greatly accelerate the pH change and hence the signal transduction process. This function of CA has already been demonstrated in the peripheral chemoreceptor response to CO_2.[3,41]

The importance of intracellular CA in the central respiratory response to CO_2 has been demonstrated by using CA inhibitors, benzolamide (which does not cross membranes), and acetazolamide (which does cross membranes) in anesthetized cats. The evidence that the ventrolateral medulla is richly vascularized might suggest that CA in the capillary endothelium is the important site of CO_2 hydration. To determine whether CA acts outside the blood–brain barrier at the luminal surface of the capillary endothelium or inside the blood–brain barrier, presumably at the chemosensitive sites, one group[12] tested the respiratory response to vertebral artery injections of CO_2-saturated saline before and after administration of intravenous acetazolamide or benzolamide. They found that the respiratory response to CO_2 was reduced only after acetazolamide treatment, indicating that (1) the central response to CO_2 is CA dependent and (2) the location of this CA is beyond the blood–brain barrier.

In a subsequent study,[11] an attempt was made to localize the physiologically important compartment more discretely by determining the effects of benzolamide or acetazolamide applied directly to the ventral surface of the medulla on the respiratory response to CO_2. The results from these studies suggested that although extracellular CA may contribute to the response, the major site of CO_2 hydration important in determining the respiratory response to CO_2 is intracellular. Additional physiological evidence in support of an intracellular location for CA in the respiratory response to CO_2 was found.[23] In this study, the effect of inhibition of CA with acetazolamide on the phrenic neurogram and ventral medullary extracellular pH responses to CO_2 forcing were assessed. It was hypothesized that if chemosensitive neurons were lacking CA, then CO_2 would diffuse to the chemosensitive neurons, and the dynamics of phrenic neurogram response to CO_2 would be unaltered but the changes in extracellular pH would be markedly delayed because of its dependence on CA. Contrary to this prediction, the dynamics of both the phrenic neurogram and extracellular pH were significantly delayed, strongly suggesting that intracellular CA is important in the CO_2 transduction function of central chemosensitive cells.

The first histochemical suggestion that CA may be present in CNS neurons was the work of Ridderstråle and Hanson.[29] Sections of the medulla, cerebellum, and parietal cortex were histochemically stained for the presence of CA. Although they found that most neuronal cell bodies and axons were unstained, many neurons and axons were stained, with great variation in the intensity of staining. Especially striking to these investigators was a small area close to the ventral surface of the medulla in which the neurophil was intensely stained for CA. The specific location of these stained neurons corresponded closely to the previously defined CO_2-chemosensitive areas. Previous studies using histochemical staining in tissue slices had failed to identify CA in CNS neurons. The difficulty, however, may be related the very high level of CA in oligodendrocytes, which may make it extremely difficult to detect cells with significantly lower concentrations, i.e., neurons and astrocytes.

In an effort to substantiate that neurons in the chemosensitive regions of the ventrolateral medulla contain intracellular CA, we have used immunocytochemical techniques to localize CA in neurons in dissociated cultures of central chemosensitive regions of the ventral medulla.[38] Dissociated cell cultures were used so that the sensitivity of the immunocytochemical technique could be enhanced, since culturing in this manner results in the growth of isolated neurons on a principally astrocytic substratum. Several regions in addition to the ventral medulla were cultured, including dorsal root ganglia (positive neuronal control), optic nerve oligodendrocytes (positive glial control), dorsal medulla, frontal cortex (sensorimotor), and hypothalamus (anterior preoptic region). Immunocytochemical staining for CA in dissociated cell cultures of the ventral medulla revealed that there were neurons in this region that stained intensely for CA. Although only 25% of the ventral medullary neurons stained for CA, if these neurons are the chemosensitive cells, this observation would be consistent with the percentage of chemosensitive neurons reported previously.[28] However, this region was not unique in containing neurons stained for CA. The dorsal medulla contained a smaller population (18%) of intensely stained neurons, the cortex contained a greater population (35%) of lightly stained neurons, and the anterior hypothalamus (59%) contained the greatest population of intensely stained neurons. This large degree of CA-positive neurons in the anterior hypothalamus may be consistent with the importance of CA in many different sensory functions, including thermosensitivity, osmosensitivity, and perhaps even CO_2 sensitivity, as suggested recently.[39]

In summary, intracellular CA II has been found in neurons as well as glia cells in both the central and peripheral nervous system. The preponderance of CA is present in oligodendrocytes, being found in all of these cells in very high concentrations. Since oligodendrocytes are the source of myelin, the primary importance of CA in these cells is perhaps its function in supplying the HCO_3^- needed for fatty acid synthesis. A much lower concentration of CA has been found in astrocytes and has been proposed to be important for the ion transport function of these cells. Finally, intracellular CA is found in retinal neurons, in sensory but not motor neurons in ganglia of the peripheral nervous system, and in a small population of CNS neurons, where it may also be related to specialized sensory functions.

References

1. Ashby, W., 1944, *J. Biol. Chem.* **155**:671–679.
2. Barakat, I., Kazimierczak, J., and Droz, B., 1986, *Cell Tissue Res.* **245**:497–505.

3. Black, A. M. S., McCloskey, D. I., and Torrance, R. W., 1971, *Respir. Physiol.* **13**:36–49.
4. Cammer, W., and Tansey, F. A., 1988, *J. Neurochem.* **50**:319–322.
4a. Carter, N. D., Fryer, A., Grant, A. G., Hume, R., Strange, R. G., and Wistrand, P. J., 1990, *Biochim. Biophys. Acta* **1026**:113–116.
5. Church, G. A., Kimelberg, H. K., and Sapirstein, V. S., 1980, *J. Neurochem.* **34**:873–879.
6. Cragg, P., Patterson, L., and Purves, M. J., 1977, *J. Physiol.* (London) **272**:137–166.
7. Delaunoy, J. P., Hog, F., Devilliers, G., Bansart, M., Mandel, P., and Sensenbrenner, M., 1980, *Cell Mol. Biol.* **26**:235–240.
8. Filippi, D., Sciaby, M., Limozin, N., and Laurent, G., 1978, *Biochimie* **60**:99–102.
9. Ghandour, M. S., Langley, O. K., Vincendon, G., and Gombos, G., 1979 *J. Histochem. Cytochem.* **27**:1634–1637.
10. Giacobini, E., 1962, *J. Neurochem.* **9**:169–177.
11. Hanson, M. A., Holman, R. B., and McCooke, H. B., 1983, in: *The Peripheral Arterial Chemoreceptors* (D. J. Pallot, ed.), Croan Helm, London, pp. 409–414.
12. Hanson, M. A., Nye, P. C. G., and Torrance, R. W., 1981, *J. Physiol.* (London) **320**:113–125.
13. Hertz, L., and Sapirstein, V., 1978, *In Vitro* **14**:376.
14. Kazimierczak, J., Sommer, E. W., Philippe, E., and Droz, B., 1986, *Cell Tissue Res.* **245**:487–495.
15. Kimelberg, H. K., Narumi, S., and Bourke, R. S., 1978, *Brain Res.* **153**:55–77.
16. Kimelberg, H. K., Steig, P. E., and Mazurkiewicz, J. E., 1982, *J. Neurochem.* **39**:734–742.
17. Kumpulainen, T., and Korhonen, L. K., 1982, *J. Histochem. Cytochem.* **30**:283–292.
18. Kumpulainen, T., and Nystrom, S. H. M., 1981, *Brain Res.* **220**:220–226.
19. Langley, O. K., Ghandour, M. S., Vincendon, G., and Gombos, G., 1980, *Histochem. J.* **12**:473–483.
20. Leusen, I., 1954, *Am. J. Physiol.* **176**:39–44.
21. Linser, P. J., and Cohen, J. S., this volume, Chapter 26.
22. McCarthy, J. D., and De Vellis, J., 1980, *J. Cell Biol.* **85**:89–902.
23. Mishra, J., Neubauer, J. A., Li, J. K. J., and Edelman, N. H., 1985, *Fed. Proc.* **44**:1583.
24. Mitchell, R. A., Loeschcke, H. H., Massion, W. H., and Severinghaus, J. W., 1963, *J. Appl. Physiol.* **18**:523–533.
25. Oswald, T., and Riley, D. A., 1987, *Brain Res.* **406**:379–384.
26. Parthe, V., 1981, *J. Neurosci. Res.* **6**:119–131.
27. Philippe, E., Omlin, F. X., and Droz, B., 1985, *J. Neurochem. Suppl.* **44**:43.
28. Pokorski, M., 1976, *Am. J. Physiol.* **230**:1288–1295.
29. Ridderstråle, Y., and Hanson, M., 1985, *Acta Physiol. Scand.* **124**:557–564.
30. Riley, D. A., Ellis, S., and Bain, J. L. W., 1984, *Neuroscience* **13**:189–206.
31. Roussel, G., Delaunoy, J. P., Nussbaum, J. L., and Mandel, P., 1979, *Brain Res.* **160**:47–55.
32. Sapirstein, V. S., Strocchi, P., and Gilbert, J. M., 1984, *Ann. N.Y. Acad. Sci.* **429**:481–493.
33. Sapirstein, V. S., Strocchi, P., Wesolowski, M., and Gilbert, J. M., 1983, *J. Neurochem.* **40**:1251–1261.
34. Schlaefke, M. E., and Loeschcke, H. H., 1967, *Pfluegers Arch. Ges. Physiol.* **297**:201–220.
35. Schlaefke, M. E., See, W. R., and Loeschcke, H. H., 1970, *Respir. Physiol.* **10**:198–212.
36. Shams, H., 1985, *J. Appl. Physiol.* **58**:357–364.
37. Snyder, D. S., Zimmerman, T. R., Jr., Farooq, M., Norton, W. T., and Cammer, W., 1983, *J. Neurochem.* **40**:120–127.
38. Sterbenz, G. C., Neubauer, J. A., Geller, H. M., and Edelman, N. H., 1988, *FASEB J.* **2**:1295.
39. Tamaki, Y., Nakayama, T., and Matsumura, K., 1986, *Pfluegers Arch.* **407**:8–13.
40. Teppema, L. J., Barts, P. W. J. A., Folgering, H. Th., and Evers, J. A. M., 1983, *Respir. Physiol.* **53**:379–395.
41. Travis, D. M., 1971, *J. Pharmacol. Exp. Ther.* **178**:529–540.
42. Trouth, C. O., Odek-Ogunde, M., and Holloway, J. A., 1982, *Brain Res.* **246**:35–45.
43. Wong, V., Barrett, C. P., Donati, E. J., Eng, L. F., and Guth, L., 1983, *J. Histochem. Cytochem.* **31**:293–300.
44. Wong, V., Barrett, C. P., Donati, E. J., and Guth, L., 1987, *J. Comp. Neurol.* **257**:122–129.

CHAPTER 28

Carbonic Anhydrase in Myelin and Glial Cells in the Mammalian Central Nervous System

WENDY CAMMER

During the past 10 years, immunocytochemical methods have been used for localization of carbonic anhydrase (CA) in glial cells and myelin in the mammalian central nervous system. Those immunocytochemical studies were, however, preceded by several decades of biochemical research, and the possible localization in the oligodendrocytes, which comprise one glial cell type, was first suggested over 45 years ago, after CA was assayed in homogenates of spinal cords and cerebra from seven species.[1] At that time, large numbers of oligodendrocytes had been observed predominantly in myelinated regions, where they were often found in rows between myelinated axons. Later, the oligodendrocytes were shown to be the cells that myelinate axons in the central nervous system (reviewed in reference 50). The astrocytes, which are the other major glial cells of brain and spinal cord, are quite diverse in their structures and putative functions, which include the transport of ions, water, and other compounds and the formation of scar tissue within the brain after injury.[23,33,40] In this chapter, the evidence for CA in myelin and in glial cells of both types will be discussed. Possible functions for the CA in myelin and glial cells will also be suggested. In another chapter, there is some discussion of localization in the peripheral nervous system and exceptions to the localization in glial cells in the mammalian central nervous system.[36] The high levels of CA in the choroid plexus[14] should also be noted.

WENDY CAMMER • Department of Neurology, Albert Einstein College of Medicine, Bronx, New York 10461.

1. Biochemical Studies of CA in Glial Cells and Myelin

The earliest reports of CA activity in glial cells were based on the results of dissecting neurons and glial cell clumps from rat brainstem. The glial cells had sixfold the CA activity per cell as did the neurons, and it was suggested that the oligodendrocytes might be the loci of CA.[18,19] Similarly, in an early histochemical study, CA activity was shown to be highest in areas of mouse brain that were rich in myelinated fibers and glial cells.[26] In three subsequent studies, where glia and neurons were separated by bulk isolation, the glia were confirmed to have the higher CA activities[24,35,45]; however, the glial cell types were not distinguished.

After methods were devised for the purification of myelin and each type of glial cell from rat brains,[13,37,47] the definitive biochemical localization of central nervous system enzymes became more feasible. Significant CA activity was observed in myelin isolated from the brains of rat, mouse, monkey, cat, and rabbit,[6,7] and the finding was confirmed.[43,51] When CA was assayed in bulk-isolated cells, the highest specific activities were observed in oligodendrocytes, and the specific activities in neurons and astrocytes were low but measurable.[46] Significant CA activities were also observed in cultured astrocytes.[25,44]

2. Immunocytochemical Localization in Glial Cells and Myelin

When sections from rodent and human brains and spinal cords were immunostained with antisera against red blood cell CA II, the most striking positive reactions were consistently found in oligodendrocytes.[4,8,15,16,28,29,41] However, for several reasons, intense CA immunoreactivity in myelinated fibers was observed in only two of the studies.[4,8] First, unless the tissue was delipidated and embedded in paraffin, the immunocytochemical reagents were unable to permeate the myelinated fibers. Second, in some of the studies in which paraffin sections were immunostained, the tissue was used after fixation with Carnoy's solution, which was later shown to remove myelin proteins.[8]

The technique used originally to distinguish whether there was CA in astrocytes, as well as oligodendrocytes, involved double immunofluorescence staining with antibodies against CA and the astrocytic marker, glial fibrillary acidic protein (GFAP).[15] Because most of the GFAP-positive ("fibrous") astrocytes in the normal brain are situated in the white matter, it was natural to focus on the white matter in those studies, and it was indeed possible to show that CA and GFAP were in different cells. However, CA-positive cells that resembled "protoplasmic" astrocytes had been observed in gray matter of brain, as distinguished from white matter.[22,41] Double immunofluorescence techniques were therefore used to show that in the gray matter there was indeed GFAP in the CA-positive cells that resembled protoplasmic astrocytes.[9,10] The finding of CA in astrocytes remains controversial and was recently attributed, for example, to possible contaminants in the antisera to CA II.[17] However, the anti-CA antibodies used to immunostain astrocytes had actually consisted of one antiserum and a monoclonal ascites fluid,[9] and it is unlikely that contamination could account for the staining with each of these. More recently, CA has been observed in reactive astrocytes of severely gliotic white matter in the jimpy mutant mouse and in rats with experimental autoimmune encephalomyelitis[10,11] as well as in astrocytes in normal gray matter of brain. However, in normal spinal cords and normal white matter of brain, CA has been detected only in oligodendrocytes. It should be noted that in the avian central nervous system CA is found in astrocytes and neurons.[34]

3. Postnatal Development of CA in Rodent Brains

In the rat forebrain, the specific activities of CA increase steadily between ages 15 and 60 days.[2,6,27,28] During this time, myelin accumulates, the rate of aerobic metabolism increases, and the CA specific activity in oligodendrocytes increases.[2,37,42,46] Sapirstein[42] has shown that the membrane-bound CA, much of which is in myelin, as well as the cytoplasmic CA in brain increase during development and that the CA specific activities reach their adult levels earliest in the more caudal regions of brain than in the more rostral.

Immunocytochemical studies have localized CA in rodent brains as early as embryonic day 16. During the last embryonic week and first 2 postnatal weeks, CA has been detected in radial-appearing glial cells[3] and subsequently in large, rounded glial cells, which are followed by cells with extensive processes.[30,31] Cells of these various types, which also can be immunostained with antibodies against markers for immature glial cells, probably are precursors of oligodendrocytes or astrocytes or common precursors of both[3,30,31] (our unpublished data).

4. CA in Glial Cell Cultures

In cultures prepared from brain under various experimental conditions, CA has been detected in both oligodendrocytes and astrocytes. When cultures were prepared from neonatal rat brains and brain extracts used to stimulate the growth of oligodendrocytes, the highest CA activities were observed between 13 and 23 days in culture.[38] In cultures from mouse brains, CA was expressed progressively in oligodendrocytes at 6–20 days.[20] These results are similar in timing to the appearance of 2′3′-cyclic nucleotide-3′-phosphohydrolase,[39] which is found in oligodendrocytes and myelin *in vivo*. Most recently, DeVitry *et al.*[12] detected CA II gene transcripts in oligodendrocytes in mouse brain cultures, beginning about 5 days earlier than CA II protein.

In primary cultures of astrocytes from neonatal rodent brains, the CA specific activities, after ~21 days in culture and treatment with dibutyryl cyclic AMP (dBcAMP), were similar to those in brain homogenates from 3-week-old rats.[44]

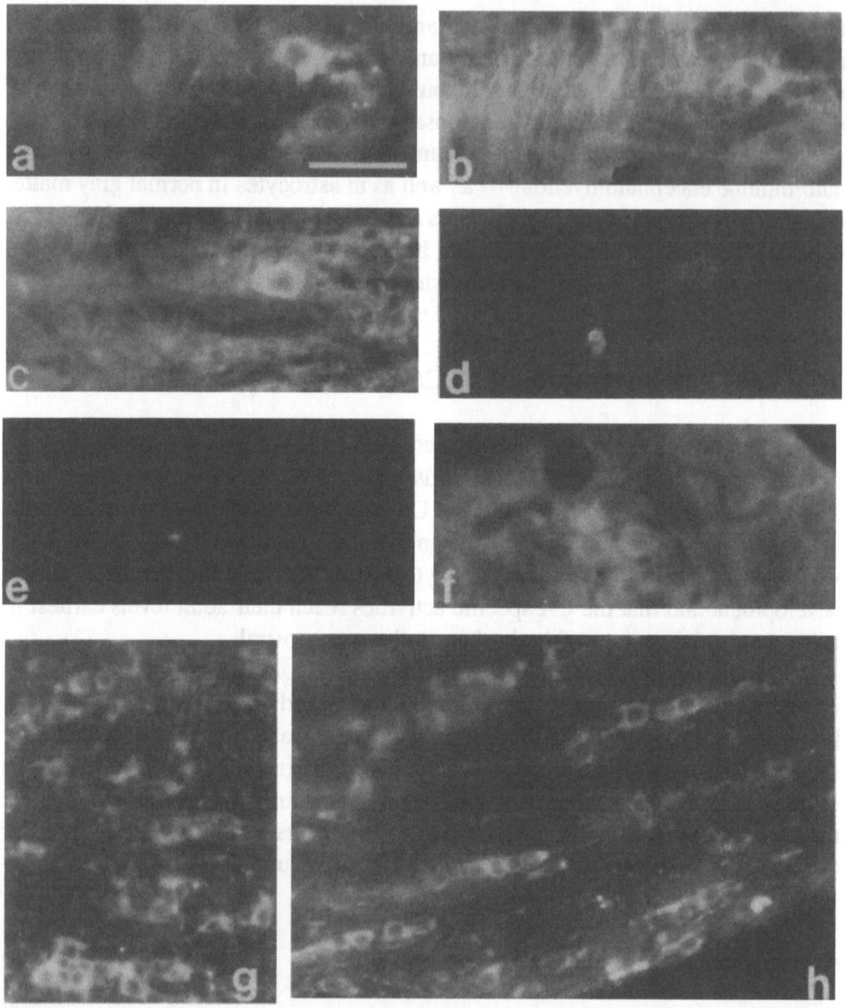

Figure 1. Immunofluorescence staining of CA and acetyl-CoA carboxylase (ACC) in brainstems of 3-week-old rats. Tissue was fixed, and vibratome sections were prepared and immunostained as described previously.[8,9] (a) Rabbit anti-ACC (1:2000), rhodamine; (b) same field as in panel a, mouse anti-CA monoclonal antibody (1:100), fluorescein; (c) rabbit anti-ACC, rhodamine; (d), same field in as panel c, no primary antibody, fluorescein; (e) normal rabbit serum, rhodamine; (f), same field as in panel e, mouse anti-CA monoclonal antibody, fluorescein; (g and h) rabbit anti-ACC with rhodamine, one-half the magnification used for panels a–f. Distance marker represents 25 μm.

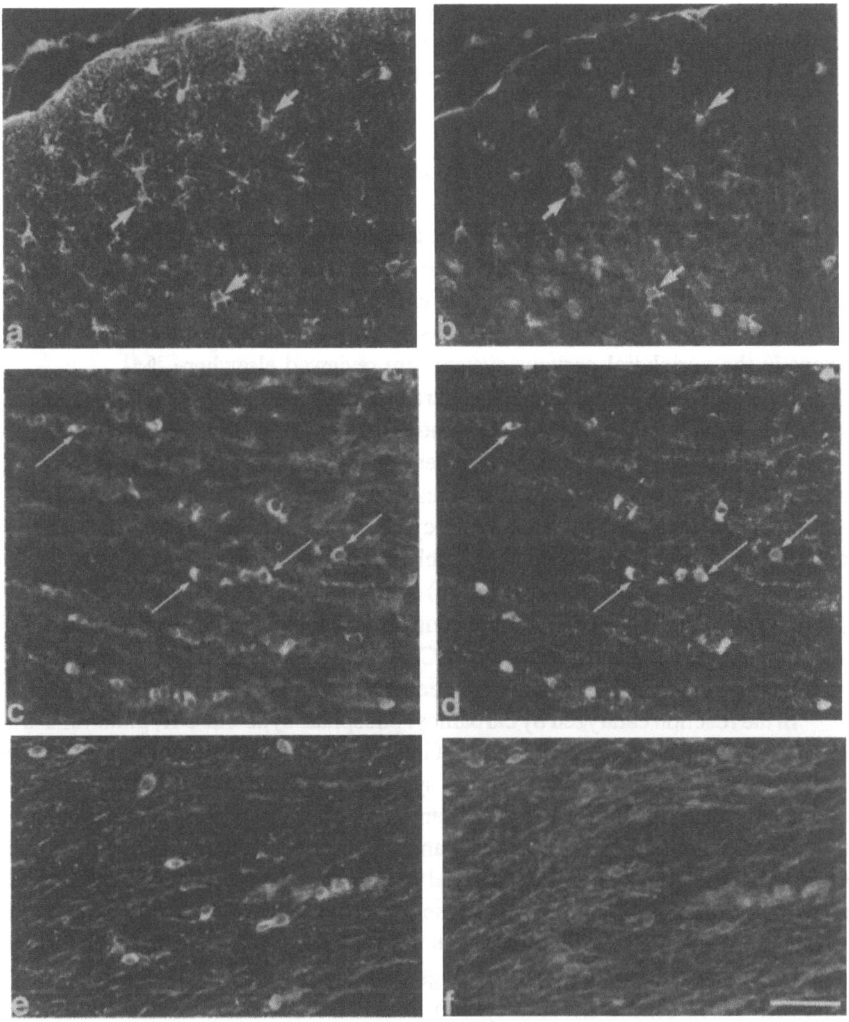

Figure 2. Double immunofluorescence staining of CA and glutamine synthetase (GS) in astrocytes in gray matter and oligodendrocytes in white matter in the mouse central nervous system. Tissue was fixed, and paraffin sections were prepared and immunostained as described previously.[8,9] (a and b) Cerebral cortex of 3-week-old mouse; (a) rabbit anti-GS (1:150), rhodamine; b, same field, mouse anti-CA (1:100), fluorescein. (c and d) White matter in spinal cord from adult mouse; c, rabbit anti-GS, rhodamine; d, same field, mouse anti-CA, fluorescein. (e and f) Control, same spinal cord as shown in panels c and d; e, rabbit anti-GS, rhodamine; f, same field, no monoclonal antibody, fluorescein isothiocyanate. Additional controls, for panels a and b, are shown in panels c and d; e, rabbit anti-GS, rhodamine; f, same field, no monoclonal antibody, fluorescein isothiocyanate. Additional controls, for panels a and b, are shown in reference 10. Thick arrows point to some examples of astrocytes in gray matter that contain both GS and CA; thin arrows point to some examples of oligodendrocytes in white matter that contain both GS and CA. The distance marker represents 50 μm.

Kimelberg and co-workers[25] have demonstrated positive immunocytochemical staining of CA in dBcAMP-treated, cultured astrocytes. Thus, the findings *in vitro* and *in vivo* consistently show CA in both oligodendrocytes and astrocytes.

5. Hypothetical Functions of CA in Glial Cells and Myelin

CA activity can contribute to ion transport and can provide bicarbonate for biochemical reactions, and more than one function can be served in the same cells. The possible roles of CA in ion transport across myelin sheaths or through astrocyte plasma membranes in the central nervous system, and in neuronal cell bodies and axons in the peripheral nervous system, are reviewed elsewhere.[36,42]

It has been suggested that CA generates bicarbonate to serve as substrate for acetyl-CoA carboxylase and for carbamoyl phosphate synthetase II.[21] These enzymes participate in fatty acid biosynthesis and pyrimidine biosynthesis, respectively. Consistent with the need to synthesize myelin lipids, acetyl-CoA carboxylase has been localized in oligodendrocytes,[48,49] as has biotin,[32] which is the cofactor for this enzyme. By using double immunofluorescence, it is possible to stain acetyl-CoA carboxylase (Fig. 1a) and CA (Fig. 1b) in the same oligodendrocytes. Panels c–f of Fig. 1 are controls for the double immunofluorescence staining, and panels g and h show acetyl-CoA carboxylase in typical interfascicular oligodendrocytes lined up in rows between myelinated axons in the white matter.

In the reaction catalyzed by carbamoyl phosphate synthetase II, glutamine is a substrate, along with bicarbonate. It is of interest that CA and glutamine synthetase occur together in the rat central nervous system in astrocytes in gray matter and in oligodendrocytes in white matter.[5,9] Figure 2, for example, shows double immunofluorescence staining of CA (panel b) and glutamine synthetase (panel a) in the same astrocytes (thick arrows) in cortical gray matter of mouse brain. Panels c and d show CA (d) and glutamine synthetase (c) in the same oligiodendrocytes (thin arrows) in white matter of mouse spinal cord. Panels e and f are controls for the double immunofluorescence. The findings suggest important biosynthetic roles for glial-cell CA in central nervous system function.

ACKNOWLEDGMENTS. I thank Drs. P. J. Linser and K. G. Thampy for antibodies used to obtain the data shown in Figs. 1 and 2, Dr. F. A. Tansey for performing those experiments, and Ms. R. Sasso for typing the manuscript. The experiments shown in Figs. 1 and 2 were supported by Public Health Service grants NS-12890 and NS-23705.

References

1. Ashby, W., 1944, *J. Biol. Chem.* **152**:235–240.
2. Ashby, W., and Schuster, E. M., 1950, *J. Biol. Chem.* **184**:109–116.

3. Benjelloun, S., Delaunoy, J. P., Gomes, D., DeVitry, F., Langui, D., and Dupouey, P., 1986, *Dev. Neurosci.* **8**:150–159.
4. Cammer, W., 1984, *Ann. N.Y. Acad. Sci.* **429**:494–497.
5. Cammer, W., 1990, *J. Neuroimmunol.* **26**:173–178.
6. Cammer, W., Bieler, L., Fredman, T., and Norton, W. T., 1977, *Brain Res.* **138**:17–28.
7. Cammer, W., Fredman, T., Rose, A. L., and Norton, W. T., 1976, *J. Neurochem.* **27**:165–171.
8. Cammer, W., Sacchi, R., and Sapirstein, V., 1985, *J. Histochem. Cytochem.* **33**:45–54.
9. Cammer, W., and Tansey, F. A., 1988, *J. Neurochem.* **50**:319–322.
10. Cammer, W., and Tansey, F. A., 1988, *J. Comp. Neurol.* **275**:65–75.
11. Cammer, W., Tansey, F. A., and Brosnan, C. F., 1989, Glia **2**:223–230.
12. DiVitry, F., Gomes, D., Rataboul, P., Dumas, S., Hillion, J., Catelon, J., Delaunoy, J. P., Tixier-Vidal, A., and Dupouey, P., 1989, *J. Neurosci. Res.* **22**:120–129.
13. Farooq, M., and Norton, W. T., 1978, *J. Neurochem.* **31**:887–894.
14. Fisher, R. G., and Copenhaver, J. H., 1959, *J. Neurosurg.* **16**:167–176.
15. Ghandour, M. S., Langley, O. K., Vincendon, G., and Gombos, G., 1979, *J. Histochem. Cytochem.* **27**:1634–1637.
16. Ghandour, M. S., Langley, O. K., Vincendon, G., and Gombos, G., 1980, *Neuroscience* **5**:559–571.
17. Ghandour, M. S., Skoff, R. P., Venta, P. J., and Tashian, R. E., 1989, *J. Neurosci. Res.* **23**: 180–190.
18. Giacobini, E., 1961, *Science* **134**:1524–1525.
19. Giacobini, E., 1982, *J. Neurochem.* **9**:169–177.
20. Griot, C., and Vandervelde, M., 1988, *J. Neuroimmunol.* **18**:333–340.
21. Gros, G., and Dodgson, S. J., 1988, *Annu. Rev. Physiol.* **50**:669–694.
22. Kahn, S., Tansey, F. A., and Cammer, W., 1986, *J. Neurochem.* **47**:1061–1065.
23. Kimelberg, H. K., and Ransom, B. R., 1986, in: *Astrocytes—Cell Biology and Pathology of Astrocytes*, Volume 3 (S. Fedoroff and A. Vernadakis, eds.), Academic Press, New York, pp. 129–166.
24. Kimelberg, H. K., Braddlecome, S., Narumi, S., and Bourke, R. S., 1978, *Brain Res.* **141**: 305–323.
25. Kimelberg, H. K., Stieg, P. E., and Mazurkiewicz, J. E., 1982, *J. Neurochem.* **39**:734–742.
26. Korhonen, L. K., Naatanen, E., and Hyyppa, M., 1964, *Acta Histochem.* **18**:336–347.
27. Koul, O., and Kanungo, M. S., 1975, *Exp. Gerontol.* **10**:273–278.
28. Kumpulainen, T., and Korhonen, L. K., 1982, *J. Histochem. Cytochem.* **30**:283–292.
29. Langley, O. K., Ghandour, M. S., Vincendon, G., and Gombos, G., 1980, *Histochem. J.* **12**: 473–483.
30. LeVine, S. M., and Goldman, J. E., 1988, *J. Neurosci.* **8**:3992–4006.
31. LeVine, S. M., and Goldman, J. E., 1988, *J. Comp. Neurol.* **277**:441–455.
32. Levine, S. M., and Macklin, W. B., 1988, *Brain Res.* **444**:199–203.
33. Lindsay, R. M., 1986, in: *Astrocytes—Cell Biology and Pathology of Astrocytes*, Volume 3 (S. Fedoroff and A. Vernadakis, eds.), Academic Press, New York, pp. 231–262.
34. Linser, P. J., 1985, *J. Neurosci.* **5**:2388–2396.
35. Nagata, Y., Mikoshiba, K., and Tsukada, Y., 1974, *J. Neurochem.* **22**:493–503.
36. Neubauer, J., this volume, Chapter 27.
37. Norton, W. T., and Poduslo, S. E., 1973, *J. Neurochem.* **21**:749–757.
38. Pettman, B., Delaunoy, J. P., Couraget, J., Devilliers, G., and Sensenbrenner, M., 1980, *Dev. Biol.* **75**:278–287.
39. Pfeiffer, S. E., 1984, in: *Oligodendroglia* (W. T. Norton, ed.), Plenum Press, New York, pp. 233–298.
40. Privat, A., and Rataboul, P., 1986, in: *Astrocytes—Development, Morphology, and Regional Specialization of Astrocytes*, Volume 1 (S. Fedoroff and A. Vernadakis, eds.), Academic Press, New York, pp. 105–129.
41. Roussel, G., Delaunoy, J. P., Nussbaum, J. L., and Mandel, P., 1979, *Brain Res.* **160**:47–55.

42. Sapirstein, V., 1983, in: *Handbook of Neurochemistry*, Volume 4 (A. Lajtha, ed.), Plenum Press, New York, pp. 385–402.
43. Sapirstein, V. S., and Lees, M. B., 1978, *J. Neurochem*. **31**:505–511.
44. Schousboe, A., Nissen, C., Bock, E., Sapirstein, V. S., Juurlink, B. H. J., and Hertz, L., 1980, in: *Tissue Culture in Neurobiology* (E. Giacobini, A. Vernadakis, and A. Shahar, eds.), Raven Press, New York, pp. 397–409.
45. Sinha, A. K., and Rose, S. P. R., 1971, *Brain Res*. **33**:205–217.
46. Snyder, D. S., Zimmerman, T. R., Jr., Farooq, M., Norton, W. T., and Cammer, W., 1983, *J. Neurochem*. **40**:120–127.
47. Snyder, D. S., Raine, C. S., Farooq, M., and Norton, W. T., 1980, *J. Neurochem*. **34**:1614–1621.
48. Tansey, F. A., and Cammer, W., 1988, *Dev. Brain Res*. **43**:123–130.
49. Tansey, F. A., Thampy, K. G., and Cammer, W., 1988, *Dev. Brain Res*. **43**:131–138.
50. Wood, P., and Bunge, R. P., 1984, in: *Oligodendroglia* (W. T. Norton, ed.), Plenum Press, New York, pp. 1–46.
51. Yandrasitz, J. R., Ernst, S. A., and Salganicoff, L., 1976, *J. Neurochem*. **27**:707–715.

CHAPTER 29

Lung Carbonic Anhydrase

SHOKO NIOKA and ROBERT E. FORSTER II

1. Carbonic Anhydrase in the Lung as a Whole

In 1946, there was considered to be no carbonic anhydrase (CA) in the lung except that in the blood.[6] Since 1952, CA has been reported to be present in lung tissue,[1,4,14,25,27,29] and it was noted[7] that alveolar CO_2 appeared to be rapidly equilibrated with lung tissue, implying the presence of CA.

A homogenate of blood-free lung tissue less major airways and vessels of the rat,[23,35] calf,[36] ox,[32] and rabbit[32] has CA activity. Average acceleration, k_{enz}/k_u, has been reported to be 122 at 37°C[23] and about 70 (recalculated).[35] More recently, we have used an ^{18}O-labeled CO_2 exchange method[16,30] to measure the average tissue CA activity in an isolated ventilated blood-free saline perfused guinea pig lung at 22°C, finding the catalyzed hydration velocity constant to be about 50 times the uncatalyzed, reasonable agreement considering the different species, techniques, and temperature.

Certainly CA is present in the lung, much of it in cytosol (Table I). Two isozymes on nondenaturing gel of rat lung homogenate at a molecular weight of 29,000 correspond to CA I and CA II in the red blood cells.[23] The histochemical methods have been of only limited help, largely because most studies have depended on the Hansson method,[21] which does not have the resolution needed. Functional methods that involve CO_2–HCO_3^- reactions or transport have provided most of the information. However, interpretation of these results depends on assumptions regarding the cellular or subcellular localization of CO_2, of HCO_3^-, and, most important, of assorted CA inhibitors, which must be borne in mind. Ethoxzolamide is assumed to penetrate all membranes rapidly, whereas benzol-

SHOKO NIOKA • Department of Biochemistry and Biophysics, University of Pennsylvania School of Medicine, Philadelphia, Pennsylvania 19104-6085. ***ROBERT E. FORSTER II*** • Department of Physiology, University of Pennsylvania School of Medicine, Philadelphia, Pennsylvania 19104-6085.

TABLE I
Lung CA[a]

Location	Isozyme	Comments[b]
Whole lung		
Parenchyma or cytosol	CA I and II	CA activity in homogenate supernatant accelerates k_u 50–120 times at 37°C.[16,23,30,32,35]
Cells		
Epithelium	?CA II	Disagreement as to presence of CA (yes[19,35]; no[28]). Mitochondria and lamellar bodies had no CA, but nuclei and microvilli did.[35]
Endothelium	?CA I and II	Hansson histochemical technique shows CA in amphibia, calf, human, monkeys, rat, reptiles.[12,28,32,35] No CA by dansylamide staining method.[19]
	?CA IV	On surface of endothelium.[32,36]
Subcellular particles		
Mitochondria	CA II	Activity in mitochondrial pellet from whole lung.[23]

[a]$k_{enz} = d[CO_2]/(dt[CO_2]) - k_u$. This is an operational definition; in units, k_u is the reaction velocity constant for hydration of CO_2 in seconds^{-1}.
[b]The references are not intended to be exhaustive but only to establish the presence or absence of CA.

amide, with an equal inhibition equilibrium constant, K_i, is relatively impermeable. Acetazolamide's properties lie in between. HCO_3^- permeates membranes some 4 orders of magnitude less rapidly than CO_2, and its charge must be compensated by anion exchange or cation pairing.

2. Postcapillary Disequilibrium Studies

In the mid 1970s, it was determined experimentally, following theoretical predictions,[15,31] that after CO_2 exchanges with a red cell suspension, the necessary pH readjustment of extracellular fluid is slow, with a half-time of seconds, because of the lack of CA, and that the same should be true of pulmonary capillary plasma.[15,24,31,34] This stimulated experiments on isolated, ventilated, and perfused blood-free lungs in which CO_2 was exchanged. These conditions were necessary to eliminate the dominant effect of the red blood cell enzyme. The pH of the postcapillary, pulmonary venous perfusate was found to vary with time, or distance, from the capillaries, except when a CA inhibitor was added to the perfusate.[2,5,17,22] The conclusion was that there is normally a CA available to the pulmonary capillary contents which produces chemical equilibrium within the blood plasma. When it is inhibited, the plasma pH continues to readjust after the fluid leaves the capillary bed.[18]

A CA inhibitor of molecular weight 100,000, dextran prontosil (the active group is sulfanilamide), which cannot leave the vascular bed, was observed to

TABLE II
Effects of CA Inhibition on Lung CO_2 Excretion and Postcapillary pH Readjustment[a]

Location of lung CA	CO_2 excretion (%)	Postcapillary pH readjustment	Inhibitor	[Inhibitor] in multiples of K_i[b]
None (control)	100	0	None	0
Capillary lumen	102	0.015	DP 100,000	1000
Capillary lumen plus interstitial space	98 + 2	0.03 + 0.004	DP 5000	1000
Whole lung: intravascular, interstitial, and intracellular	84 + 1	0.026 + 0.0002	Prontosil	2000
	87 + 1	0.29 + 0.005	Acetazolamide	1400

[a] CO_2 excretion is normalized to the control value. DP is dextran prontosil.
[b] The concentrations of inhibitors were adjusted to produce approximately the same inhibition in terms of the inhibition equilibrium constant, K_i.

produce postcapillary disequilibrium (Table II).[17] This finding suggests the presence of CA on the surface of the endothelium, in agreement with an immunohistochemical study which has localized CA on the surface of cultured calf pulmonary arterial endothelium.[32] To make matters more complicated, dextran prontosil of molecular weight 5000, which can permeate the capillary wall but still cannot permeate cell membranes, caused an amplification in the pH disequilibrium. A possible explanation is that there is additional CA in the intercellular space available through interendothelial passage.

3. Net CO_2 Exchange and Multiple Indicator Dilution Studies

Isolated, ventilated blood-free rabbit lungs were perfused at 37°C with CA-free electrolyte solution. CO_2 excretion was decreased about 80% on adding 50× the inhibition equilibrium constant, K_i, of acetazolamide to the perfusate.[26] The drug should have inhibited intracellular as well as endothelial CA, and the effects cannot be distinguished.

In another series of experiments on a similar preparation, two boluses of fluid were injected simultaneously into the pulmonary artery. One volume contained ^{13}C-labeled bicarbonate that was (1) acid and therefore largely CO_2 or (2) alkaline and therefore HCO_3^-, or (3) equilibrated at pH 7.4. The second bolus was a neutralizing solution chosen to titrate the mixed injectate to pH 7.4. The excretion of labeled CO_2 was approximately the same whether CO_2 or HCO_3^- was injected.[8] After addition of a large concentration of acetazolamide to the perfusate, the CO_2 bolus produced increased CO_2 excretion, but the HCO_3^- and equilibrated bolus produced less. It was concluded that there was sufficient CA available to the capillary contents to produce equilibrium.

In the reverse of the above experiments, 0.5 ml of ^{14}C-labeled HCO_3^- solution was introduced into the distal airways of the same preparation, and the concentration of ^{14}C in the venous outflow was determined.[10] Adding acetazolamide into the airways caused no change in the findings, but adding it to the perfusate reduced ^{14}C transport. Adding CA to the perfusate had no effect, but adding it to the airways increased ^{14}C in the outflowing blood. It was considered that labeled CO_2 was formed in the airways at an uncatalyzed rate, except when CA was added and diffused into capillary bed, where CA caused it to hydrate rapidly.

In a similar preparation ventilated with 5% CO_2, an injected bolus of alkaline buffer resulted in a transient fall in end expiratory CO_2, which was decreased by adding sulfonamide inhibitor to the perfusate and increased by including CA in the bolus. Again, endothelial as well as intracellular CA were affected, and their functions cannot be separated.[20]

The uptake of radioactive CO_2 from alveolar gas was measured in a similar isolated dog lung lobe preparation in which the lung intracellular CA was inhibited but the intracapillary CA was active.[11] The perfusate contained phosphate buffer at pH 7.7 and enough spinach CA to accelerate the hydration of CO_2 at 22°C 100-fold. When acetazolamide was added to the perfusate, which should have inhibited all intracellular CA but only 15% of the intravascular spinach CA because it is relatively insensitive, the CO_2 uptake, expressed as a diffusing capacity of the lung for CO_2, fell 40%. This must have been caused by the fall in intracellular CA activity alone.

In a mirror image of the above experiment, more than $1000 \times K_i$ of prontosil or dextran prontosil was added to the perfusate to inhibit endothelial or endothelial plus extracellular (interstitial) CA, respectively, without affecting intracellular CA, but neither altered CO_2 exchange (Table II).[22] In contrast, acetazolamide and prontosil, which are relatively permeable and presumably inhibited intracellular as well as extracellular CA, reduced CO_2 exchange without reaching a saturation plateau even when concentration of inhibitor was raised to over $5000 \times K_i$. Multiple indicator dilution experiments on the same preparation[9] using radioactively labeled acetazolamide plus tracers distributed in lung water and tracers limited to the extracellular space demonstrated that there were binding sites for the drug accessible to the vascular contents.

4. Histological Studies

There is general agreement that the lung capillary endothelium contains CA as stained by the Hansson technique in amphibia and reptiles,[12] in rats,[28,35] and in monkeys and humans.[28] CA in cultured bovine endothelial cells was demonstrated immunologically with electron microscopic stain,[32] whereas none was apparent with dansylamide, a fluorescent CA inhibitor.[19]

There is less agreement about CA in pulmonary epithelium. Lönnerholm found none,[28] whereas Sugai *et al.* found it in both type I and type II cells.[35] By dansylamide staining, CA was found in some epithelial cells.[19] Information about CA in subcellular particles is very sketchy. Small vesicles in the cytosol of endothelial cells reportedly contain CA.[32] Nuclei, cytosol, and the cores of microvilli of type II cells have high concentrations of CA.[35]

5. Separation of Cells and Subcellular Particles

The supernatant after low-speed centrifugation of a homogenate of blood-free rat lung in mannitol–sucrose contains 67% of the total lung CA, as measured at 4°C using a glass electrode in stirred vessel.[16] The remainder is bound tightly to the particulate components which sediment. When the pellet is resuspended in a solution of 0.3 M Tris buffer at pH 9.0 and washed three times, all but 8% of the total CA goes into solution. This last fraction is probably membrane bound CA on the endothelial surface. The average catalyzed hydration velocity constant (k_{enz}) of the homogenate corrected to 37°C was 100 sec^{-1}.[23] There is a membrane-bound CA of molecular weight 52,000 in rat, rabbit, ox, and human lung, which is stable in SDS–gel with activity comparable to that of CA II but slightly less sensitive to sulfonamides.[36] The increased molecular weight results from a carbohydrate moiety possibly involved in the attachment of the enzyme to the membrane. This CA represents a very small fraction of total lung CA and presumably corresponds to the 8% bound to large particles.[23] Membrane CA (CA IV) has in general these characteristics.[3,32,33] CA activity is associated with the microsomal, membrane fraction of a homogenate of cultured endothelial cells from bovine pulmonary artery.[32]

It is difficult from the studies described above to determine which of the many cell types in the lung contain the CA and its subcellular localization. However, cultured pulmonary artery endothelium contains CA, and subcellular particles were labeled by immunofluorescence.[32] Isolated epithelial type II cells (unpublished data of A. Chander, L. Lin, and S. J. Dodgson) contain CA.

Mitochondrial and microsomal fractions of rat lung homogenate contained 8 and 7%, respectively, of lung CA activity.[23] However, the lack of staining of CA in mitochondria of type II cells[35] suggests that the enzyme is present in some mitochondria and microsomes.

6. Physiological Functions of Lung Carbonic Anhydrases

Although several authors have suggested that the capillary-available CA increases CO_2 output from the blood, the amount of CO_2 produced in the plasma during the readjustment of postcapillary disequilibrium should be small compared with the CO_2 produced in the red cell by normal HCO_3^- exchange with plasma.

Postcapillary pH disequilibrium and CO_2 exchange appear to be independent functions insofar as their dependence on CA (Table II). Inhibition of intravascular CA produces a postcapillary pH disequilibrium but does not decrease CO_2 exchange.[22] Klocke[26] found a CA acceleration of fivefold in pulmonary capillary contents, which works out to be only several percent of the total lung CA.

Enns and Hill[11] demonstrated an acetazolamide-inhibitable fraction, 33%, of CO_2 transport through a 2-mm-thick layer of compressed dog lung, presumed to be facilitation by the mechanism diagrammed in Fig. 1. However, the flux of CO_2 compared with CO, chosen as a chemically inert control in the absence of hemoglobin, was considerably less than expected on the basis of physical solubility alone. Facilitation has been invoked[11,17] to explain the increased alveolar–capillary CO_2 exchange produced by intracellular CA in the lung. At the same time, facilitation should be proportionally less in the alveolar membrane because it is so thin.[17] CA inhibitors did not reduce the diffusion of CO_2 out of the lung through

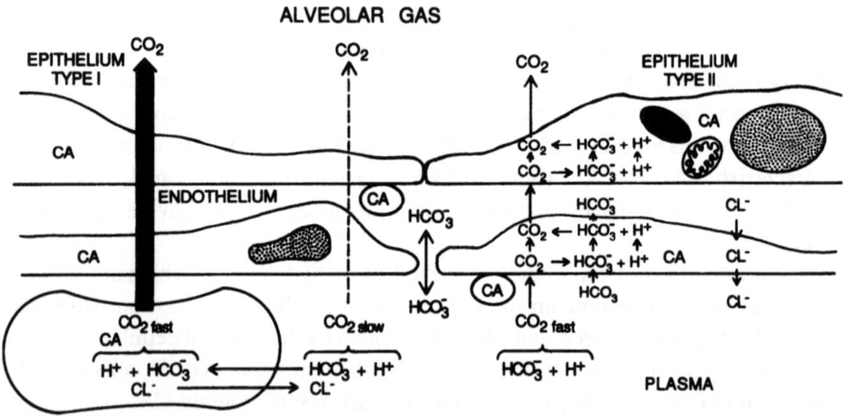

Figure 1. The thick solid arrow indicates the major flux of CO_2 from capillary blood to alveolar gas, i.e., that formed in the red blood cells by the CA accelerated dehydration of HCO_3^-, largely that which has entered from the plasma in exchange for chloride ion. The dashed arrow indicates the minimal flux of CO_2 formed by the uncatalyzed dehydration of plasma HCO_3^-; the serial arrows represent that formed under the influence of CA on the surface of the endothelium. Facilitated flux of CO_2 is diagrammed for the latter path in both endothelium and epithelium and applies equally well to the two other streams of $CO_2 \cdot HCO_3^-$ can exchange for chloride across the endothelium but not across the epithelium because there is no return path for chloride as there is for the endothelium (although it is not completed in the diagram for the latter). The CA in the interstitial space is that available to 5000-dalton dextran prontosil but not to 100,000-dalton prontosil, which implies that the interendothelial passage is permeable to the former but not the latter. This CA contributes to CO_2 formation from HCO_3^- that is available to it from the plasma. Although the CA available to the plasma and the interstitial fluid is important in producing equilibration of H^+, HCO_3^-, and CO_2 in plasma, it is doubtful that it contributes to CO_2 exchange.

the pleura,[13] although it had to pass through alveolar septa and any effect of facilitation should have been observed.

HCO_3^- can diffuse or exchange into lung tissue cells, albeit slowly, which would provide access to intracellular CA and produce CO_2 which would add to that produced in the red cells. However, this would require a steady flux of counter ions to balance the charge movement. In the studies that measured labeled HCO_3^- movement, self-exchange of the anion and production of labeled CO_2 intracellularly would have been much more rapid because there is no charge exchange. However, this does not necessarily mean that the intracellular CA augments net CO_2 exchange.

Since the effective solubility for CO_2 of the various perfusates was much less than that of blood, the arteriovenous pCO_2 difference for the same CO_2 exchange is much greater than *in vivo*. This increases the fraction of CO_2 output provided by dissolved CO_2 in the perfusate and might exaggerate the importance of intracellular CA, which is available to CO_2, in gas exchange in the *in vitro* lung preparations.

The most obvious origin of lung CA is synthesis in the lung cells. However, the finding that there are two isozymes of CA in rat lung which correspond to the two in red blood cells[23] raises the question of whether some lung CA derives from red cells.

7. Conclusions

CA is available to the intravascular fluid in the lung, is probably on the endothelial wall, and produces chemical equilibrium in the plasma, which otherwise would take many seconds to achieve. This CA is unimportant in CO_2 exchange in the lungs. A part of this CA is in the interstitial space, again on the cell membrane, or on the endothelium but behind a diffusion barrier to molecules larger than about 5000 daltons.

Most of the lung CA is intracellular, but the distribution among cell types is not clear. Certainly endothelial cells contain CA; the amount in epithelial cells is less certain, although type II cells contains some. Subcellular particles separated from homogenates of the lung contain CA, but the types of particles and the cells of their origin are unclear.

The function of the large amount of intracellular CA is also unclear. It buffers swings in alveolar pCO_2 during breathing, but the importance of this is not obvious. The cyclic variations are only some 10–15%. Inhibition of intracellular CA reduces CO_2 exchange by an undetermined mechanism(s), possibly because of a reduction in facilitated transport of CO_2 in the cytosol. However, the alveolar–capillary membrane is normally very thin and dissolved CO_2 is very permeable, so facilitation would not increase CO_2 transport much. The *in vitro* experiments using blood-free perfusate exaggerate the arteriovenous pCO_2 difference and thus the relative importance of dissolved CO_2 in gas transport in the lung.

References

1. Berfenstam, R., 1952, *Acta Paediatr.* **41**:310–315.
2. Bidani, A., and Crandall, E. D., 1978, *J. App. Physiol.* **45**:565–573.
3. Bruns, W., Dermietzel, R., and Gros, G., 1986, *J. Physiol.* **371**:351–364.
4. Chinard, F. P., Enns, T., and Nolan, M. F., 1960, *Am. J. Physiol.* **198**:78–88.
5. Crandall, E. D., and O'Brasky, J. E., 1978, *J. Clin. Invest.* **62**:618–622.
6. Davenport, H. W., 1946, *Physiol. Rev.* **26**:560–573.
7. DuBois, A. B., 1968, in: *CO_2: Chemical, Biochemical and Physiological Aspects.* (R. E. Forster, J. T. Edsall, A. B. Otis, and F. J. W. Roughton, eds.), National Aeronautics and Space Administration, Washington, D. C., pp. 257–259.
8. Effros, R. M., Chang, R. S. Y., and Silverman, P., 1978, *Science* **199**:427–429.
9. Effros, R. M., Shapiro, L., and Silverman, P., 1980, *J. Appl. Physiol.* **49**:589–600.
10. Effros, R. M., Mason, G., and Silverman, P., 1981, *Physiology* **51**:190–193.
11. Enns, T., and Hill, E. P., 1983, *J. Appl. Physiol.* **54**:483–490.
12. Fain, W., and Rosen, S., 1973, *Histochem. J.* **5**:519–528.
13. Farhi, L. E., Plewes, J. L., and Olszowka, A. J., 1976, *Ciba Found. Symp.* **38**:235–249.
14. Fisher, D. A., 1961, *Proc. Soc. Exp. Biol. Med.* **107**:359–363.
15. Forster, R. E., and Crandall, E. D., 1975, *J. Appl. Physiol.* **38**:710–718.
16. Forster, R. E., II, this volume, Chapter 6.
17. Geers, C., Heming, T. A., Gros, G., Bidani, A., and Crandall, E. D., 1986, *Prog. Respir. Res.* **21**:26–29.
18. Gros., G., Forster, R. E., and Dodgson, S. J., 1988, in: *pH Homeostatis: Mechanisms and Control* (D. Häussinger, ed.), Academic Press, New York, pp. 203–231.
19. Gros., G., Geers, C., and Dermietzel, R., 1988, *Funktionsanal. Biol. Syst.* **18**:45–52.
20. Hanson, M. A., Nye, P. C. G., and Torrance, R. W., 1981, *J. Physiol.* **319**:93–109.
21. Hansson, H. P. J., 1967, *Histochemie* **11**:112–128.
22. Heming, T. A., Geers, C., Gros, G., Bidani, A., and Crandall, E. D., 1986, *J. Appl. Physiol.* **61**:1849–1856.
23. Henry, R. P., Dodgson, S. J., Forster, R. E., and Storey, B. T., 1986, *J. Appl. Physiol.* **60**:638–645.
24. Hill, E. P., Power, G. G., and Gilbert, R. D., 1977, *J. Appl. Physiol.* **42**:928–934.
25. Hodgen, G. D., and Falk, R. J., 1971, *Int. J. Appl. Radiat. Isot.* **22**:492–495.
26. Klocke, R. A., 1978, *J. Appl. Physiol.* **44**:882–888.
27. Longmuir, I. S., Forster, R. E., Woo, C., 1966, *Nature* (London) **209**:393–394.
28. Lonnerholm, G., 1982, *J. Appl. Physiol.* **52**:352–356.
29. Maren, T. H., 1967, *Physiol. Rev.* **47**:594–781.
30. Nioka, S., Henry, R. P., and Forster, R. E., 1988, *J. Appl. Physiol.* **65**:2236–2244.
31. Roughton, F. J. W., 1935, *Physiol. Rev.* **15**:241–296.
32. Ryan, U. S., Whitney, P. L., and Ryan, J. W., 1982, *J. Appl. Physiol.* **53**:914–919.
33. Sanyal, G., and Maren, T. H., 1981, *J. Biol. Chem.* **256**:608–612.
34. Sirs, J. A., 1958, *Trans. Faraday. Soc.* **54**:207–212.
35. Sugai, N., Ninomiya, Y., and Oosaki, T., 1981, *Histochemistry* **72**:415–424.
36. Whitney, P. L., and Briggle, T. V., 1982, *J. Biol. Chem.* **257**:12056–12059.

CHAPTER 30

Carbonic Anhydrase and Chemoreception in Carotid and Aortic Bodies

SUKHAMAY LAHIRI

1. Carbonic Anhydrase and Peripheral Chemoreceptors

Intracellular pH in eukaryotic cells is maintained more acid than extracellular pH (6.8 vs. 7.4), but it is more alkaline than predicted by electrochemical gradient at equilibrium (reviewed in reference 14). The intracellular pH is therefore maintained by some active process. What critical role the carbonic anhydrases (CAs) play in the maintenance of steady-state cellular cytosolic pH in the glomus cells of carotid and aortic bodies is unclear. It is clear, however, that in dynamic physiological states, where speed of reaction matters, CAs are important.

Several identified mechanisms are used for cellular pH regulation: Na^+/H^+ antiport; Cl^-/HCO_3^- antiport; Na^+/HCO_3^- symport; and H^+-translocating ATPase. The CAs could influence these mechanisms, particularly in specialized cells. Peripheral chemoreceptors, carotid and aortic bodies, are specialized tissues. They continuously monitor CO_2 and O_2 partial pressures in the arterial blood and send appropriate neural signals to the central nervous system for the reflex control particularly of respiration, autonomic nervous system, and blood circulation. Without these signals, the reflexes would not originate and survival of the organism would be in jeopardy; integrated control system for CO_2 and O_2 transport would be in chaos without the strategic presence and function of CAs in the cells.

SUKHAMAY LAHIRI • Department of Physiology, University of Pennsylvania School of Medicine, Philadelphia, Pennsylvania 19104-6085.

1.1. Localization of Carbonic Anhydrase

Both CA and the arterial peripheral chemoreceptors were discovered in the third decade of this century (for reviews, see references 3 and 18), but the presence of CA and its physiological role in the arterial chemoreceptors were not studied until three decades later. The enzyme has been localized in the glomus cells of the cat carotid body[12,17] and in the isolated glomus cells of the rat carotid body (C. A. Nurse, personal communication). Since CA distribution in the cells varies,[4,20] identification of its localization in the glomus cells would be useful. The isozyme character of CA in the carotid body is not known.

1.2. Effects of Carbonic Anhydrase Inhibitors

Intravascular administration of a CA inhibitor (methazolamide or acetazolamide) is promptly followed by a decrease in the carotid and aortic chemoreceptor activities[5,7] (S. Lahiri, unpublished observations). Assuming that these inhibitors actually inhibited CA, only a part of this immediate effect is certainly due to decreases in arterial pCO_2 and $[H^+]$. Inhibitors of CA decrease alveolar pCO_2 partly because HCO_3^- is slow to give rise to CO_2 in the lungs and partly because the animal begins to hyperventilate as a result of the rise of pCO_2 in the brain and central chemoreceptor tissue. It seems that the chemosensory activity remains low even if alveolar pCO_2 is artificially maintained, raising arterial pCO_2 at the carotid body.[5,7] The problem is that after CA inhibition, the reversible CA hydration reaction is never complete in the flowing blood; the arterial blood gas values measured *in vitro* overestimate pCO_2 and $[H^+]$ and are not useful. It is possible that $[H^+]$ at the intracellular receptor site is actually low after CA inhibition.

Unlike central chemoreceptors in the brainstem after CA inhibition, metabolic CO_2 of the tiny carotid body (less than 1 mg) may not raise its tissue pCO_2 because of the high tissue blood flow (reviewed in reference 3).

1.3. Carbonic Anhydrase and Chemosensory Response to CO_2–H^+

Carotid chemoreceptors are exquisitely CO_2 sensitive, the gain of which is augmented by hypoxia.[6] With each breath, these blood gases reciprocally oscillate; pCO_2 rises during expiration and falls with inspiration. Blockade of carotid body CA blunts these signal transductions.[7,15,21] The explanation is that the uncatalyzed reaction, $CO_2 + H_2O \rightleftharpoons H_2CO_3 \rightleftharpoons H^+ + HCO_3^-$, is slow, and the H^+ stimulus does not rise sufficiently; hence, the pH-sensitive receptors are not stimulated as rapidly. The steady-state response to CO_2 is also diminished in vivo.[5,7] More recently, Iturriaga and Lahiri (unpublished observations) confirmed the steady-state results in vitro.

The chemoreceptor afferents from the aortic bodies are less responsive to CO_2 for unknown reason.[9] However, blockade of aortic body CA also attenuates its response to CO_2.

Unlike the case for CO_2, extracellular application of H^+ does not generate a comparable chemoreceptor response, presumably because the H^+ cannot enter the cells as rapidly as CO_2 generating intracellular H^+ and the receptors are essentially intracellular.

1.4. Carbonic Anhydrase and CO_2–O_2 Stimulus Interaction

Hypoxia augments the chemosensory response to CO_2 both in transient[10] and steady[6] states and in stimulus threshold.[8] These stimulus interactions are reminiscent of the cooperative phenomenon: at a lower CO_2 stimulus, a greater hypoxic stimulus is needed to elicit a given response. However, the hypoxic response can be blocked by appropriate doses of metabolic inhibitors[16] and by chronic hyperoxia.[11] Under these conditions, the response to hypercapnia is augmented, mimicking the effect of hypoxia. Inhibition of CA by acetazolamide attenuates the response to the onset of hypercapnia.[2] Because the effect of hypoxia is dependent on the response to CO_2,[6] a decrease in the CO_2 response can lead to a decrease in the hypoxic response. However, there are other explanations. For example, after administration of acetazolamide, neuromuscular transmission was blocked in a nerve–muscle preparation of the frog.[19] The mechanism of this effect is unclear, but a similar effect may occur in the chemosensory transduction process. Whether CA directly affects the chemosensory response to hypoxia has not been settled. For example, the response to cyanide is often not attenuated after CA inhibition by methoxzolamide (unpublished observations).

1.5. Carbonic Anhydrase and H^+ Pump

Because CA is present in the glomus cells, one may speculate that CA helps in regulating cellular [H^+]. Obviously, H^+ will be generated at a greater rate in the compartment where CA is present, such as glomus cells. The meaning of this greater rate of H^+ increase may be that the stimulus elicits reflex hyperventilation, which would reduce the CO_2 level. A decreased cellular pH might decrease Ca^{2+}-dependent K^+ conductance[13,14] and increase the chemosensory discharge. It has recently been reported that hypoxia decreases the conductance of some K^+ channel, which is seen as the pO_2 sensor.[13] Whether the same channel is affected by CO_2 has not been reported. It is unlikely, however, that a K^+ channel which is common to many membrane phenomena could be the key to a specific stimulus response.

2. Physiological Significance

Regardless of the mechanism of effects, a functional significance of cellular CA in the carotid and aortic bodies is that the chemoreceptor and chemoreflex responses would lose the speed that is compatible with the normal respiratory and circulatory oscillations. The physiological reflexes, respiratory and cardiovascular, through the autonomic nervous system will be delayed and perhaps attenuated. Also, the chemoreceptor input to the central nervous system will arrive at an inappropriate time of the rhythmic sensitivity cycle in the brainstem. Consequently, the responses will not be fully useful to the organism for the feedback control for which the strategic placement of the CA was originally designed.

ACKNOWLEDGMENT. This research was supported in part by grants HL-19737 and NS-21068.

References

1. Black, A. M. S., and Torrance, R. W., 1971, *Respir. Physiol.* **13**:221–237.
2. Erhan, B., Mulligan, E., and Lahiri, S., 1981, *Neuroscience Lett.* **24**:143–147.
3. Fitzgerald, R. S., and Lahiri, S., 1986, in: *Handbook of Physiology; The Respiratory System*, Volume II (A. P. Fishman, ed.), American Physiological Society, Bethesda, Md., pp. 313–362.
4. Gros, G., and Dodgson, S. J., 1988, *Annu. Rev. Physiol.* **50**:669–694.
5. Hays, M. W., Maini, B. K., and Torrance, R. W., 1976, in: *Morphology and Mechanisms of Chemoreceptors* (A. S. Paintal, ed.), University of Delhi, Delhi, India, pp. 36–45.
6. Lahiri, S., and DeLaney, R. G., 1975, *Respir. Physiol.* **24**:249–266.
7. Lahiri, S., DeLaney, R. G., and Fishman, A. P., 1976, *Physiologist* **19**:261.
8. Lahiri, S., Mokashi, A., DeLaney, R. G., and Fishman, A. P., 1978, *Respir. Physiol.* **34**:359–375.
9. Lahiri, S., Mulligan, E., Nishino, T., and Mokashi, A., 1979, *J. Appl. Physiol.* **47**:858–866.
10. Lahiri, S., Mulligan, E., and Mokashi, A., 1982, *Brain Res.* **234**:137–147.
11. Lahiri, S., Mulligan, E., Andronikou, S., Shirahata, M., and Mokashi, A., 1987, *J. Appl. Physiol.* **62**:1924–1931.
12. Lee, K. D., and Mattenheimer, H., 1964, *Enzymol. Biol. Clin.* **4**:199–216.
13. Lopez-Barneo, J., Lopez-Lopez, J. R., Urena T., and Gonzalez, C., 1988, *Science* **241**:580–582.
14. Madshus, H., 1988, *Biochem. J.* **250**:1–8.
15. McCloskey, D. I., 1968, in: *Arterial Chemoreceptors* (R. W. Torrance, ed.), Blackwell, Oxford, pp. 279–295.
16. Mulligan, E., S. Lahiri, and B. Storey, 1981, *J. Appl. Physiol.* **51**:438–446.
17. Ridderstråle, Y., and Hanson, M. A., 1984, *Ann. N.Y. Acad. Sci.* **429**:398–400.
18. Roughton, F. J. W., 1954, in: *Respiratory Physiology in Aviation* (W. M. Boothbey, ed.), Air University, Randolph Field, Texas, pp. 51–102.
19. Scheid, P., and Siffert, W., 1985, *J. Physiol.* (London) **361**:91–101.
20. Tashian, R. E., 1989, *BioEssays* **10**:186–192.
21. Torrance, R. W., 1968, in: *Arterial Chemoreceptors* (R. W. Torrance, ed.), Blackwell, Oxford, pp. 1–40.

CHAPTER 31

Carbonic Anhydrases in the Kidney

SUSANNA J. DODGSON

1. Carbonic Anhydrases of the Nephron

There have been several recent excellent reviews in which localization of the carbonic anhydrase (CA) (carbonate hydro-lyase; EC 4.2.1.1.) isozymes throughout the mammalian nephron have been considered. These include histochemical studies by Ridderstråle's technique,[32] which have mapped the CA activity throughout the nephron,[21,22] immunohistochemical studies,[19] and functional studies in which the membrane-bound CA IV is considered in regulation of acid–base balance and H^+ secretion.[13] Several laboratories have reported the absence of CA activity in renal papilla; however, recently it has been reported that inner medullary collecting duct cells from rat papilla in culture have CA activity equivalent per milligram of protein to the activity in erythrocytes.[2] The CA inhibitors and their uses in the renal clinic and laboratory have been reviewed recently.[30]

The reader is particularly directed to work from the laboratory of Dr. Per Wistrand. His lucid understanding of the function of CA in the body led to his postulation, and later discovery, of the existence of a CA in the ciliary processes of the eye.[40] His reasoning that there must be a CA in the renal tubular cytosol as well as in the membrane to assist with HCO_3^- reabsorption led also to the discovery of CA isozymes in several renal cell compartments (reviewed in reference 42).

CA activity in the mammalian kidney was first described in 1941.[7] In the past 15 years, this activity has been found to be due to isozymes II, IV, and V. The isozymes are not distributed evenly throughout the nephron, and there is great variation between species.[33] CA activity has been found in the intercalated cells of the collecting duct of the cortex and medulla but not of the papilla.[3] The CAs of the developing human fetal kidney were studied by Ridderstråle's technique[20]; stage I

SUSANNA J. DODGSON • Department of Physiology, University of Pennsylvania School of Medicine, Philadelphia, Pennsylvania 19104-6085.

nephrons (renal vesicles) had little CA activity, which contrasted with the activity found in the cortical collecting tubules at the luminal cell surface. Stage II nephrons (S-shaped tubules) had CA activity, whereas there was none in the glomerulus. The authors concluded that there may be functions for CA isozymes in the developing kidney which have disappeared in the adult.

1.1. CA II

The high-activity isozyme CA II is present in the cytoplasm of renal tubular cells.[41,45] This isozyme has also been characterized in the human renal medulla.[37]

It was proposed in 1945 that a cytosolic CA would function by facilitating the secretion of H^+ from the proximal tubule cells.[29] It has been clearly established that the proximal tubule is responsible for reabsorption of 80–85% of filtered bicarbonate (reviewed in reference 13). The connection between filtered bicarbonate and CA has been studied in several laboratories[4,5,23,24]; the connection between CA and Cl^- transport is not so obvious, and there is considerable discussion about whether the CA isozymes influence Cl^- transport in the proximal tubule (reviewed in reference 34).

In a study with rabbit renal cortex basolateral membrane vesicles, Na^+ uptake stimulated by an imposed HCO_3^- gradient was unaffected by acetazolamide.[35] When the vesicles were equilibrated in a bicarbonate–CO_2 buffer, Na^+ influx was stimulated 500% by an outward NH_4^+ gradient; this stimulation was 75% inhibited by 600 μM acetazolamide. With an inward HCO_3^- gradient, Na^+ influx was inhibited 70% by an inward NH_4^+ gradient but only 35% when 600 μM acetazolamide was also included. This is evidence that cytosolic CA II indirectly provides HCO_3^- for Na^+–HCO_3^- transport.

With microelectrodes, the intracellular pH in renal tubular epithelial cells was determined to be 7.10, with a mean membrane potential of −51.8 mV. The pH increased to 7.35 after acetazolamide treatment of the rats, although the membrane potential was unchanged.[17] This points to a function of CA II in maintaining a low intracellular pH.

1.2. CA IV

In 1957, Dr. Robert Berliner suggested that CA on the luminal outer side of the proximal tubule cells would catalyze the rate of dehydration of luminal H_2CO_3 formed from titration of filtered bicarbonate by secreted H^+.[1] Thus, a function for membrane-bound CA was proposed before it was known whether this enzyme existed. A significant disequilibrium pH in rat proximal tubules was found only when CA was inhibited, as was also the case in the distal tubules where CA was absent.[31] Later studies have concluded that this finding is functional evidence for the existence of membrane-bound CA IV.[13]

CA activity associated with the membrane fraction of homogenates of human kidney was first described in 1976.[25] A study in which CA activity of isolated brush border and plasma membranes from kidney cortex were quantitated came to the conclusion that the membrane-associated CA was CA II.[14] That this was a different isozyme was reported first in a study with bovine lung homogenates with partial characterization of the membrane-associated CA.[38] After the physiological importance of this membrane-bound CA was examined[43] and the enzyme was classified biochemically, it was designated CA IV.[42] CA IV is found in homogeneous preparations of microvilli and basal infoldings of rat renal tubular cells on the brush border and basolateral membranes.[43] The molecular weight of human CA IV is 34,400.[44] The instability of the isozyme had previously led to an greater estimate.[42]

CA IV is thought to be primarily responsible for bicarbonate resorption in the proximal tubule; it is these cells that have the highest concentration of this isozyme. An early report from Dr. Wistrand's laboratory indicated that when CA IV is reconstituted into lipid bilayer membranes, it forms channels selective for small ions.[8] This report was possibly premature in that a native, pure preparation to CA IV had not been achieved at that time.[44] There is now also evidence for a CA IV which is attached by a carbohydrate tail to the membrane.[28,46]

1.3. CA V

A study in 1959 gave some evidence of CA activity associated with the mitochondrial fraction of rat kidney[6]; this finding was tentatively affirmed by two independent groups,[18,27] yet in all three studies the data were equivocal, and the investigators all suggested that the CA activity might be due to contamination from other parts of the kidney or from erythrocytes.[16] Histochemical studies in which localization of CA isozymes was investigated by Ridderstråle's technique[32] failed to detect rat renal CA V.[22] Thus, a decade ago when the author first was given test tubes containing mitochondria prepared from several guinea pig organs, the question of whether there was CA activity in kidney mitochondria was unanswered. The observation that there was CA in liver and skeletal muscle mitochondria[12] but none in mitochondria prepared from decapsulated kidneys led this author to ignore the kidney for several years. The finding that CA inhibitors decrease the rates of production of intermediates for ureagenesis by guinea pig liver mitochondria led to the hypothesis that the mitochondrial CA, CA V provides the HCO_3^- substrate for the first enzyme of urea synthesis; subsequent experiments expanded this hypothesis to include the gluconeogenic enzyme, pyruvate carboxylase (reviewed in references 9 and 16). The rat kidney cortex also synthesizes glucose from pyruvate, leading to the puzzled query in this laboratory as to why the liver has CA V but the kidney does not. The riddle was solved simply: CA V is indeed present in rat kidney mitochondria,[11] and rat renal proximal tubules synthesize from pyruvate at a rate 5–10 times faster than do guinea pig proximal tubules, which lack CA V.

1.4. Total CA Activity in the Rat and Guinea Pig Kidney Proximal Tubule

CA activity of isolated, disrupted renal cortical proximal tubules prepared from 48-hr starved rats and guinea pigs has been measured with 25 mM $NaHCO_3$ (1% labeled with ^{18}O) at 37°C, pH 7.4, by the isotope ratio mass spectrometric CA assay[10] (see reference 15 for technique). In units of k_{enz} (milliliters per second per milligram of protein) total activity for rat was as follows: proximal tubules, 2.5 ± 0.8 ($n = 3$); mitochondria, 0.15 ± 0.4 ($n = 3$). Assuming that between 20 and 30% of total proximal tubule protein is mitochondrial, then it is calculated that CA V contributes approximately 2% of the total CA activity in the proximal tubule. These data contrast with the values of k_{enz} determined for guinea pig proximal tubule, 0.5 ± 0.2 ($n = 3$), and mitochondria, <0.01 ($n = 2$).

By the same technique, CA activities were compared of mitochondria prepared from rats that were either normally fed, starved for 48 hr, or fed with only dilute acid as drinking water.[11] CA activity of kidney mitochondria from the mildly acidotic rats doubled compared with activity of the starved rats.

2. Renal Proximal Tubular Mitochondrial Carbonic Anhydrase and Glucose Synthesis

It was reported in 1983 from another laboratory that the CA inhibitor acetazolamide decreased rat tubular glucose synthesis; the authors were unaware of the existence of CA V and concluded that this was an example of a nonspecific effect of acetazolamide on the proximal tubule.[36]

Recent studies from this laboratory have been aimed at examining the hypothesis that renal CA V is involved in glucose synthesis in proximal convoluted tubules when pyruvate is the sole added substrate. Figure 1 is a metabolic scheme delineating the possible fates of pyruvate. In the first study, experiments were designed to test the hypothesis that CA V is needed to provide the substrate HCO_3^- for pyruvate carboxylase. Rats were starved for 48 hr so that all endogenous glucose was removed from the proximal convoluted tubules and also because this stress maximizes the activity of the gluconeogenic enzymes. The lipophilic CA inhibitor ethoxzolamide was shown to readily inhibit CA V activity of intact mitochondria. It was then determined that glucose synthesis was a linear function of time in both the presence and absence of ethoxzolamide. When proximal convoluted tubules were incubated in Krebs–Henseleit buffer (which contains 25 mM $NaHCO_3$) and 10 mM pyruvate with 5% CO_2 and 95% O_2, glucose synthesis was decreased increasingly by increasing the concentrations of ethoxzolamide; 50% was decreased by 0.6 μM. It was not possible to reduce the rate of glucose synthesis to less than 30% with ethoxzolamide, indicating that some HCO_3^- can be provided for pyruvate carboxylase in the absence of any CA activity. Doubling the concen-

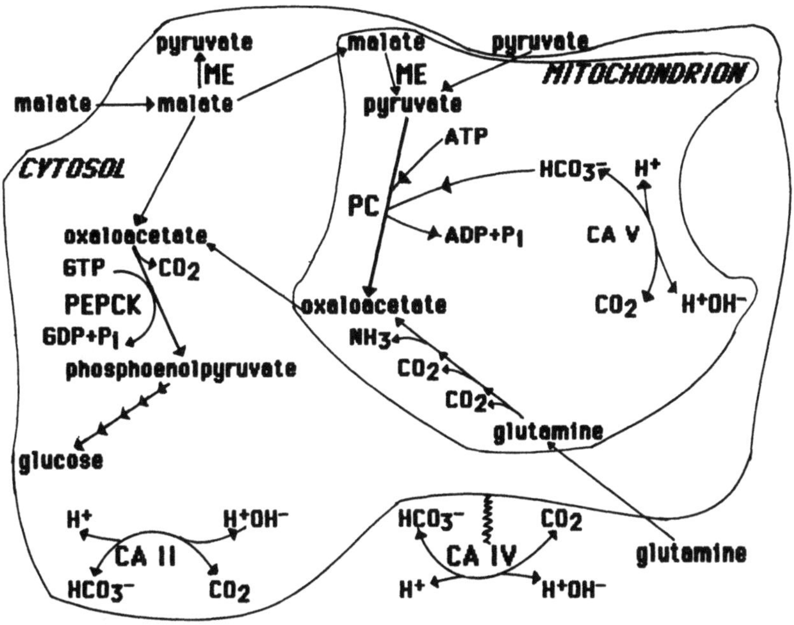

Figure 1. Pathway for glucose production by rat renal proximal tubules when pyruvate is the only added substrate. CA IV is located on both the brush border and the luminal basolateral membranes.

tration of $NaHCO_3$ and pCO_2 did not alter the rate of glucose synthesis; however, in these conditions, ethoxzolamide no longer inhibited glucose synthesis.[10]

In a subsequent study, acetazolamide (100 μM) and benzolamide (600 μM) mediated decrease of glucose synthesis could not be overridden by doubled pCO_2, leading to the conclusion that a nonmitochondrial CA participates in glucose synthesis.[47]

ACKNOWLEDGMENTS. The work done in this laboratory by the author, Miss Lisa C. Contino, and Mrs. Kay Cherian was funded by National Institutes of Health grant DK-38041. The mass spectrometer facility was funded by grants to Dr. Robert E. Forster II in HL PO19737.

References

1. Berliner, R. W., and Davidson, D. G., 1957, *J. Clin. Invest.* **36**:1416–1421.
2. Brion, L. R., Schwartz, J. H., Zavilowitz, B. J., and Schwartz, G. J., 1988, *Anal. Biochem.* **175**:289–297.
3. Brown, D., and Kumpulainen, T., 1985, *Histochem.* **83**:153–158.

4. Chou, S., Porush, J. G., Slater, P. A., Flombaum, C. D., Shafi, T., and Fein, P. A., 1977, *J. Clin. Invest.* **60**:162–170.
5. Clapp, J. R., Watson, J. F., and Berliner, R. W., 1977, *Am. J. Physiol.* **233**:F307–F314.
6. Datta, P. K., and Shepherd, T. H., 1959, *Arch. Biochem. Biophys.* **81**:124–129.
7. Davenport, H. W., and Wilelmi, A. E., 1941, *Proc. Soc. Exp. Biol. Med.* **48**:53–56.
8. Diaz, E., Sandblom, J. P., and Wistrand, P. J., 1982, *Acta Physiol. Scand.* **116**:461–463.
9. Dodgson, S. J., this volume, Chapter 25.
10. Dodgson, S. J., and Cherian, K., 1989, *Am. J. Physiol.* **257**:E791–E796.
11. Dodgson, S. J., and Contino, L. C., 1988, *Arch. Biochem. Biophys.* **260**:334–341.
12. Dodgson, S. J., Forster, R. E., II, Storey, B. T., and Mela, L., 1980, *Proc. Natl. Acad. Sci. USA* **77**:5562–5566.
13. Dubose, T. D., Jr., and Bidhani, A., 1988, *Annu. Rev. Physiol.* **50**:653–667.
14. Eveloff, J. E., Swenson, E. R., and Maren, T. H., 1979, *Biochem. Pharmacol.* **28**:1434–1437.
15. Forster, R. E., II, this volume, Chapter 6.
16. Gros, G., and Dodgson, S. J., 1988, *Annu. Rev. Physiol.* **50**:669–694.
17. Henderson, R. M., Bell, P. B., Cohen, R. D., Browning, C., and Iles, R. A., 1986, *Am. J. Physiol.* **250**:F203–209.
18. Karler, R., and Woodbury, D. M., 1960, *Biochem. J.* **75**:538–543.
19. Kumpulainen, T., 1984, *Ann. N.Y. Acad. Sci.* **429**:359–368.
20. Larsson, L., and Lonnerholm, G., 1985, *Biol. Neonate* **48**:168–171.
21. Lönnerholm, G., 1984, *Ann. N. Y. Acad. Sci.* **429**:369–381.
22. Lönnerholm, G., and Ridderstråle, Y., 1980, *Kidney Int.* **17**:162–174.
23. Lönnerholm, G., Wistrand, P. J., and Barany, E., 1986, *Acta Physiol. Scand.* **126**:51–60.
24. Lucci, M. S., Warnock, D. G., Rector, F. C., Jr., 1979, *Am. J. Physiol.* F58–F65.
25. McKinley, D. N., and Whitney, P. L., 1976, *Biochim. Biophys. Acta* **445**:780–790.
26. Maren, T. H., 1985, *N. Engl. J. Med.* **313**:179–181.
27. Maren, T. H., and Ellison, A. C., 1967, *Mol. Pharmacol.* **3**:503–508.
28. Murakami, H., Marelich, G. P., Grubs, J. H., Kyle, J. W. , and Sly, W. S., 1987, *Genomics* **1**: 159–166.
29. Pitts, R. F., and Alexander, R. S., 1945, *Am. J. Physiol.* **144**:239–254.
30. Preisig, P. A., Toto, R. D., and Alpern, R. J., 1987, *Renal Physiol.* **10**:136–159.
31. Rector, F. C., Jr., Seldin, D. W., Roberts, A. D., Jr., and Smith, J. S., 1960, *J. Clin. Invest.* **39**:1706.
32. Ridderstråle, Y., this volume, Chapter 10.
33. Ridderstråle, Y., Kashgarian, M., Koepper, B., Giebisch, G., Stetson, D., Ardito, T., and Stanton, B., 1988, *Kidney Int.* **34**:655–670.
34. Sasaki, S., and Marumi, F., 1989, *News Physiol. Sci.* **4**:18–22.
35. Soleimani, M., and Aronson, P. S., 1989, *J. Clin. Invest.* **83**:945–951.
36. Tannen, R. L., and Ross, B. L., 1983, *J. Lab. Clin. Med.* **102**:536–542.
37. Wåhlstrand, T., and Wistrand, P. J., 1980, *Uppsala J. Med. Sci.* **85**:7–17.
38. Whitney, P. L., and Briggle, T. V., 1982, *J. Biol. Chem.* **257**:12056–12059.
39. Wirthensohn, G., and Guder, W. G., 1986, *Physiol. Rev.* **66**:469–497.
40. Wistrand, P. J., 1951, *Acta Physiol. Scand.* **24**:144–148.
41. Wistrand, P. J., 1980, *Acta Physiol. Scand.* **109**:239–248.
42. Wistrand, P. J., 1984, *Ann. N.Y. Acad. Sci.* **429**:195–206.
43. Wistrand, P. J., and Kinne, R., 1977, *Pfluegers Arch.* **370**:121–126.
44. Wistrand, P. J., and Knuuttila, K.-G., 1989, *Kidney Int.* **35**:851–859.
45. Wistrand, P. J., Lindahl, S., and Wåhlstrand, T., 1975, *Eur. J. Biochem.* **57**:189–195.
46. Zhu, X. L., and Sly, W. S., 1990, *J. Biol. Chem.* **265**:8795–8801.
47. Dodgson, S. J., and Cherian, K., 1990, *Arch. Biochem. Biophys.* **282**:1–7.

CHAPTER 32

Localization of Carbonic Anhydrase Isozymes in Calcified Tissues

H. KALERVO VÄÄNÄNEN and EEVA-KAISA PARVINEN

Carbonic anhydrase (CA) has been shown to be associated with the calcification process in invertebrates as well as in vertebrates. In several invertebrate species, including molluscs, calcareous sponges, corals, crabs, echinoderms, and barnacles, CA activity has been directly connected to calcification.[3,4,13,14,17,18,19,24] The function of the CAs in the mineralization of different invertebrates as well as in egg shell calcification is supported by several studies demonstrating decrease or inhibition of the calcification rate by CA inhibitors.[6,9,16,21,38] CA may also function in decalcifying hen egg shells.[32]

In mammalian bone and cartilage, CA has been associated with both endochondral bone growth and bone resorption. During endochondral bone growth, CA has been suggested to have a role in cartilage calcification. However, in contrast to calcification in invertebrates, no clear evidence of the effects of CA inhibition on cartilage calcification or bone mineralization and bone growth in mammals is available.

Most of the data on the localization of CA isozymes in different calcifying tissues has been obtained by immunocytochemistry. Hard tissue processing for immunocytochemistry has several difficulties in comparison with analysis of soft tissues.[34] Most of these difficulties can be avoided by using fetal tissues or young animals in which mineralization of hard tissues is incomplete. Histochemical localization of CA isozyme activity by using cobalt salt techniques could face additional problems in the calcified tissues because of the high possibility of nonspecific precipitation of metal salts to hydroxyapatite. This problem has further encouraged the use of immunohistochemical techniques in studies of CA localization in mineralized tissues.

H. KALERVO VÄÄNÄNEN and EEVA-KAISA PARVINEN • Department of Anatomy, University of Oulu, Kajaanintie 52 A, SF-90220 Oulu, Finland.

Distribution of soluble CA isozymes in rat and human bone and cartilage is summarized in Table I.

CA activity in the extracellular fluid of calcifying cartilage has been demonstrated, leading to the suggestion of the presence of CA in the extracellular matrix.[7] Later immunohistochemical studies showed that hypertrophic chondrocytes are the main cellular origin of CA in growth plate cartilage.[11,25,33] Extracellular localization of CA in growth plate cartilage has been confirmed recently by enzyme histochemistry.[26] On the basis of data obtained by these three methods, it can be concluded that CA, and more specifically CA II, is secreted somehow from hypertrophic chondrocytes into extracellular matrix just before during the calcification of the cartilage matrix. It is unknown whether CA II in cartilage matrix is in the extracellular fluid or is packed into matrix vesicles, which are small membrane particles, where the first crystals of hydroxyapatite appear. CA in growth plate cartilage may facilitate CO_2 transport and may explain the high pH of cartilage extracellular fluid, which favors the precipitation of calcium phosphate.

In contrast to hypertrophic chondrocytes and calcifying cartilage, neither bone-forming cells nor newly mineralized osteoid contains detectable CA. CA in developing hamster molars was localized histochemically in the stratum intermedium and stellate reticulum layers.[8] Some activity was observed also in odontoblasts and ameloblasts.

In vitro as well as *in vivo* experiments with acetazolamide clearly demonstrate that bone resorption can be partly blocked by inhibiting CA activity.[15,22,23,27,29,36,37]

Several immunohistochemical and enzyme histochemical studies have confirmed the presence of CA II in osteoclasts.[2,10,12,20,26,31,32,35] CA II antiserum labels both multinucleated and mononuclear osteoclasts (Fig. 1). These cells are

TABLE I
Distribution of Soluble CAs in Mammalian Calcified Tissues Based on Immunohistochemical Staining Reaction

	Reaction[a]		
Tissue	CA I	CA II	CA III
Bone			
Osteoclast	−	++	−
Osteoblast	−	(+)	−
Osteocytes	−	−	−
Bone matrix	−	−	−
Cartilage			
Articular	−	−	−
Growth plate	−	++	−

[a] ++, Strong staining reaction; +, moderate staining reaction; (+), weak staining reaction; −, negative.

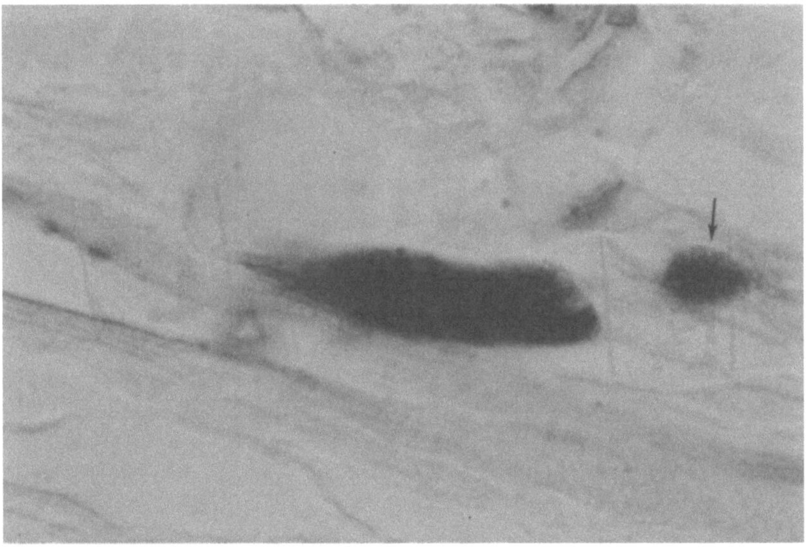

Figure 1. Localization of CA II in rat bone by using the peroxidase-antiperoxidase (PAP) method reveals strongly stained multinuclear and mononuclear (arrow) osteoclasts. ×640.

responsible for bone resorption by secreting protons and proteinases into resorption lacuna through the ruffled border membrane. On the basis of immunocytochemistry, the content of CA II in osteoclasts is rather high, in agreement with recent measurement of CA activity in isolated osteoclasts.[1,30] The function of CA in osteoclasts is probably to supply protons to proton-transporting membrane ATPase. This hypothesis is also supported by the observations that some of the CA II molecules in parathyroid hormone-stimulated osteoclasts are associated with the ruffled border membrane and that this association is somehow regulated by calcitonin.[5] In our recent immunogold study, we were not able to confirm this membrane association of CA II.[28] However, this could be due to the fact that the described type of distribution is found only in hormone-activated osteoclasts. Further studies are clearly needed to clarify the mechanism of possible membrane association of CA II in all acid-secreting cells.

Immunogold labeling of CA II shows that it is distributed diffusively through the osteoclasts (Fig. 2). Clear nuclear labeling with CA II antiserum has been noted in several types of cells, including osteoclasts. We originally thought that this was due to a diffusion artifact related to tissue processing. However, after obtaining more experience with different tissues and with several antisera, we have come to the conclusion that nuclear staining is not an artifact but most probably reflects physiological situation in which soluble CA enters the nucleus through nuclear pores. In fact, on several occasions gold labels were seen within nuclear pores.

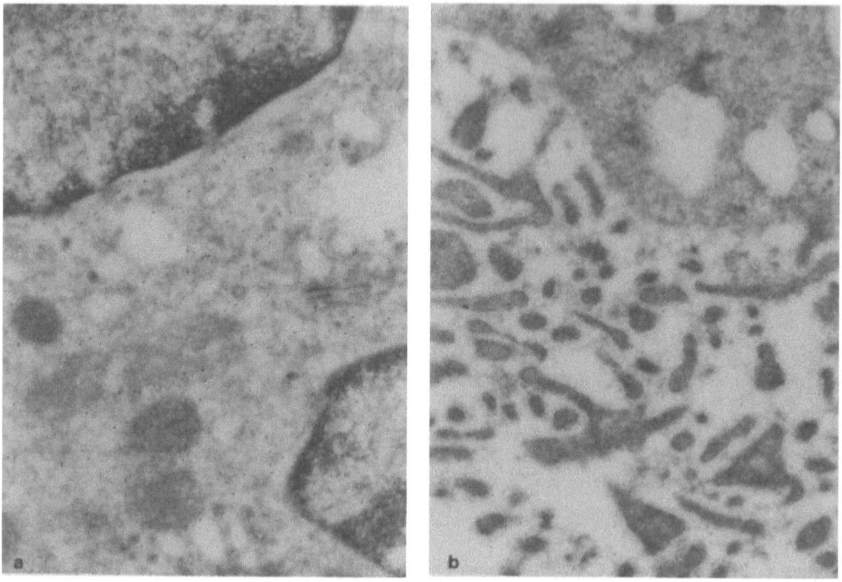

Figure 2. Immunogold localization CA II shows diffuse nucleoplasmic and cytoplasmic labeling (a). (b) Ruffled border area. ×7300.

The molecular weight of CA is also small enough to allow the transport of soluble cytoplasmic enzyme protein into nucleoplasm. The function, if any, of CA in the nucleus remains an open question.

REFERENCES

1. Anderson, R. E., Jee, W. S. S., and Woodbury, D. M., 1985, *Calcif. Tissue Int.* **37**:646–650.
2. Anderson, R. E., Schraer, H., and Gay, C. V., 1982, *Anat. Rec.* **204**:9–20.
3. Boer, H. H., and Witteveen, J., 1980, *Calcif. Tissue Res.* **209**:383–390.
4. Cameron, J. N., and Wood, C. M., 1985, *J. Exp. Biol.* **114**:181–196.
5. Cao, H., and Gay, C. V., 1985, *Experientia* **41**:1472–1474.
6. Costlow, J. D., 1959, *Physiol. Zool.* **32**:177–184.
7. Cuervo, L. A., Pita, J. C., and Howell, D. S., 1981, *Calcif. Tissue Res.* **7**:220–231.
8. Dogterom, A. A., Vervoets, T. J. M., Lyaruu, D. M., and Wöltgens, J. H. M., 1983, *J. Biol. Buccale* **11**:15–21.
9. Freeman, J. A., 1960, *Biol. Bull.* **118**:412–418.
10. Gay, C. V., and Müller, W. J. , 1974, *Science* **183**:432–434.
11. Gay, C. V., Anderson, R. E., Schraer, H., and Howell, D. S., 1982, *J. Histochem. Cytochem.* **30**:391–394.
12. Gay, C. V., Ito, M. B., and Schraer, H., 1983, *Metab. Bone Dis. Rel. Res.* **5**:33–39.
13. Giraud, M. M., 1981, *Comp. Biochem. Physiol.* **69**:381–387.

14. Goreau, T. F., 1959, *Biol. Bull.* **116**:59–75.
15. Hall, G. E., and Kenny, A. D., 1987, *Calcif. Tissue Int.* **40**:212–218.
16. Heatfield, B. M., 1970, *Biol. Bull.* **139**:151–163.
17. Henry, R. P., 1984, *Ann. N.Y. Acad. Sci.* **429**:544–546.
18. Henry, R. P., and Kormanik, G. A., 1985, *J. Crust. Biol.* **5**:234–241.
19. Henry, R. P., and Saintsing, D. G., 1983, *Physiol. Zool.* **56**:274–280.
20. Jilka, R. L., Rogers, J. I., Khalifah, R. G., and Väänänen, H. K., 1985, *Bone* **6**:445–449.
21. Jones, W. C., and Ledger, P. W., 1986, *Comp. Biochem. Physiol.* **84**:149–158.
22. Kenny, A. D., 1972, *Proc. Soc. Exp. Biol. Med.* **140**:135–139.
23. Kenny, A. D., 1985, *Calcif. Tissue Int.* **37**:126–133.
24. Kingsley, R. J., and Watabe, N., 1987, *J. Exp. Zool.* **241**:171–180.
25. Kumpulainen, T., and Väänänen, H. K., 1982, *Calcif. Tissue Int.* **34**:428–430.
26. Marie, J. P., and Hott, M., 1987, *J. Histochem. Cytochem.* **35**:245–250.
27. Minkin, C., and Jennings, J. M., 1972, *Science* **176**:1031–1033.
28. Parvinen, E. K., 1990, *Acta Histochemica*, (in press).
29. Raisz, L. G., Simmons, H. A., Thompson, W. J., Shepard, K. L., Anderson, P. S., and Rodan, G. A., 1988, *Endocrinology* **122**:1083–1086.
30. Silverton, S. F., Dodgson, S. J., Fallon, M. D., and Forster, R. E., III, 1987, *Am. J. Physiol.* **253**:E670–674.
31. Sundquist, K. T., Leppilampi, M., Järvelin, K., Kumpulainen, T., and Väänänen, H. K., 1987, *Bone* **8**:33–38.
32. Tuan, R. S., 1984, *Ann. N.Y. Acad. Sci.* **429**:459–472.
33. Väänänen, H. K., 1984, *Histochemistry* **81**:485–487.
34. Väänänen, H. K., and Korhonen, L. K., 1984, in: *Methods of Calcified Tissue Preparation* (G. Dickson, ed.) Elsevier, Amsterdam, pp. 309–332.
35. Väänänen, H. K., and Parvinen, E.-K., 1983, *Histochemistry* **78**:481–485.
36. Väänänen, H. K., and Tuukkanen, J., 1986, in: *Cell Mediated Calcification and Matrix Vesicles* (S. Y. Ali, ed.), Elsevier Science Publishers, B. V. (Biomedial Division), Amsterdam, pp. 143–146.
37. Waite, L. C., Volkert, W. A., and Kenny, A. D., 1970, *Endocrinology* **87**:1129–1139.
38. Yule, A. B., Crisp, D. J., and Cotton, I. H., 1982, *Mar. Biol. Lett.* **3**:273–288.

CHAPTER 33

Carbonic Anhydrase and Skeletogenesis

SUSAN F. SILVERTON

In a variety of species, the biochemical pathways that control Ca^{2+} accumulation in the skeleton have been closely related to carbonic anhydrase (CA) activity. In addition, because bone is the most important reservoir for Ca^{2+} homeostasis, Ca^{2+} is reclaimed by dissolution of deposited mineral and matrix.[5,10,22,47] This resorptive process also requires CA activity. While the more primitive exoskeletons rely on a mineral phase composed of $CaCO_3$,[6,55] vertebrates employ phosphate as the counter ion for mineralization.[52–54] Localization of CA isoenzymes in invertebrate and vertebrate tissue is reviewed elsewhere.[7,8,46]

1. Invertebrates

The invertebrate species that use $CaCO_3$ as an exoskeleton have evolved an efficient Ca^{2+} transport mechanism.[20,49] Studies of several of these marine organisms, including certain species of coral, the oyster, and sea urchin embryos, have demonstrated CA involvement in forming the shell or exoskeleton,[4,9] with Ca^{2+} obtained from seawater. Uptake of this cation is linked to bicarbonate accumulation[55]; both processes are inhibited by CA inhibitors.[33] In the sea urchin larva, sequestered $CaCO_3$ is transported within specialized pseudopodial cables that are extended from the primary mesenchymal cells toward the area where the skeleton will be formed. The pseudopods bud off small vesicles containing the $CaCO_3$ and other minerals that form the skeletal substance, magnesian calcite. Vesicles coalesce and form the spicules of the sea urchin. With the appearance of CA activity, the embryo simultaneously develops the ability to transport Ca^{2+}. Functional Ca^{2+}

SUSAN F. SILVERTON • Department of Oral Medicine, University of Pennsylvania School of Dental Medicine, Philadelphia, Pennsylvania 19104-6003.

translocation is dependent on CA activity.[9] During the critical period of skeletal formation, the Ca^{2+} transport mechanism is also inhibited by agents which interfere with chloride transport, suggesting that the HCO_3^-/Cl^- exchange is part of the Ca^{2+} accumulation mechanism. A relationship of HCO_3^-/Cl^- handling to CA function is repeated in higher species (see below).

2. Avian Skeletogenesis

Ca^{2+} for skeletogenesis in the chick is acquired indirectly from the hen via an egg shell composed mainly of $CaCO_3$. CA activity is linked to the process of Ca^{2+} deposition in the egg shell as well as to the subsequent dissolution of $CaCO_3$ from the shell for chick bone mineralization.[43]

Ca^{2+} for the egg shell originates in the bones of the hen. During the egg-laying cycle, large amounts of Ca^{2+} are transferred from the hen to the egg shell.[35] The Ca^{2+} pool tapped for this process was laid down during non-egg-laying periods as medullary bone, an extension of trabecular bone. Trabecular bone is sensitive to the body's demands for Ca^{2+} homeostasis, turnover is rapid, and hormones strongly influence the trabecular bone metabolic rates. In contrast, cortical bone, which makes up the major part of the bone shaft, fulfills an architectural role.

The non-laying hen stores massive amounts of Ca^{2+} as medullary bone in the marrow canals of the long bones. When estrogens increase and the Ca^{2+} demands for laying begin, numerous bone-resorbing cells, osteoclasts, cover the surface of the medullary bone. These multinucleate cells dissolve the mineral and degrade the bone matrix. The Ca^{2+} released into the bloodstream by osteoclastic bone resorption is taken up by the shell gland and deposited in the shell as $CaCO_3$. CA inhibitors impede shell–gland Ca^{2+} transport and result in a noncalcified shell.[44]

When the egg is first laid, the shell serves as a protective environment for the embryo. As the chick develops, Ca^{2+} is required for skeletal mineralization.[28] The importance of the shell in providing Ca^{2+} to the embryo has been graphically illustrated by raising chick embryos without shells.[43] This is accomplished by maintaining the chick *in ovo* only for the first 3 days of the 21-day developmental period. The shell is then removed, and the egg is cultured in a plastic wrap cup. The resulting severe Ca^{2+} deficiency causes skeletal deformities of the limbs and beak. Similar deformities occur when the shell is present and the embryo is exposed to CA inhibitors during development.

Studies by Tuan and colleagues[43,44] have implicated CA in the Ca^{2+} transport process of the chorioallantoic membrane of the chick egg shell. In the presence of a shell, the Ca^{2+} transport function in the ectodermal cells appears on the same time scale as CA activity. Inhibition of CA by sulfonamide CA inhibitors significantly decreases Ca^{2+} transport. In the shell-less embryo, CA activity develops with its usual time course, but the Ca^{2+} transport process is significantly impaired. To compensate for the reduced Ca^{2+} availability, the amount of Ca^{2+} binding protein increases. These results suggest that Ca^{2+} transport in the chick chorioallantoic membrane requires at least CA, a Ca^{2+} binding protein, and an unknown Ca^{2+} pump. A possible interaction of this Ca^{2+} transport mechanism with HCO_3^-/Cl^- exchange, like Ca^{2+} bicarbonate deposition in invertebrate species, has not been explored. Conversely, in the sea urchin, the role of Ca^{2+} binding proteins in CA-mediated Ca^{2+} transport is unknown.

3. Vertebrate Skeletogenesis

The anlagen of the skeleton structure is laid out in the embryo by mesenchymal cells. The bone structure elongates toward the cartilaginous caps. There is a continuous deposition of bone tissue by bone-forming cells (osteoblasts). The result is an encroachment of bone into the areas that previously were cartilage. During growth, the cartilage cap has a specialized organ, the growth plate. At this place, columns of chondrocytes are continuously produced as bone deposition advances toward the bone ends. Defects in growth plate function result in growth retardation and dwarfism, i.e., rickets, achondroplasia, and Ca^{2+} deficiency.

CA may influence the development of the epiphyseal growth plate.[30] A relatively alkaline pH in the cartilage fluid was obtained by micropuncture and fluid withdrawal from the mineralizing portion of the growth plate.[11] CA inhibitors, given orally, abolished the pH gradient in cartilage. Howell *et al.*[25,26] postulated that the mismatch of O_2 and CO_2 tensions that had been shown by other workers in the growth plate was the result of the alkaline pH and was driven by CA-directed secretion of CO_3 ion. A similar mechanism has been discussed for acid secretion in kidney brush border.[51] Histochemical studies have localized CA isoenzymes to the growth plate.[12,16,45] Enzyme localization is increased in mature chondrocytes and is present in the extracellular matrix surrounding mature cartilage cells. The specific activity of CA in isolated matrix vesicles is fivefold that in chondrocytes and exceeds that in red cells (unpublished data). The enhancement of CA activity in matrix vesicles compared with chondrocytes parallels the observations made on alkaline phosphatase in matrix vesicles.[19] Alkaline phosphatase is presumed to generate free phosphate ion in the presence of large extracellular Ca^{2+} concentrations which would initiate mineralization. Perhaps a CA-directed bicarbonate generation in matrix vesicles may be used to produce local increases in pH. The increased alkalinity could augment matrix vesicle alkaline phosphatase activity and favor mineralization.

At skeletal maturity, the epiphyseal growth plates of the long bones cease to exist. Except for fracture repair, the transition of mineralized cartilage to bone no longer takes place. Instead, bone is laid down by osteoblasts and destroyed by osteoclasts in a remodeling cycle that continues throughout adult life. Although bone is constantly being formed as well as resorbed, after the age of 30, humans have a continuously decrementing bone mass. The minute amount of bone loss per year represents the slightly greater efficiency that the small osteoclast population exerts in destroying the bone substance, compared with a vastly greater population of osteoblasts. This loss of structural material has dramatic consequences on human mortality: many older people die from complications of hip fractures. CA appears to play a pivotal role in osteoclast function and in bone resorption. Studies in whole animals, in organ culture, and in isolated osteoclasts have implicated CA in the hormonal control of Ca^{2+} homeostasis and bone resorption.

Studies linking CA activity to bone resorption can be divided into three groups: (1) CA in Ca^{2+} homeostasis and bone resorption, (2) localization of CA, and (3) CA and osteoclast function.

The interaction of parathyroid hormone (PTH) and bone CA to regulate bone Ca^{2+} availability has been studied *in vivo* and *in vitro*. Waite[48] used a nephrectomized rat model to demonstrate the relationship of bone CA to Ca^{2+} homeostasis. When treated with intraperitoneal CA inhibitors, rats were not able to increase plasma [Ca^{2+}] either after bilateral nephrectomy or with exogenous PTH. These results suggest that Ca^{2+} generated from bone for Ca^{2+} homeostasis is dependent on intact CA function. The overall maintenance of serum [Ca^{2+}] by the bone Ca^{2+} reservoir has a somewhat controversial literature. While it is generally agreed that osteoclasts are responsible for long-term physiological Ca^{2+} losses, acute changes in Ca^{2+} levels may be mediated by decreases in osteoblastic transport of Ca^{2+} into the mineralizing bone. Thus, targeting serum Ca^{2+} changes to osteoclastic CAs may be a simplistic approach.

Several *in vitro* studies explored the interaction of bone CA with release of Ca^{2+} from bone organ cultures. These experimental models evaluate Ca^{2+} release from bone tissue by prelabeling the bone with $^{45}Ca^{2+}$ and quantitating the amount of radioactivity released into the culture medium with known resorbing agents. Since $^{45}Ca^{2+}$ is already within the bone, the effects of osteoblastic Ca^{2+} transport are minimized in this model. Mahgoub and Stern[31] and Jennings,[32] using fetal long-bone cultures and calvarial cultures, respectively, demonstrated a decrease in PTH-stimulated prelabeled bone ^{45}Ca release with CA inhibitors. In another study, increases in medium Ca^{2+} caused by dibutyryl cyclic AMP and forskolin were obliterated in the presence of CA inhibitors.[21] Equivalent inhibition of the resorptive effects of 1,25-dihydroxy D_3 and prostaglandin E_2 has also been shown.[23,24] Again, stimulation of bone resorption by these agents was prevented by inhibition of CA. Thus, CA activity has been linked to known mediators of Ca^{2+} homeostasis and bone resorption.

Another study using a rat model of disuse osteoporosis[29] showed that nonhormonal bone resorption was also dependent on CA activity. One limb was neurectomized, and the contralateral limb served as a control. Generally, there is a significant loss of bone in the immobile limb. However, with continuous subcutaneous infusion of acetazolamide, this loss was significantly decreased. These results suggest that both neurogenic and hormonal pathways of bone resorption overlap in their requirement for CA. Further experiments showed that continuous infusion of the CA inhibitor was necessary to maintain drug concentrations high enough to show an effect.

Localization of bone CA involved defining the cell of origin as well as the isoenzyme present. CA was localized in osteoclasts by using radioactively labeled acetazolamide.[14] Further studies used immunohistochemical methods and demonstrated high activity enzyme in chick and human osteoclasts.[15,45,46] The use of ultrastructural immunocytochemistry targeted the enzyme to the ruffled border membrane and to cytoplasmic vesicles.[16] Additional information on CA response to calcemic hormones was provided by Gay and co-workers.[27]

Osteoclasts have evolved as specific phagocytes that respond to signals in the bone environment and destroy bone. These large multinucleate cells are relatively difficult to isolate, are not available in large numbers, do not proliferate in culture, and appear to lose morphological integrity within a few days *in vitro*. Recent work suggests that protein factors in bone may initiate resorptive activity.[18] The osteoclast responds to cellular signals by sealing off an area on the bone and developing a specialized ruffled border. This organelle is the site of lysosomal acid hydrolase export.[2] Along with secretion of these enzymes, there is an active production of acid, which is essential for bone destruction. Osteoclast-resorptive cavity fluid has been measured at pH 5.[13] Although acidification is known to be a prominent feature of osteoclast function, the biochemical mechanism of acid production in this cell is not understood. ATPases are present in osteoclasts, but none have been localized to the ruffled border.[3] The proton pump of the osteoclast does not seem at present to be similar to that of the kidney tubule. CA II is associated with acid secretion in renal tubular cells[51] and gastric mucosa[15] and has been postulated to be involved in osteoclastic acid secretion. It has been shown that in isolated osteoclasts, PTH increases CA activity[37] and osteoclast acid production.[27] Both of these activities have been shown to be inhibited by CA inhibitors.

Families have been found with CA II deficiency and ineffective bone remodeling.[38,39] The decreased CA is manifested by an overabundance of bone and cartilage in the interior of the shaft of the long bones. With CA II deficiency, osteoclasts are still competent enough to allow some marrow space to be carved out of the long bone shafts. Although the osteoclasts are defective, no biochemical studies have been carried out to assess acid secretion. It is possible that more than one isozyme is present in osteoclasts and that localization is similar to that of the renal proximal tubule.[40,50,51] The ultrastructural demonstration of CA at the ruffled border of the chick osteoclast[1] would support this interpretation. As the mechanisms for osteoclastic acid production are better understood, the relationship of CA activity to bone resorption will become easier to investigate. On a larger plane, the understanding of the function of CAs in skeletogenesis awaits more experimentation on the link between Ca^{2+} and bicarbonate translocation.

ACKNOWLEDGMENT. This research was supported by Public Health Service grant MO1-RR-01224.

References

1. Anderson, R. E., Schraer, H., and Gay, C. V., 1982, *Anat. Rec.* **204**:9–20.
2. Baron, R., Neff, L., Louvard, D., and Courtoy, P. J., 1983, *J. Cell Biol.* **97**:108.
3. Baron, R. E., 1988, 3rd International Conference on Chemistry and Biology of Mineralized Tissue, Chatham, Mass., Abstract 1.
4. Brinkman, R., 1934, *J. Physiol.* **80**:171–173.
5. Cao, H., and Gay, C. V., 1982, *Experientia* **41**:1472–1474.
6. Carson, D. D., Farach, M. C., Earles, D. S., Decker, G. I., and Lennarz, W. J., 1985, *Cell* **41**: 639–648.
7. Carter, M. J., 1972, *Biol. Rev.* **47**:465–513.
8. Carter, N., and Jeffrey, S., 1985, *Biochem. Soc. Trans.* **13**:531–533.
9. Chow, G., and Benson, S. C., 1979, *Exp. Cell Res.* **124**:451–453.
10. Conway, H. H., Waite, L. C., and Kenny, A. D., 1973, *Calcif. Tissue Res.* **11**:323–330.
11. Cuervo, L. A., Pita, J. C., and Howell, D. S., 1971, *Calcif. Tissue Res.* **7**:220–231.
12. Ellison, A. C., 1965, *Proc. Soc. Expt. Biol.* **120**:415.
13. Fallon, M. D., 1984, in: *Endocrine Control of Bone and Calcium Metabolism* (D. V. Cohn *et al.*, eds.), Elsevier Science Publishers B. V., Amsterdam, pp. 144–146.
14. Gay, C. V., and Mueller, W. J., 1973, *J. Histochem. Cytochem.* **21**:693.
15. Gay, C. V., Faleski, E. J., Schraer, H., and Schraer, R., 1974, *J. Histochem. Cytochem.* **22**:415.
16. Gay, C. V., Anderson, R. E., Schraer, H., and Howell, D. S., 1982, *J. Histochem. Cytochem.* **30**:391–394.
17. Gay, C. V., Ito, M. B., and Schraer, H., 1983, *Metab. Bone Dis. Rel. Res.* **5**:33–39.
18. Glowacki, J., and Lian, J. B., 1987, *Cell Diff.* **21**:247–254.
19. Golub, E. E., Schattschneider, S. C., Berthold, P., Burke, A., and Shapiro, I. M., 1983, *J. Biol. Chem.* **258**:612–621.
20. Goreau, T. F., 1959, *Biol. Bull.* **116**:59.
21. Gunasekaran, S., Hall, G. E., and Kenny, A. D., 1986, *Proc. Soc. Exp. Biol. Med.* **181**:438–442.
22. Hall, G. E., and Kenny, A. D., 1986, *J. Pharmacol. Exp. Ther.* **238**:778–782.
23. Hall, G. E., and Kenny, A. D., 1985, *Pharmacology* **30**:339–347.
24. Hall, G. E., and Kenny, A. D., 1985, *Calcif. Tissue Int.* **37**:134–142.
25. Howell, D. S., Pita, J. C., Marquez, J. F., and Madruga, J. E., 1968, *J. Clin. Invest.* **47**:1121–1132.
26. Howell, D. S., Pita, J. C., Marquez, J. F., and Gatter, R. A., 1969, *J. Clin. Invest.* **48**:630–641.
27. Hunter, S. J., Schraer, H., and Gay, C. V., 1988, *J. Bone Min. Res.* **3**:297–303.
28. Johnston, P. M., and Comar, C. L., 1955, *Am. J. Physiol.* **183**:365–370.
29. Kenny, A. D., 1985, *Pharmacology* **31**:97–107.
30. Kumpulainen, T., and Väänänen, H. K., 1982, *Calcif. Tissue Res.* **34**:428–430.
31. Mahgoub, A., and Stern, P. H., 1974, *Am. J. Physiol.* **226**:1272–1275.
32. Minken, C., and Jennings, J. M., 1972, *Science* **176**:1031–1033.
33. Mitsunaga, K., Akasaka, K., Shimada, H., Fujino, Y., Yasumasu, I., and Numanoi, H., 1986, *Cell Diff.* **18**:257–262.
34. Pearson, T. W., and Goldner, A. M., 1974, *Am. J. Physiol.* **227**:465–468.
35. Pearson, T. W., Pryor, T. J., and Goldner, A. M., 1977, *Am. J. Physiol.* **232**:E437–443.
36. Romanoff, A., 1967, *Biochemistry of the Avian Embryo: A Quantitative Analysis of Prenatal Development*, Wiley-Interscience, New York.
37. Silverton, S. F., Dodgson, S. J., Fallon, M. D., and Forster, R. E., II, 1987, *Am. J. Physiol.* **253**:E670–674.
38. Sly, W. S., Hewett-Emmett, D., Whyte, M. P., Yh, Y. S., and Tashian, R. E., 1983, *Proc. Natl. Acad. Sci. USA* **80**:2751–2756.
39. Sly, W. S., Whyte, M. P., Sundaram, V., Tashian, R. E., Hewett-Emmett, D., Guibaud, P.,

Vainsel, M., Baluarte, H. J., Gruskin, A., Al-Mosawi, M., Sakati, N., and Ohlsson, A., 1985, *N. Engl. J. Med.* **313**:139–145.
40. Tashian, R. E., Hewett-Emmett, D., Dodgson, S. J., Forster, R. E., and Sly, W. S., 1984, *Ann. N.Y. Acad. Sci.* **429**:262–275.
41. Terepka, A. R., Stewart, M. E., and Merkel, N., 1969, *Exp. Cell Res.* **58**:107–117.
42. Terepka, A. R., Coleman, J. R., Garrison, J., and Spataro, R., 1971, in: *Cellular Mechanisms for Calcium Transfer and Homeostasis* (G. Nichols and R. H. Wasserman, eds.) Academic Press, New York, pp. 371–389.
43. Tuan, R. S., 1980, *Dev. Biol.* **74**:196–204.
44. Tuan, R. S., and Zrike, J., 1978, *Biochem. J.* **176**:67–74.
45. Väänänen, H. K., 1984, *Histochemistry* **81**:485–487.
46. Väänänen, H. K., and Parvinen, E. K., 1983, *Histochemistry* **78**:481–485.
47. Väänänen, H. K., and Parvinen, E. K., this volume, Chapter 32.
48. Waite, L. C., 1972, *Endocrinology* **91**:1160–1165.
49. Wilbur, K. M., and Jodrey, L. H., 1955, *Biol. Bull.* **108**:359.
50. Wistrand, P. J., 1984, *Ann. N.Y. Acad. Sci.* **429**:195–206.
51. Wistrand, P. J., and Knuuttila, K.-G., 1980, *Acta Physiol. Scand.* **109**:239–248.
52. Wuthier, R. E., 1969, *Calcif. Tissue Res.* **4**:20–38.
53. Wuthier, R. E., 1971, *Calcif. Tissue Res.* **8**:24–35.
54. Wuthier, R. E., 1977, *Calcif. Tissue Res.* **23**:125–133.
55. Yasumasu, I., Mitsunaga, K., and Fujino, Y., 1985, *Exp. Cell. Res.* **159**:80–90.

CHAPTER 34

Carbonic Anhydrases Secreted in the Saliva

ROSS T. FERNLEY

1. Introduction

Secreted carbonic anhydrases (CAs) are synthesized in the salivary glands and secreted into saliva. Although it has been known for some time that saliva contains CA activity,[17] it was not realized that the enzyme present might be very different from the other isozymes. In 1979, we reported that the CA of sheep saliva and about 40% of the enzyme in parotid gland homogenate was a high-molecular-mass protein.[6] In comparison, the cytoplasmic isozymes are relatively small proteins with a molecular mass of 30,000. The salivary enzyme was also less sensitive to sulfonamide inhibition than the high-activity CA II isozyme and did not cross-react with antibodies raised against sheep CA II. This initial study suggested that this enzyme was a distinct CA isozyme and not just a complex of CA II with some other protein. The isozyme has now been classified as CA VI.[8,15] This chapter reviews the purification, properties, distribution, and possible role of these secreted CAs.

2. Purification

To fully characterize this enzyme, a convenient purification protocol needed to be developed to prepare sufficient protein in a homogeneous form. CA isozymes are purified routinely by affinity chromatography; although CA VI is less susceptible to sulfonamide inhibition than is CA II, it still binds strongly to sulfonamide affinity resins, and this method has been used in all cases reported in the literature. The purification of CA isozymes is reviewed in Chapter 7 by W. R. Chegwidden.

ROSS T. FERNLEY • Howard Florey Institute of Experimental Physiology and Medicine, Parkville, Victoria 3052, Australia.

Feldstein and Silverman[5] obtained pure rat CA VI by using affinity purification only, as did Kadoya *et al.*[12] with human CA VI. Murakami and Sly[15] used affinity chromatography followed by ion exchange chromatography on DEAE–Sephacel to obtain homogeneous human CA VI from saliva, and we have used wheat germ lectin affinity chromatography and gel filtration on Sepharose 6B following the initial affinity purification step to obtain homogeneous sheep CA VI from parotid glands[8] or saliva.[9] The enzyme is a relatively abundant protein, constituting about 10–15% of the total proteins in sheep parotid saliva. This, and the fact that large volumes of parotid saliva (2–3 liters overnight) can be collected from sheep by cannulation of the parotid duct, means large amounts of the enzyme can be purified. About 6 mg of pure enzyme can be obtained from 2 liters of sheep parotid saliva in 3 working days. The purification procedure is summarized in Table I.

3. Properties

As mentioned previously, CA VI occurs as a high-M_r form in the native state.[6,8] On polyacrylamide gels in the presence of 0.1% SDS, the apparent M_r of CA VI is 45,000 in sheep,[7,8] 46,000 in rats,[5] and 42,000[15] or 40,000[12] in humans. When the enzyme from humans, sheep, cows, and dogs are run on the same gel, all have the same apparent M_r (about 45,000), whereas the mouse CA VI appears to be smaller, at 42,000.[10] The smaller apparent size of the mouse enzymes may be due to the presence of large amounts of proteolytic enzymes in the mouse salivary glands even though proteolytic enzyme inhibitors were included in the homogenization buffer. It has been observed[5] that purification of the rat salivary gland CA VI in the absence of protease inhibitors results in species with M_r of 32,000–42,000.

The specific activities reported for the secreted CAs vary widely. Sheep CA VI is reported to have about 70% of the activity of sheep CA II[8] and rat CA VI is reported to have about 75% of the k_{cat} value of rat CA II,[5] whereas the specific activity of human CA VI was reported as only about 1% of that of human CA II.[15] We have found a higher specific activity for human CA VI, at about 35% of that of sheep CA VI (R. T. Fernley, unpublished observations). It is possible that there are individual variations in the human enzymes. Sheep CA VI was found to have no effect on *p*-nitrophenyl acetate, which acts as a substrate for the other CA isozymes.[8]

All the secreted CAs studied are known to be glycoproteins,[5,8,10,15] and the N-linked carbohydrate can be removed enzymatically. The M_rs of the deglycosylated human[15] and sheep[8] enzymes are both 36,000. There is no evidence of any O-linked carbohydrate in CA VI.[15]

The secreted CAs have limited cross-reactivity with antibodies raised against other CA isozymes. Antibodies raised against sheep CA II and CA VI did not cross-react with one anothers' antigens,[8] whereas antibodies to the human secreted CA cross-reacted only slightly with human CA II[15] or not at all with CA I and CA

TABLE I
Purification of Sheep CA VI from Saliva

Step	Total protein (mg)	Total activity (units)[a]	Specific activity (units·mg^{-1})	Yield (%)
Saliva	181	8590	47.5	100
Sulfonamide-Sepharose	14.2	3429	241	39.8
Wheat germ lectin–Sepharose	13.1	3308	252	38.4
Sepharose 6B	6.5	2210	340	25.7

[a]One unit of enzyme activity is the amount required to complete the reaction in a time half that of the uncatalyzed reaction.

II.[12] Secreted CAs isolated from several species were found to cross-react strongly with anti-sheep CA VI antibody, the exceptions being the mouse and rat CA VIs.[10]

4. Amino Acid Sequence

The complete amino acid sequence of CA VI from sheep saliva has been determined.[9] It consists of a single polypeptide chain of 307 amino acids, compared with the 259 or 260 residues in the cytoplasmic isozymes. The calculated M_r is 35,565, which agrees well with the M_r of the deglycosylated enzyme.[8] Most of the extra sequence is contained in a hydrophilic carboxyl-terminal extension whose function is unknown. Overall, sheep CA VI has a 33% sequence identity with sheep CA II, with several gaps being introduced into the sequences to optimize alignments (Fig. 1). The low degree of identity and the number of gaps introduced in the alignment process suggest that the gene coding for this isozyme diverged from the ancestral CA gene at an early stage of evolution. The identification of this isozyme in lower vertebrates would help elucidate its evolutionary development. By analysis of cyanogen bromide peptides of the nonreduced protein, it was determined[9] that there is an intramolecular disulfide bond between Cys-25 and Cys-207. Such disulfide bonds are a common feature of secreted proteins, and they may help to stabilize the molecule. Indeed, reduction of the bond with 10 mM dithiothreitol over 25 hr abolished the carbon dioxide hydrating activity of CA VI.

Using a synthetic oligodeoxyribonucleotide whose sequence was based on the amino terminal amino acid sequence of sheep CA VI,[10] cDNA clones coding for human CA VI were obtained and the nucleotide sequence was determined.[1] These clones code for a protein of 308 amino acids, including a 17-amino-acid leader sequence typical of secreted proteins. This sequence is not present in the mature protein.[10] Human CA VI has a sequence identity of 71.5% with sheep CA VI and 35% with human CA II. The active-site residues, two cysteines, and potential glycosylation sites have been conserved. Compared with the sheep enzyme, human

```
                                                       *                          ▼
CA VI   H3N+ G H G V E W T Y S E G M L D E A H W P L E Y P K C G G R R Q S P I D L Q M K K V Q Y N P S L R A L N    50
CA II          AcS H H W G Y G E - H N G P E H W H K D F P I A D G E R Q S P V D I D T K A V V P D P A L K P L A    47

                                                                                     *  Zn  Zn
CA VI        L T G Y G L W H G E F P V T N N G H T V Q I S L P S T M S M T T S - D G - - - T Q Y L A K Q M H F H    96
CA II        L L Y E Q A A S R R - - M V N N G H S F N V E F D D S Q D K A V L K D G P L T G T Y R L V Q F H F H    95

                       * *                 *  Zn
CA VI        W G G A S S E I S G S E H T V D G M R Y V I E I H V V H Y N S K Y N S Y E E A Q K E P D G L A V L A   146
CA II        W G - - S S D D Q G S E H T V D R K K Y A A E L H L V H W N T K Y G D F G T A A Q Q P D G L A V V G   143

CA VI        A L V E V K D Y T E N A Y Y S K F I S H L E D I R Y A G Q S T V L R G L D I E D M L P G D L R Y Y Y   196
CA II        V F L K V G D A N P - - A L Q K V L D V L D S I K T K G K S A D F P N F D - P S S L L K R A L N Y W   190

               *       *           *                                            ▼
CA VI        S Y L G S L T T P P C T E N V H W F V V A D T V K L S K T Q V E K L E N S L L N H Q N K T I Q - - -   243
CA II        T Y P G S L T N P P L L E S V T W V V L K E P T S V S S Q Q M L K F R S L N F N A E G E P E L L M L   240

             *
CA VI        N D Y R R T Q P L N H R V V E A N F M S R P H Q E Y T L A S K L H F Y L N N I D Q T L E Y L R R F I   293
CA II        A N W R P A Q P L K N R Q V R V F P K-OH    259

CA VI        E Q K I P K R K K Q E N W P-OH   307
```

Figure 1. Amino acid sequences of sheep CA VI and sheep CA II. Gaps have been introduced into both sequences to maximize homology. Identical sequences are boxed. The active-site zinc-binding histidines are identified by the symbol Zn; the amino acid residues that are hydrogen bonded either directly or indirectly to these histidine residues or the zinc-bound solvent molecules are identified by an asterisk. The carbohydrate-binding asparagines are identified by arrowheads. The sequence of sheep CA II is from reference 19. (Reprinted from reference 9 with permission.)

CA VI has a three-amino-acid deletion in the carboxyl-terminal segment, and this region has a total of 29 amino acids, compared with the 45 amino acid tail of sheep CA VI. The gene encoding human CA VI (*CA6*) maps to chromosome 1 by Southern analysis of a somatic cell hybrid panel and to the tip of the short arm of this chromosome (lp 36.22-lp 36.33) by *in situ* hybridization.[18] It is therefore not linked to the cytoplasmic CA genes on chromosome 8[20] nor to *CA7* on chromosome 16.[14]

As mentioned previously, the enzyme is a glycoprotein, and two glycosylation sites have been identified at Asn-50 and Asn-239 on the sheep enzyme.[9] The function of these carbohydrate groups is not known, but they may play a role in stabilizing the enzyme. *In vitro* site-directed mutagenesis studies have been performed on the enzyme renin to eliminate its two glycosylation sites,[11] and the results have shown that the resulting enzyme is much less stable than the native glycoprotein. Similar *in vitro* site-directed mutagenesis experiments could be carried out on CA VI to ascertain the role of the disulfide bond and carbohydrate group, and expression of clones lacking the carboxyl-terminal segment of the molecule could help elucidate its role.

5. Tissue Distribution

CA VI was found originally in the parotid gland (and parotid saliva) of sheep[6] and subsequently in rat salivary glands[5] and in human saliva.[12,15] A number of tissues of sheep have been examined for the presence of CA VI by immunoblot analysis using affinity-purified anti-sheep CA VI antibodies.[10] Only the parotid and submandibular salivary gland contained CA VI, with much more CA VI in the parotid gland. None was detected in the sublingual salivary gland. These salivary glands are of different types; the parotid is a serous gland, the submandibular is a mixed serous and mucous gland, and the sublingual is a mucous gland. To investigate the site of synthesis of CA VI in the salivary gland, the technique of hybridization histochemistry as described by Penschow et al.[16] was used. An oligonucleotide probe, 56 bases long, corresponding to the amino-terminal 19 amino acids of sheep CA VI, was synthesized and 5' end labeled with [^{32}P]ATP.[10] This probe labeled the acinar cells of the parotid gland very strongly, whereas the duct cells of the gland were not labeled (Fig. 2d). By contrast, an ovine kallikrein oligonucleotide probe (Fernley, unpublished observations) labeled the duct cells but not the acinar cells in the parotid gland (Fig. 2c). Carpentier *et al.*[2] have demonstrated histochemical staining for CA in granules in the serous end pieces of the parotid gland. Kadoya *et al.*,[12] using CA VI antisera, have shown granular staining in the serous acinar cells of human salivary glands and with saliva contained in the ducts. Thus, it appears that CA VI is synthesized in the acinar cells of serous salivary glands and stored in granules before being secreted into saliva.

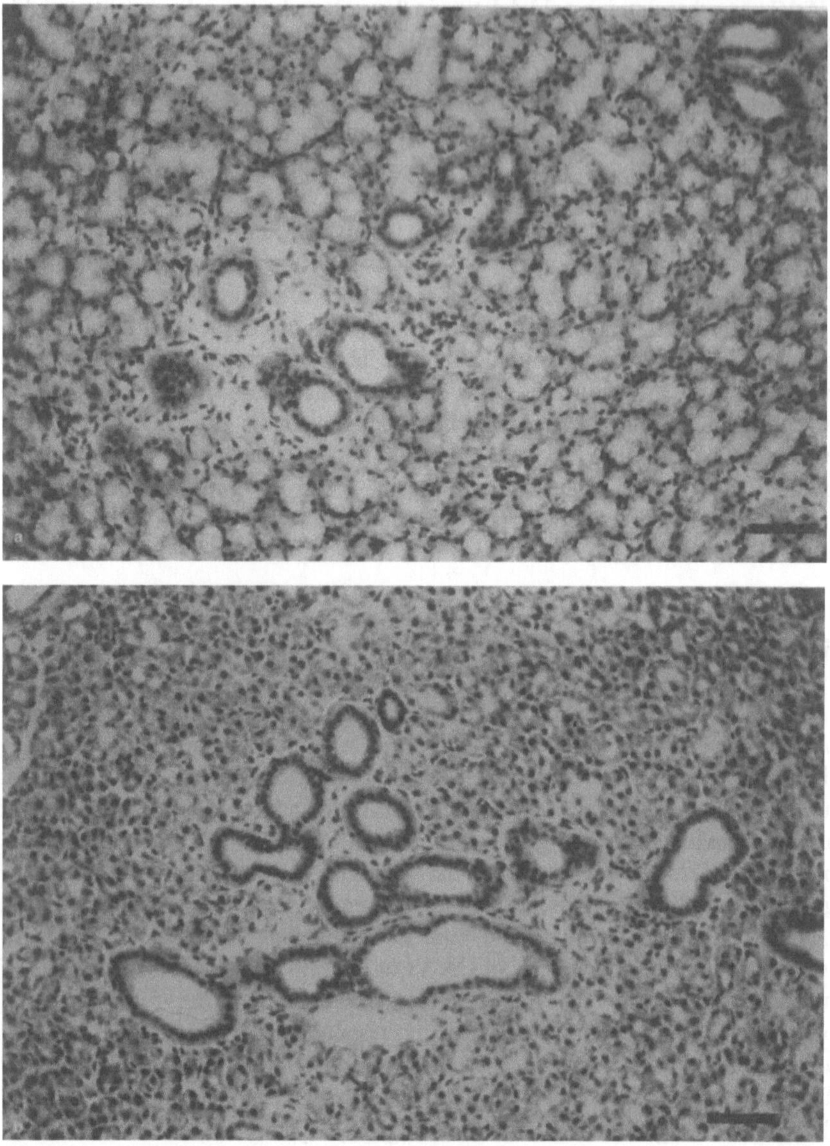

Figure 2. Autoradiographs of 5-μm frozen sections from sheep parotid gland after hybridization with ^{32}P-labeled oligodeoxyribonucleotide probes complementary to mRNA for sheep kallikrein or sheep CA VI. Panels c and d are dark-field photomicrographs showing the same fields as panels a and b, respectively, photographed under bright-field illumination. Silver grains over the striated ducts can be seen clearly under dark field in panel c; this is the site of kallikrein biosynthesis. By contrast, silver grains visible in panel d are located clearly over the acinar cells, the site of synthesis of CA VI. Stains used were hematoxylin and eosin. Bar = 50 μm.

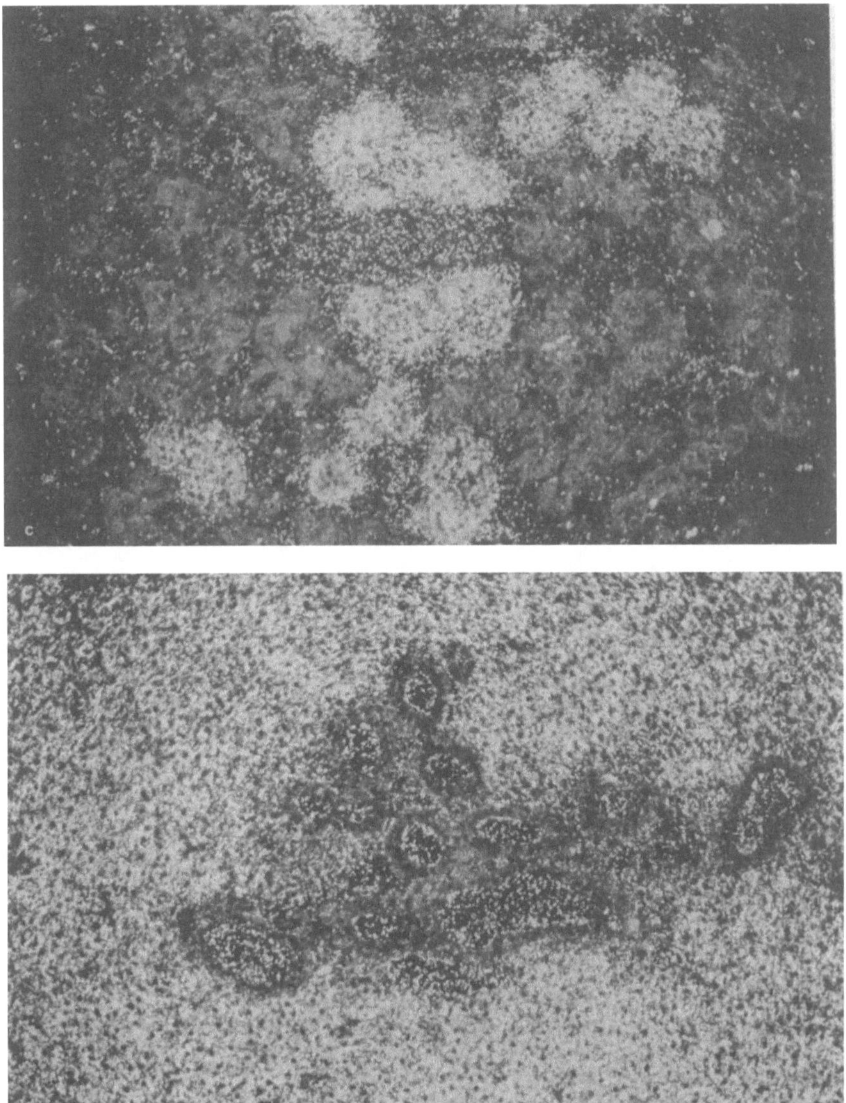

Figure 2. *(Continued)*

CA VI appears to be widespread among mammals, being found in the sheep and goat,[6] rat,[5] human,[12,15] and mouse, cow, and dog.[10] Large amounts are also present in the saliva of the kangaroo (Fernley, unpublished observations).

6. Other Salivary Proteins

A number of proteins are known to occur in saliva[13]; however, there is great variation in the proteins present and their relative amounts.[3] For example, whereas amylase is a major protein of human saliva, it appears to be absent from the saliva of ruminants[4] but present in saliva from other mammalian species. The protein profiles of human and sheep parotid salivas on polyacrylamide gels in 0.1% SDS are shown in Fig. 3; as can be seen, they are significantly different. These differences in protein profiles reflect the various functions required by salivary proteins in animals with very different diets and digestive processes. A good example is lysozyme, which in many species occurs in saliva, has a neutral pH optimum, and acts to prevent infection in the oral cavity. In ruminants it is not

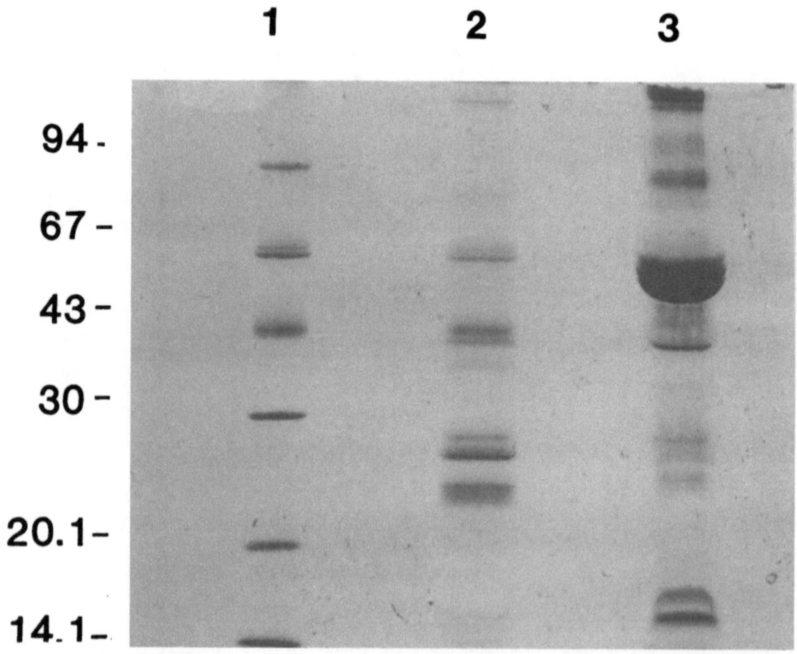

Figure 3. Polyacrylamide gel electophoresis of human and sheep parotid saliva. Samples (100 μl) of saliva and M_r standards were run on a 15% polyacrylamide gel in 0.1% SDS and stained with Coomassie blue. Sizes of the M_r standards (lane 1) are indicated on the left. Lane 2, Sheep parotid saliva; lane 3, human parotid saliva.

present in saliva, as mentioned previously, but it has evolved into a digestive enzyme of the stomach and has a pH optimum near 5.0.[4] From the results of a comparative study on mammalian salivary CAs[10] and the protein profiles shown in Fig. 3, it appears that CA VI is one of the few proteins consistently present in saliva from a wide range of mammalian species. This would indicate that it has an important basic function in saliva. It has been suggested[5] that its function is to provide a greater buffering capacity in the oral cavity. However, it is possible that it has an as yet unidentified activity. Study of patients who lack the CA II gene and who appear also to lack CA VI in saliva[15] may provide valuable insights into its function.

ACKNOWLEDGMENT. This work is supported by the National Health and Medical Research Council of Australia. I would like to thank Terri Hammill for excellent technical assistance.

References

1. Aldred, P., Fu, P., Barrett, G., Penschow, J. D., Wright, R. D., Coghlan, J. P., and Fernley, R. T., *Biochemistry*, in press.
2. Carpentier, P., Fournie, J., and Chetail, M., 1984, *Ann. N.Y. Acad. Sci.* **429**:216–218.
3. Chauncey, H. H., Henriques, B. L., and Tanzer, J. M., 1963, *Arch. Oral Biol.* **8**:615–627.
4. Dobson, D. E., Prager, E. M., and Wilson, A. C., 1982, *J. Biol. Chem.* **259**:11607–11616.
5. Feldstein, J. B., and Silverman, D. N., 1984, *J. Biol. Chem.* **259**:5447–5453.
6. Fernley, R. T., Wright, R. D., and Coghlan, J. P., 1979, *FEBS Lett.* **105**:299–302.
7. Fernley, R. T., Congiu, M., Wright, R. D., and Coghlan, J. P., 1984, *Ann. N.Y. Acad. Sci.* **429**:212–213.
8. Fernley, R. T., Coghlan, J. P., and Wright, R. D., 1988, *Biochem. J.* **249**:210–207.
9. Fernley, R. T., Wright, R. D., and Coghlan, J. P., 1988, *Biochemistry* **27**:2815–2820.
10. Fernley, R. T., Darling, P. E., Aldred, P., Wright, R. D., and Coghlan, J. P., 1989, *Biochem. J.* **259**:91–96.
11. Hori, H., Yoshino, Y., Ishizuka, Y., Yamauchi, T., and Murakami, K., 1988, *FEBS Lett.* **232**:391–394.
12. Kadoya, Y., Kuwahara, H., Shimazaki, M., Ogawa, Y., and Yagi, T., 1987, *Osaka City Med. J.* **33**:99–109.
13. Mason, D. K., and Chisholm, D. M., 1975, *Salivary Glands in Health and Disease*, W. B. Saunders, London.
14. Montgomery, J. C., Shows, T. B., Venta, P. J., and Tashain, R. E., 1987, *Am. J. Hum. Genet.* **41**:A229.
15. Murakami, H., and Sly, W. S., 1987, *J. Biol. Chem.* **262**:1382–1388.
16. Penschow, J. D., Haralambidis, J., Aldred, P., Tregear, G. W., and Coghlan, J. P., 1986, *Methods Enzymol.* **124**:534–548.
17. Rapp, G. W., 1946, *J. Am. Dent. Assoc.* **33**:191–194.
18. Sutherland, G. R., Baker, E., Fernandez, K. E. W., Callen, D. F., Aldred, P., Coghlan, J. P., Wright, R. D., and Fernley, R. T., *Cytogenet. Cell Genet.* **50**:149–150.
19. Tanis, R. J., Ferrell, R. E., and Tashian, R. E., 1974, *Biochim. Biophys. Acta* **371**:534–548.
20. Venta, P. J., Montgomery, J. C., and Tashian, R. E., 1987, *Curr. Top. Biol. Med. Res.* **14**:59–72.

Index

MIX
Papier aus verantwortungsvollen Quellen
Paper from responsible sources
FSC® C105338

If you have any concerns about our products,
you can contact us on
ProductSafety@springernature.com

In case Publisher is established outside the EU,
the EU authorized representative is:
Springer Nature Customer Service Center GmbH
Europaplatz 3, 69115 Heidelberg, Germany

Printed by Libri Plureos GmbH
in Hamburg, Germany